Thinkwell Geometry

Worksheets and Answer Keys

Thinkwell Geometry: Worksheets and Answer Keys

Worksheets

Worksheets

Worksheets

Answer Keys

Answer Keys

Answer Keys

Answer Keys

Worksheets

Understanding Points, Lines, and Planes

Example 1

Use the figure to name each of the following.

1. five points
2. two lines
3. two planes
4. point on $\overleftrightarrow{BD}$

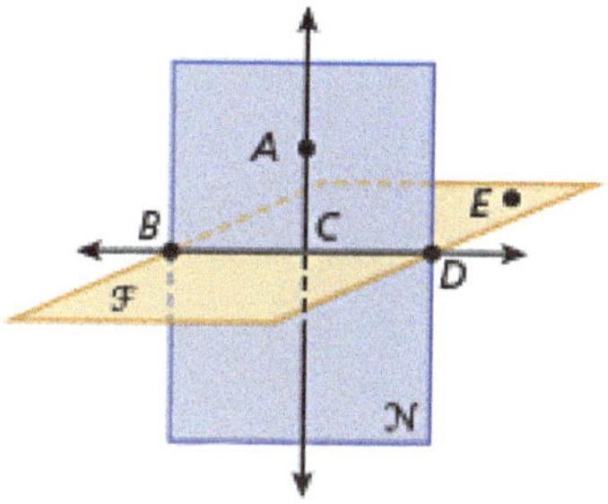

Example 2

Draw and label each of the following.

5. a segment with endpoints *M* and *N*

6. a ray with endpoint *F* that passes through *G*

Example 3

Use the figure to name each of the following.

7. a line that contains *A* and *C*

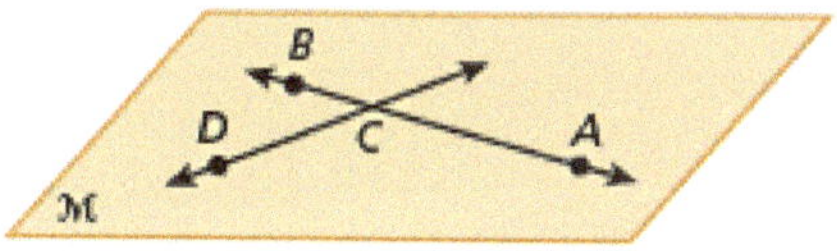

8. a plane that contains *A*, *D*, and *C*

Example 4

Sketch a figure that shows each of the following.

9. three coplanar lines that intersect in a common point

10. two lines that do not intersect

Fundamentals of Geometry

Points, Lines, Planes, and Angles

Measuring and Constructing Segments

Example 1

Find each length.

1. *AB*

2. *BC*

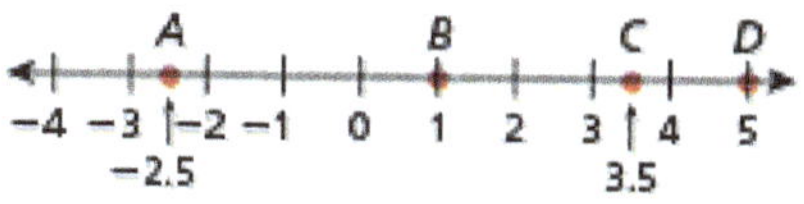

Example 2

3. Sketch, draw, and construct a segment congruent to $\overline{RS}$.

Example 3

4. *B* is between *A* and *C*, $AC = 15.8$, and $AB = 9.9$. Find *BC*.

5. Find *MP*.

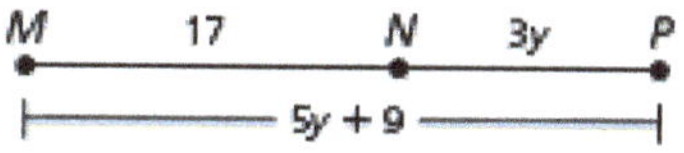

Example 4

6. **Travel** If a picnic area is located at the midpoint between Lubbock and Amarillo, find the distance to the picnic area from the sign.

Example 5

7. **Multi-Step** K is the midpoint of $\overline{JL}$, $JL = 4x - 2$, and $JK = 7$. Find x, KL, and JL.

8. E bisects $\overline{DF}$, $DE = 2y$, and $EF = 8y - 3$. Find DE, EF, and DF.

Example 1

1. **Music** Musicians use a metronome to keep time as they play. The metronome's needle swings back and forth in a fixed amount of time. Name all of the angles in the diagram.

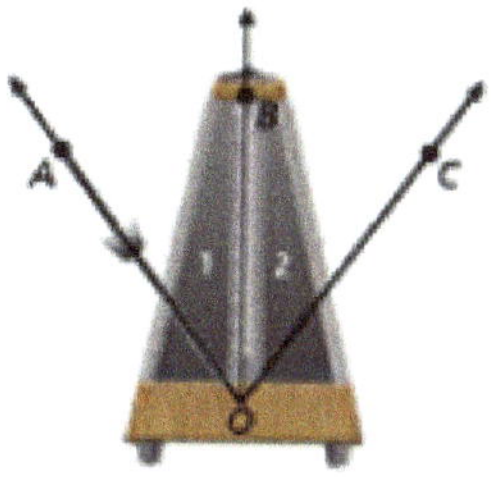

Example 2

Use the protractor to find the measure of each angle. Then classify each as acute, right, or obtuse.

2. $\angle VXW$

3. $\angle TXW$

4. $\angle RXU$

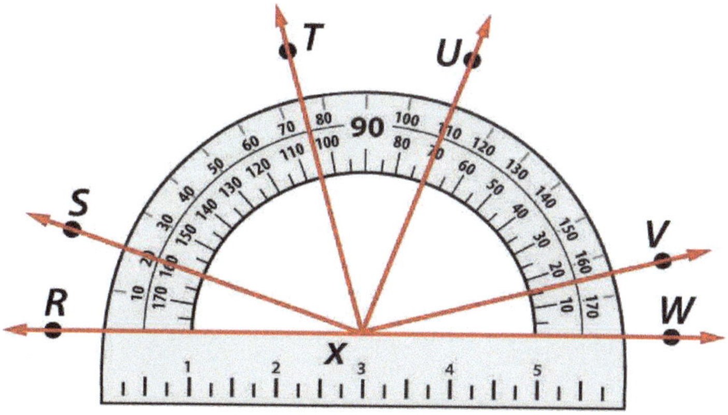

Example 3

L is in the interior of $\angle JKM$. Find each of the following.

5. $m\angle JKM$ if $m\angle JKL = 42°$ and $m\angle LKM = 28°$

6. $m\angle LKM$ if $m\angle JKL = 56.4°$ and $m\angle JKM = 82.5°$

Example 4

Multi-Step $\overrightarrow{BD}$ **bisects** $\angle ABC$. **Find each of the following.**

7. $m\angle ABD$ if $m\angle ABD = (6x + 4)^\circ$ and $m\angle DBC = (8x - 4)^\circ$

8. $m\angle ABC$ if $m\angle ABD = (5y - 3)^\circ$ and $m\angle DBC = (3y + 15)^\circ$

Pairs of Angles

Example 1

Tell whether the angles are only adjacent, adjacent and form a linear pair, or not adjacent.

1. $\angle 1$ and $\angle 2$

2. $\angle 1$ and $\angle 3$

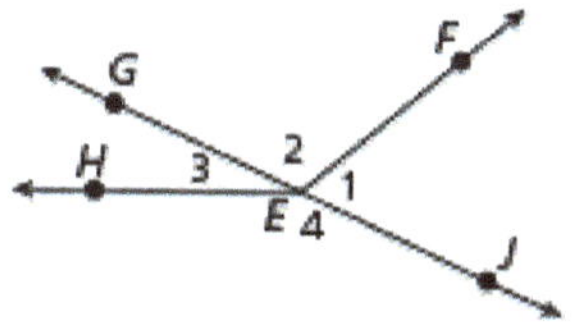

3. $\angle 2$ and $\angle 4$

4. $\angle 2$ and $\angle 3$

Example 2

Find the measure of each of the following.

5. supplement of $\angle A$

6. complement of $\angle A$

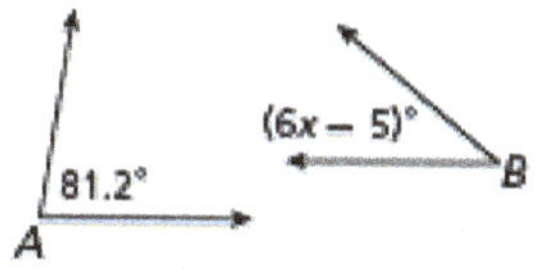

7. supplement of $\angle B$

8. complement of $\angle B$

Example 3

9. **Multi-Step** An angle's measure is 6 degrees more than 3 times the measure of its complement. Find the measure of the angle.

Example 4

10. **Landscaping** A sprinkler swings back and forth between A and B in such a way that $\angle 1 \cong \angle 2$. $\angle 1$ and $\angle 3$ are complementary, and $\angle 2$ and $\angle 4$ are complementary. If $m\angle 1 = 18.5°$, find $m\angle 2$, $m\angle 3$, and $m\angle 4$.

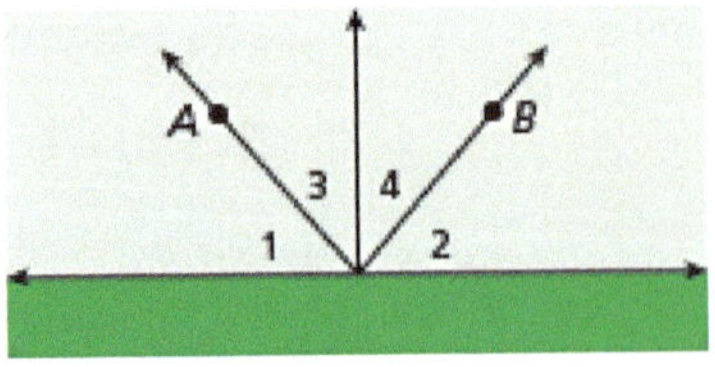

Example 5

11. Name each pair of vertical angles.

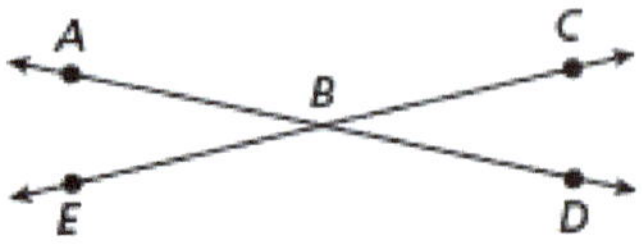

Fundamentals of Geometry

Review 1.1 / Points, Lines, Planes, and Angles

1.1.1 Understanding Points, Lines, and Planes

Use the figure to name each of the following.

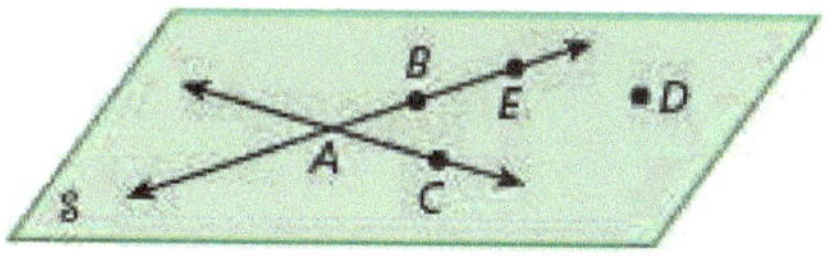

1. three collinear points

2. four coplanar points

3. a plane containing *E*

Draw and label each of the following.

4. a line containing *X* and *Y*

5. a pair of opposite rays that both contain *R*

Use the figure to name each of the following.

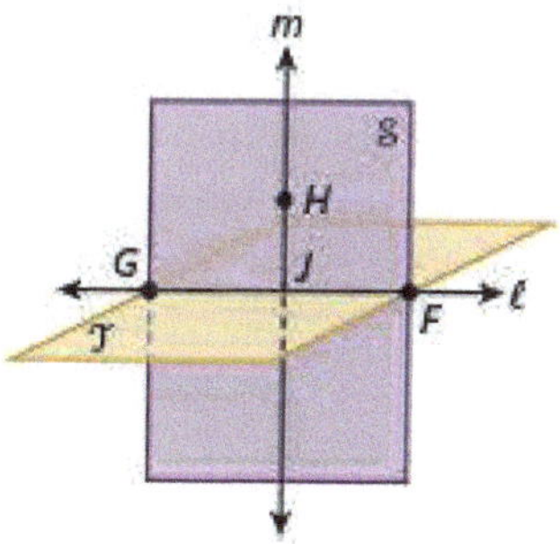

6. two points and a line that lie in plane $\mathcal{T}$

7. two planes that contain ℓ

Sketch a figure that shows each of the following.

8. a line that intersects two nonintersecting planes

9. three coplanar lines that intersect in three different points

1.1.2 Measuring and Constructing Segments

Find each length.

10. *DB*

11. *CD*

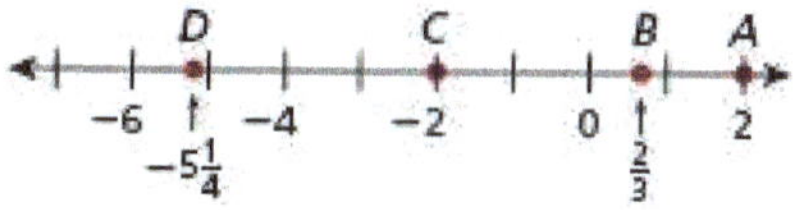

12. Sketch, draw, and construct a segment twice the length of $\overline{AB}$.

13. D is between C and E, $CE = 17.1$, and $DE = 8$. Find CD.

14. Find MN.

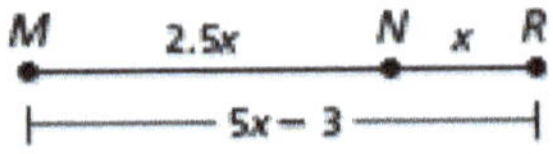

15. **Sports** During a football game, a quarterback standing at the 9-yard line passes the ball to a receiver at the 24-yard line. The receiver then runs with the ball halfway to the 50-yard line. How many total yards (passing plus running) did the team gain on the play?

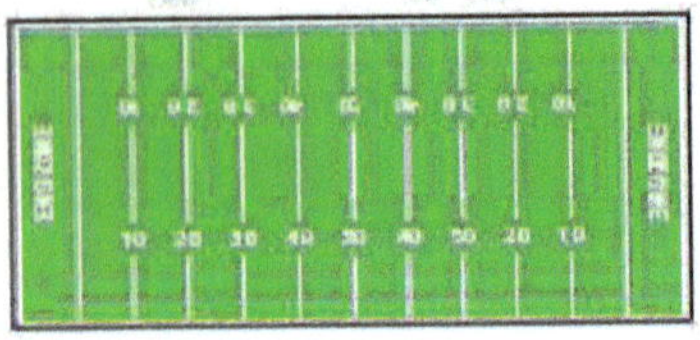

16. **Multi-Step** E is the midpoint of $\overline{DF}$, $DE = 2x + 4$, and $EF = 3x - 1$. Find DE, EF, and DF.

17. Q bisects $\overline{PR}$, $PQ = 3y$, and $PR = 42$. Find y and QR.

1.1.3 Measuring and Constructing Angles

18. **Physics** Pendulum clocks have been used since 1656 to keep time. The pendulum swings back and forth once or twice per second. Name all of the angles in the diagram.

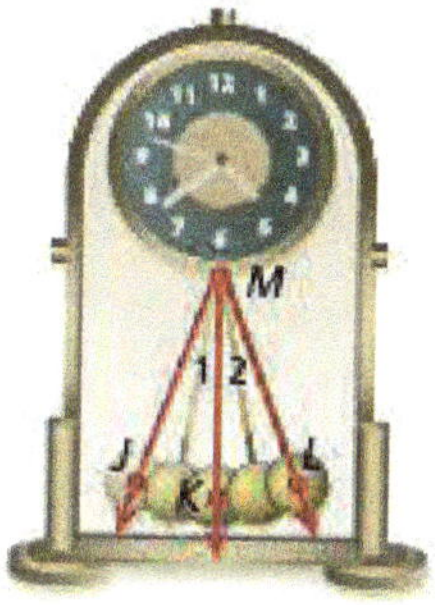

Use the protractor to find the measure of each angle. Then classify each as acute, right, or obtuse.

19. $\angle CGE$

20. $\angle BGD$

21. $\angle AGB$

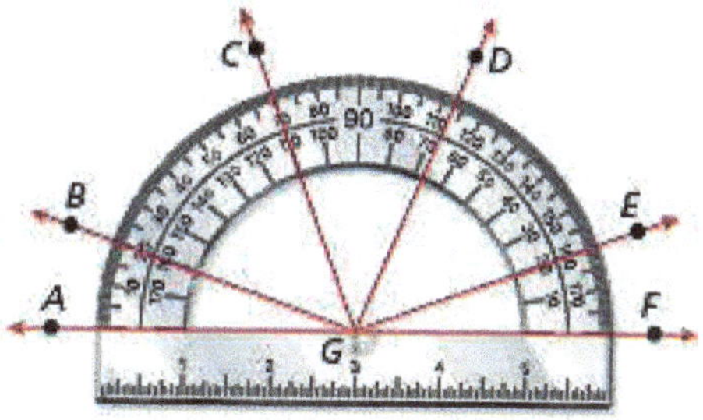

T is in the interior of $\angle RSU$. Find each of the following.

22. $m\angle RSU$ if $m\angle RST = 38°$ and $m\angle TSU = 28.6°$

23. $m\angle RST$ if $m\angle TSU = 46.7°$ and $m\angle RSU = 83.5°$

Multi-Step $\overrightarrow{SP}$ **bisects $\angle RST$. Find each of the following.**

24. $m\angle RST$ if $m\angle RSP = (3x - 2)°$ and $m\angle PST = (9x - 26)°$

25. $m\angle RSP$ if $m\angle RST = \frac{5}{2}y°$ and $m\angle PST = (y + 5)°$

1.1.4 Pairs of Angles

Tell whether the angles are only adjacent, adjacent and form a linear pair, or not adjacent.

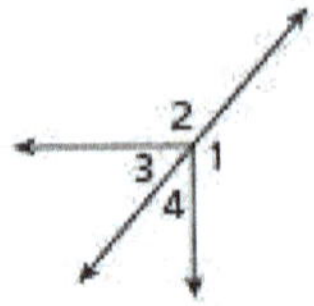

26. $\angle 1$ and $\angle 4$

27. $\angle 2$ and $\angle 3$

28. $\angle 3$ and $\angle 4$

29. $\angle 3$ and $\angle 1$

Given $m\angle A = 56.4°$ and $m\angle B = (2x - 4)°$, find the measure of each of the following.

30. supplement of $\angle A$

31. complement of $\angle A$

32. supplement of $\angle B$

33. complement of $\angle B$

34. **Multi-Step** An angle's measure is 3 times the measure of its complement. Find the measure of the angle and the measure of its complement.

35. **Art** In the stained glass pattern, $\angle 1 \cong \angle 2$. $\angle 1$ and $\angle 3$ are complementary, and $\angle 2$ and $\angle 4$ are complementary. If $m\angle 1 = 22.3°$, find $m\angle 2$, $m\angle 3$, and $m\angle 4$.

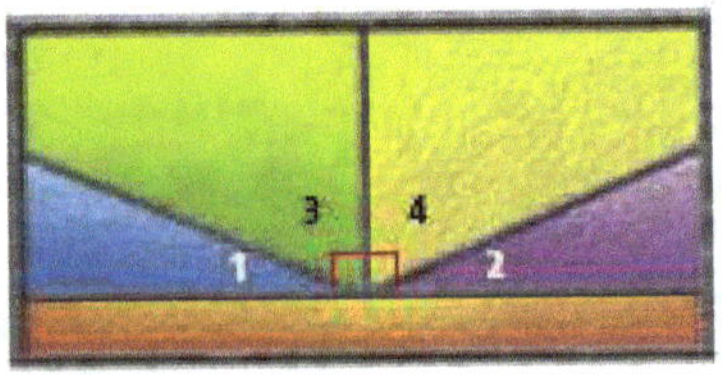

36. Name the pairs of vertical angles.

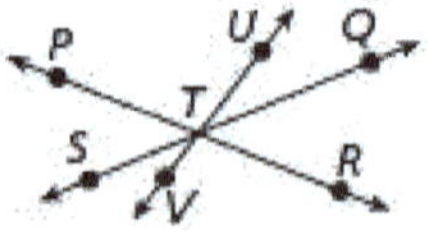

Fundamentals of Geometry

Coordinate and Transformation Tools

Using Formulas in Geometry

Example 1

Find the perimeter and area of each figure.

1. 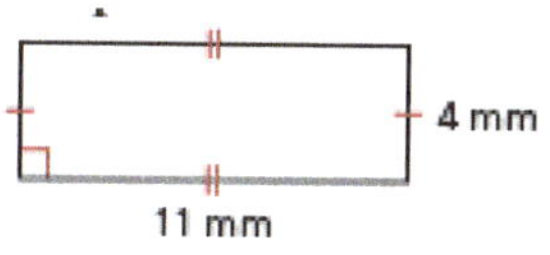

2. 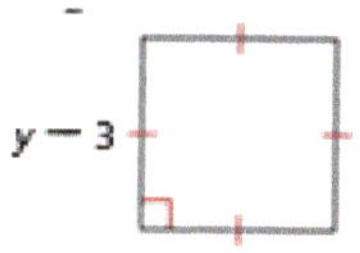

3. 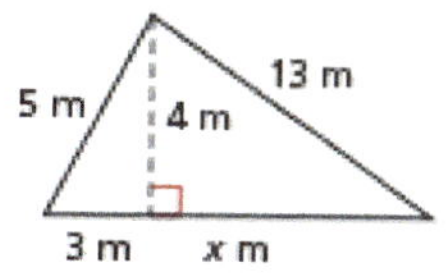

Example 2

4. **Manufacturing** A puzzle contains a triangular piece with a base of 3 in. and a height of 4 in. A manufacturer wants to make 80 puzzles. Find the amount of wood used if each puzzle contains 20 triangular pieces.

Example 3

Find the circumference and area of each circle. Use the π key on your calculator. Round to the nearest tenth.

5. 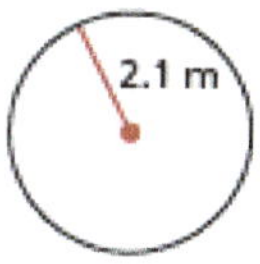

6. 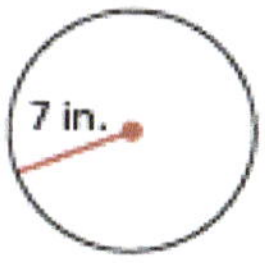

7. 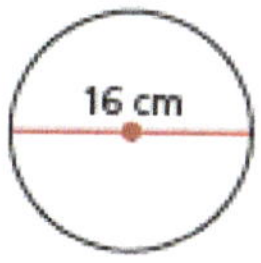

Example 1

Find the coordinates of the midpoint of each segment.

1. $\overline{AB}$ with endpoints $A(4, -6)$ and $B(-4, 2)$

2. $\overline{CD}$ with endpoints $C(0, -8)$ and $D(3, 0)$

Example 2

3. M is the midpoint of $\overline{LN}$. L has coordinates $(-3, -1)$, and M has coordinates $(0, 1)$. Find the coordinates of N.

4. B is the midpoint of $\overline{AC}$. A has coordinates $(-3, 4)$, and B has coordinates $\left(-1\frac{1}{2}, 1\right)$. Find the coordinates of C.

Example 3

Multi-Step **Find the length of the given segments and determine if they are congruent.**

5. $\overline{JK}$ and $\overline{FG}$

6. $\overline{JK}$ and $\overline{RS}$

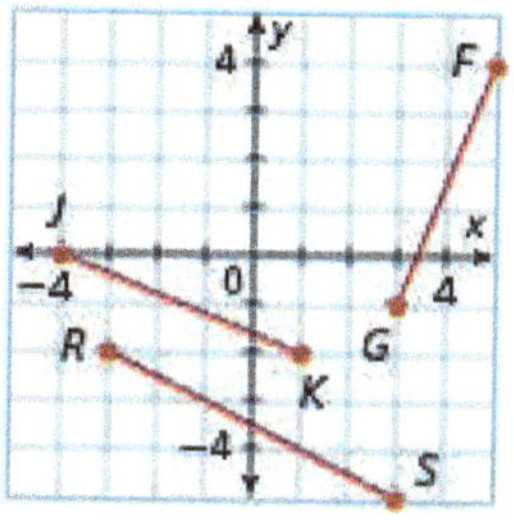

Example 4

Use the Distance Formula and the Pythagorean Theorem to find the distance, to the nearest tenth, between each pair of points.

7. $A(1, -2)$ and $B(-4, -4)$

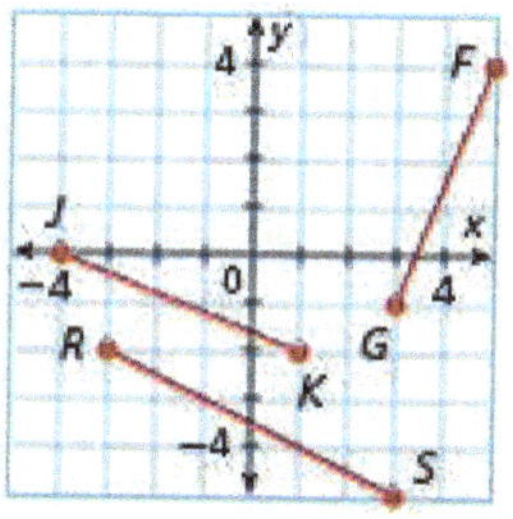

8. $X(-2, 7)$ and $Y(-2, -8)$

9. $V(2, -1)$ and $W(-4, 8)$

Example 5

10. **Architecture** The plan for a rectangular living room shows electrical wiring will be run in a straight line from the entrance E to a light L at the opposite corner of the room. What is the length of the wire to the nearest tenth?

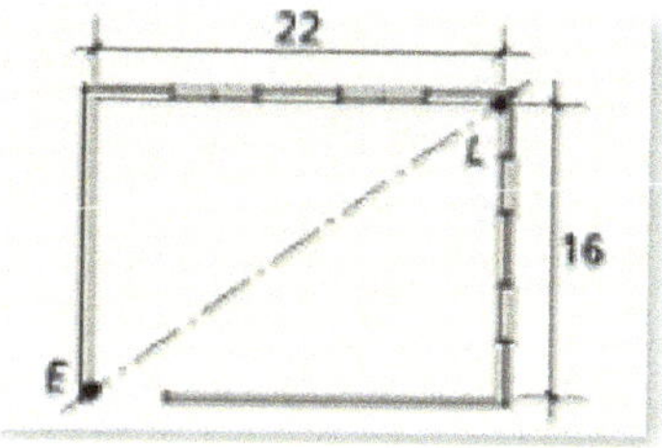

Fundamentals of Geometry

Coordinate and Transformation Tools

Transformations in the Coordinate Plane

Example 1

Identify each transformation. Then use arrow notation to describe the transformation.

1. 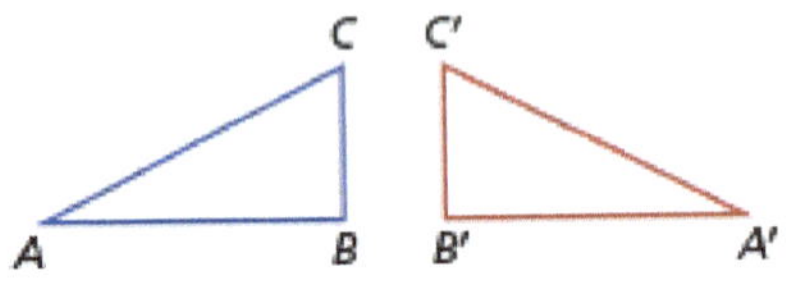

2. 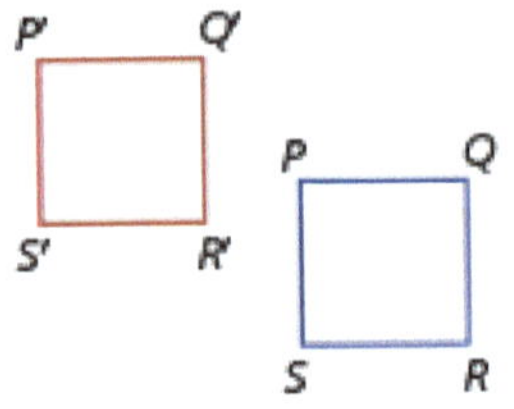

Example 2

3. A figure has vertices at $A(-3, 2)$, $B(-1, -1)$, and $C(-4, -2)$. After a transformation, the image of the figure has vertices at $A'(3, 2)$, $B'(1, -1)$, and $C'(4, -2)$. Draw the preimage and image. Then identify the transformation.

Example 3

4. **Multi-Step** The coordinates of the vertices of $\triangle DEF$ are $D(2, 3)$, $E(1, 1)$, and $F(4, 0)$. Find the coordinates for the image of $\triangle DEF$ after the translation $(x, y) \rightarrow (x - 3, y - 2)$. Draw the preimage and image.

Example 4

5. **Animation** In an animated film, a simple scene can be created by translating a figure against a still background. Write a rule for the translation that maps the rocket from position 1 to position 2.

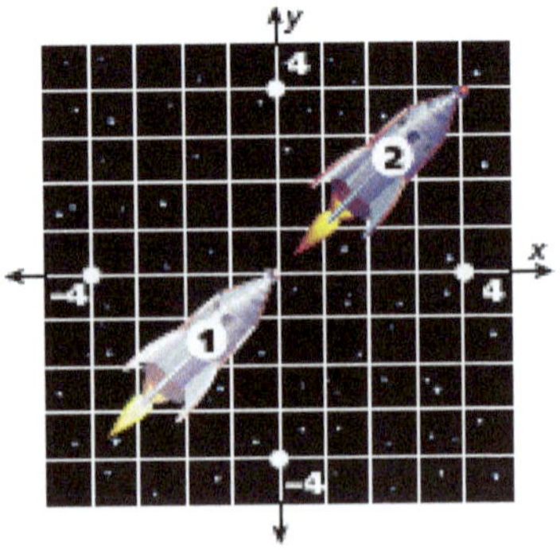

1.2.1 Using Formulas in Geometry

Find the perimeter and area of each figure.

1. 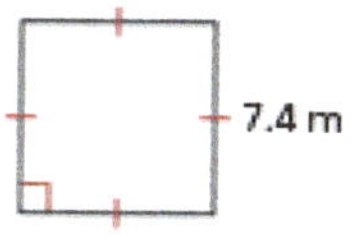

2. 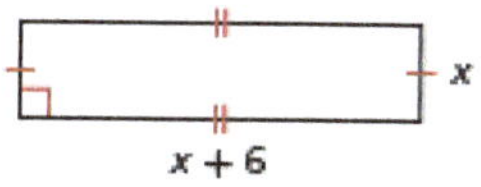

3. 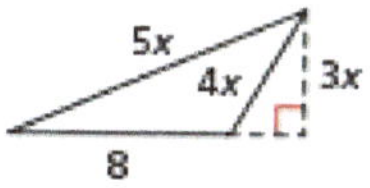

4. **Crafts** The quilt pattern includes 32 small triangles. Each has a base of 3 in. and a height of 1.5 in. Find the amount of fabric used to make the 32 triangles.

Find the circumference and area of each circle with the given radius or diameter. Use the π key on your calculator. Round to the nearest tenth.

5. $r = 12$ m

6. $d = 12.5$ ft

7. $d = \frac{1}{2}$ mi

1.2.2 Midpoint and Distance in the Coordinate Plane

Find the coordinates of the midpoint of each segment.

8. $\overline{XY}$ with endpoints $X(-3, -7)$ and $Y(-1, 1)$

9. $\overline{MN}$ with endpoints $M(12, -7)$ and $N(-5, -2)$

10. M is the midpoint of $\overline{QR}$. Q has coordinates $(-3, 5)$, and M has coordinates $(7, -9)$. Find the coordinates of R.

11. D is the midpoint of $\overline{CE}$. E has coordinates $(-3, -2)$, and D has coordinates $\left(2\frac{1}{2}, 1\right)$. Find the coordinates of C.

Multi-Step **Find the length of the given segments and determine if they are congruent.**

12. $\overline{DE}$ and $\overline{FG}$

13. $\overline{DE}$ and $\overline{RS}$

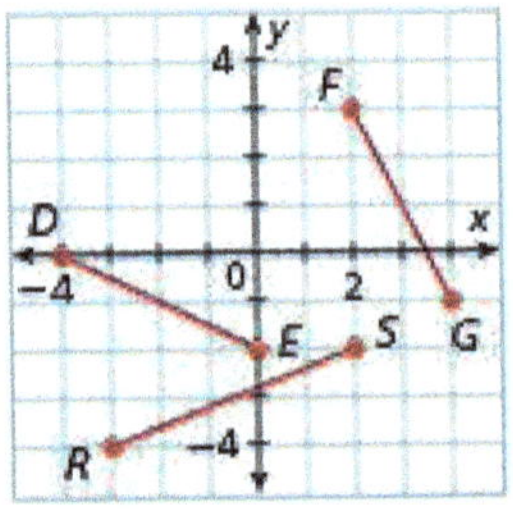

Use the Distance Formula and the Pythagorean Theorem to find the distance, to the nearest tenth, between each pair of points.

14. $U(0, 1)$ and $V(-3, -9)$

15. $M(10, -1)$ and $N(2, -5)$

16. $P(-10, 1)$ and $Q(5, 5)$

17. **Consumer Application** Televisions and computer screens are usually advertised based on the length of their diagonals. If the height of a computer screen is 11 in. and the width is 14 in., what is the length of the diagonal? Round to the nearest inch.

1.2.3 Transformations in the Coordinate Plane

Identify each transformation. Then use arrow notation to describe the transformation.

18.

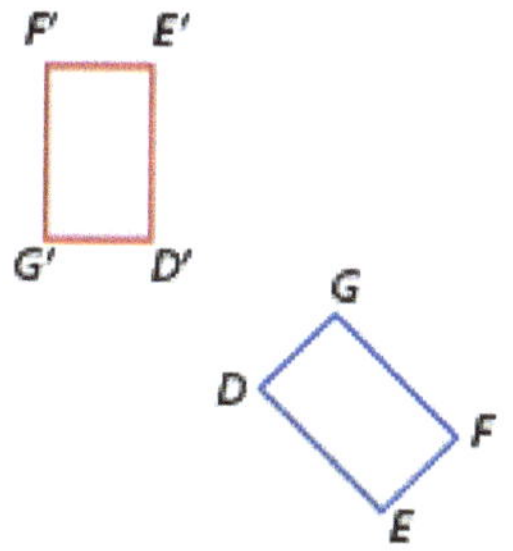

19.

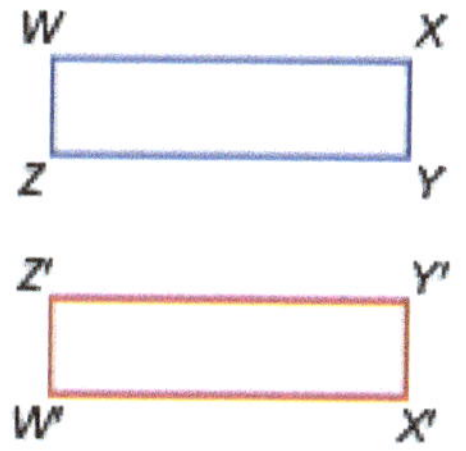

20. A figure has vertices at $J(-2, 3)$, $K(0, 3)$, $L(0, 1)$, and $M(-2, 1)$. After a transformation, the image of the figure has vertices at $J'(2, 1)$, $K'(4, 1)$, $L'(4, -1)$, and $M'(2, -1)$. Draw the preimage and image. Then identify the transformation.

21. **Multi-Step** The coordinates of the vertices of rectangle $ABCD$ are $A(-4, 1)$, $B(1, 1)$, $C(1, -2)$, and $D(-4, -2)$. Find the coordinates for the image of rectangle $ABCD$ after the translation $(x, y) \rightarrow (x + 3, y - 2)$. Draw the preimage and the image.

22. **Travel** Write a rule for the translation that maps the descent of the hot air balloon.

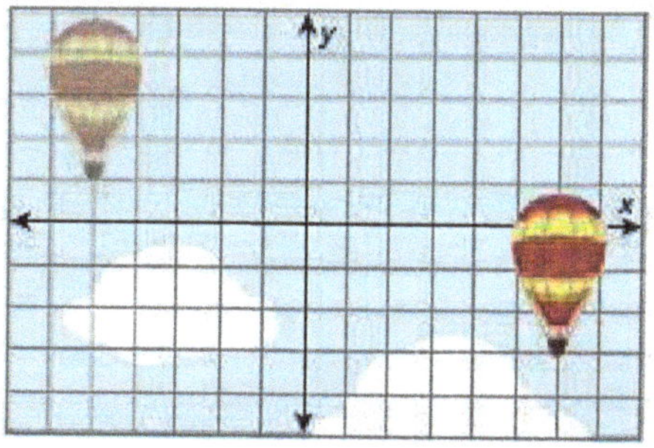

Reasoning and Writing Geometric Proofs

Inductive and Deductive Reasoning

Using Inductive Reasoning to Make Conjectures

Example 1

Find the next item in each pattern.

1. March, May, July, ...
2. $\frac{1}{3}, \frac{2}{4}, \frac{3}{5}, \ldots$
3. 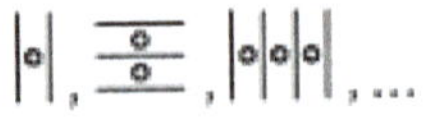

Example 2

Complete each conjecture.

4. The product of two even numbers is __?__.

5. A rule in terms of n for the sum of the first n odd positive integers is __?__.

Example 3

6. **Biology** A laboratory culture contains 150 bacteria. After twenty minutes, the culture contains 300 bacteria. After one hour, the culture contains 1200 bacteria. Make a conjecture about the rate at which the bacteria increases.

Example 4

Show that each conjecture is false by finding a counterexample.

7.

President	Age at Inauguration
Washington	57
T. Roosevelt	42
Truman	60
Kennedy	43
Clinton	46

Kennedy is the youngest U.S. president to be inaugurated.

8. Three points on a plane always form a triangle.

9. For any real number x, if $x^2 \geq 1$, then $x \geq 1$.

Example 1

Identify the hypothesis and conclusion of each conditional.

1. If a person is at least 16 years old, then the person can drive a car.

2. A figure is a parallelogram if it is a rectangle.

3. The statement $a - b < a$ implies that b is a positive number.

Example 2

Write a conditional statement from each of the following.

4. Eighteen-year-olds are eligible to vote.

5. $\left(\frac{a}{b}\right)^2 < \frac{a}{b}$ when $0 < a < b$.

Example 3

Determine if each conditional is true. If false, give a counterexample.

6. If three points form a triangle, then they lie in the same plane.

7. If $x > y$, then $|x| > |y|$.

8. If the season is spring, then the month is March.

Example 4

9. **Travel** Write the converse, inverse, and contrapositive of the following conditional statement. Find the truth value of each.

   If Brielle drives at exactly 30 mi/h, then she travels 10 mi in 20 min.

Using Deductive Reasoning to Verify Conjectures

Example 1

Does each conclusion use inductive or deductive reasoning?

1. At Bell High School, students must take Biology before they take Chemistry. Sam is in Chemistry, so Marcia concludes that he has taken Biology.

2. A detective learns that his main suspect was out of town the day of the crime. He concludes that the suspect is innocent.

Example 2

Determine if each conjecture is valid by the Law of Detachment.

3. Given: If you want to go on a field trip, you must have a signed permission slip.
   Zola has a signed permission slip.
   Conjecture: Zola wants to go on a field trip.

4. Given: If the side lengths of a rectangle are 3 ft and 4 ft, then its area is 12 ft^2.
   A rectangle has side lengths of 3 ft and 4 ft.
   Conjecture: The area of the rectangle is 12 ft^2.

Example 3

Determine if each conjecture is valid by the Law of Syllogism.

5. Given: If you fly from Texas to California, you travel from the central to the Pacific time zone. If you travel from the central to the Pacific time zone, then you gain two hours.
   Conjecture: If you fly from Texas to California, you gain two hours.

6. Given: If a figure is a square, then the figure is a rectangle. If a figure is a square, then it is a parallelogram.
Conjecture: If a figure is a parallelogram, then it is a rectangle.

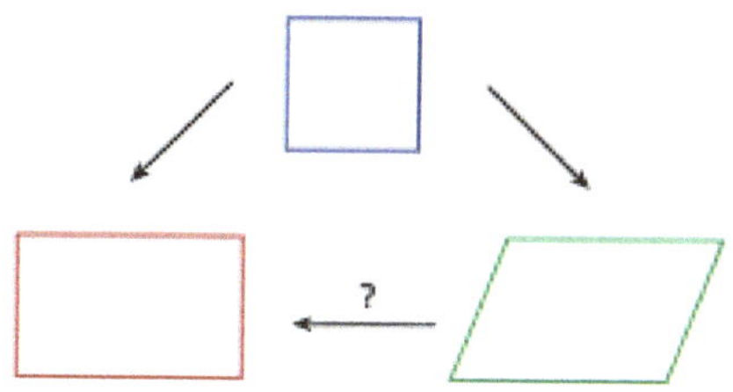

Example 4

7. Draw a conclusion from the given information.
Given: If you leave your car lights on overnight, then your car battery will drain. If your battery is drained, your car might not start. Alex left his car lights on last night.

Biconditional Statements and Definitions

Example 1

Write the conditional statement and converse within each biconditional.

1. Perry can paint the entire living room if and only if he has enough paint.

2. Your medicine will be ready by 5 P.M. if and only if you drop your prescription off by 8 A.M.

Example 2

For each conditional, write the converse and a biconditional statement.

3. If a student is a sophomore, then the student is in the tenth grade.

4. If two segments have the same length, then they are congruent.

Example 3

Multi-Step Determine if each biconditional is true.
If false, give a counterexample.

5. $xy = 0 \leftrightarrow x = 0$ or $y = 0$.

6. A figure is a quadrilateral if and only if it is a polygon.

Example 4

Write each definition as a biconditional.

7. Parallel lines are two coplanar lines that never intersect.

8. A hummingbird is a tiny, brightly colored bird with narrow wings, a slender bill, and a long tongue.

2.1.1 Using Inductive Reasoning to Make Conjectures

Find the next item in each pattern.

1. 8 A.M., 11 A.M., 2 P.M., ...
2. 75, 64, 53, ...
3. △, □, ⬠, ...

Complete each conjecture.

4. A rule in terms of n for the sum of the first n even positive integers is __?__.

5. The number of nonoverlapping segments formed by n collinear points is __?__.

6. **Industrial Arts** About 5% of the students at Lubbock High School usually participate in the robotics competition. There are 526 students in the school this year. Make a conjecture about the number of students who will participate in the robotics competition this year.

Show that each conjecture is false by finding a counterexample.

7. If $1 - y > 0$, then $0 < y < 1$.

8. For any real number x, $x^3 \geq x^2$.

9. Every pair of supplementary angles includes one obtuse angle.

2.1.2 Conditional Statements

Identify the hypothesis and conclusion of each conditional.

10. If an animal is a tabby, then it is a cat.

11. Four angles are formed if two lines intersect.

12. If 8 ounces of cereal cost \$2.99, then 16 ounces of cereal cost \$5.98.

Write a conditional statement from each sentence.

13. You should monitor the heart rate of a patient who is ill.

14. After three strikes, the batter is out.

15. Congruent segments have equal measures.

Determine if each conditional is true. If false, give a counterexample.

16. If you subtract -2 from -6, then the result is -4.

17. If two planes intersect, then they intersect in exactly one point.

18. If a cat is a bird, then today is Friday.

Write the converse, inverse, and contrapositive of each conditional statement. Find the truth value of each.

19. **Probability** If the probability of an event is 0.1, then the event is unlikely to occur.

20. **Meteorology** If freezing rain is falling, then the air temperature is 32°F or less. (*Hint:* The freezing point of water is 32°F.)

2.1.3 Using Deductive Reasoning to Verify Conjectures

Does each conclusion use inductive or deductive reasoning?

21. The sum of the angle measures of a triangle is 180°. Two angles of a triangle measure 40° and 60°, so Kandy concludes that the third angle measures 80°.

22. All of the students in Henry's Geometry class are juniors. Alexander takes Geometry, but has another teacher. Henry concludes that Alexander is also a junior.

23. Determine if the conjecture is valid by the Law of Detachment.
Given: If one integer is odd and another integer is even, their product is even.
The product of two integers is 24.
Conjecture: One of the two integers is odd.

24. **Science** Determine if the conjecture is valid by the Law of Syllogism.
Given: If an element is an alkali metal, then it reacts with water. If an element is in the first column of the periodic table, then it is an alkali metal.
Conjecture: If an element is in the first column of the periodic table, then it reacts with water.

25. Draw a conclusion from the given information.
Given: If Dakota watches the news, she is informed about current events.
If Dakota knows about current events, she gets better grades in Social Studies.
Dakota watches the news.

2.1.4 Biconditional Statements and Definitions

Write the conditional statement and converse within each biconditional.

26. Three points are coplanar if and only if they lie in the same plane.

27. A parallelogram is a rectangle if and only if it has four right angles.

28. A lunar eclipse occurs if and only if Earth is between the Sun and the Moon.

For each conditional, write the converse and a biconditional statement.

29. If today is Saturday or Sunday, then it is the weekend.

30. If Greg has the fastest time, then he wins the race.

31. If a triangle contains a right angle, then it is a right triangle.

Multi-Step **Determine if each biconditional is true. If false, give a counterexample.**

32. Felipe is a swimmer if and only if he is an athlete.

33. The number $2n$ is even if and only if n is an integer.

Write each definition as a biconditional.

34. A circle is the set of all points in a plane that are a fixed distance from a given point.

35. A catcher is a baseball player who is positioned behind home plate and who catches throws from the pitcher.

Example 1

Multi-Step **Solve each equation. Write a justification for each step.**

1. $y + 1 = 5$

2. $t - 3.2 = -8.3$

3. $2p - 30 = -4p + 6$

4. $\frac{x+3}{-2} = 8$

5. $\frac{1}{2}n = \frac{3}{4}$

6. $0 = 2(r - 3) + 4$

Example 2

7. **Nutrition** Amy's favorite breakfast cereal has 102 Calories per serving. The equation $C = 9f + 90$ relates the grams of fat f in one serving to the Calories C in one serving. How many grams of fat are in one serving of the cereal? Solve the equation for f and justify each step.

8. **Movie Rentals** The equation $C = \$5.75 + \$0.89m$ relates the number of movie rentals m to the monthly cost C of a movie club membership. How many movies did Elias rent this month if his membership cost $11.98? Solve the equation for m and justify each step.

Example 3

Write a justification for each step.

9.

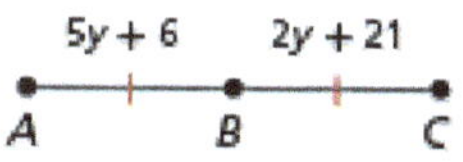

$AB = BC$

$5y + 6 = 2y + 21$

$3y + 6 = 21$

$3y = 15$

$y = 5$

10.

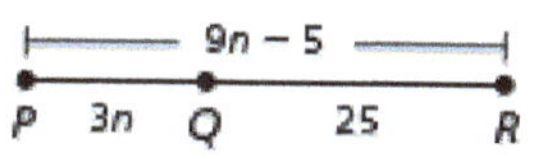

$PQ + QR = PR$

$3n + 25 = 9n - 5$

$25 = 6n - 5$

$30 = 6n$

$5 = n$

Example 4

Identify the property that justifies each statement.

11. $\overline{AB} \cong \overline{AB}$

12. $m\angle 1 = m\angle 2$, and $m\angle 2 = m\angle 4$. So $m\angle 1 = m\angle 4$.

13. $x = y$, so $y = x$.

14. $\overline{ST} \cong \overline{YZ}$, and $\overline{YZ} \cong \overline{PR}$. So $\overline{ST} \cong \overline{PR}$.

Reasoning and Writing Geometric Proofs

Mathematical Proof

Geometric Proof

Example 1

1. Write a justification for each step, given that $m\angle A = 60°$ and $m\angle B = 2m\angle A$.

1. $m\angle A = 60°$, $m\angle B = 2m\angle A$
2. $m\angle B = 2(60°)$
3. $m\angle B = 120°$
4. $m\angle A + m\angle B = 60° + 120°$
5. $m\angle A + m\angle B = 180°$
6. $\angle A$ and $\angle B$ are supplementary.

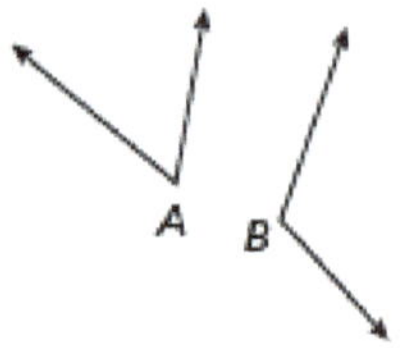

Example 2

2. Fill in the blanks to complete the two-column proof.

Given: $\angle 2 \cong \angle 3$

Prove: $\angle 1$ and $\angle 3$ are supplementary.

Proof:

Statements	Reasons
1. $\angle 2 \cong \angle 3$	1. Given
2. $m\angle 2 = m\angle 3$	2. a. ___?___
3. b. ___?___	3. Lin. Pair Thm.
4. $m\angle 1 + m\angle 2 = 180°$	4. Def. of supp. ∡
5. $m\angle 1 + m\angle 3 = 180°$	5. c. ___?___ *Steps 2, 4*
6. d. ___?___	6. Def. of supp. ∡

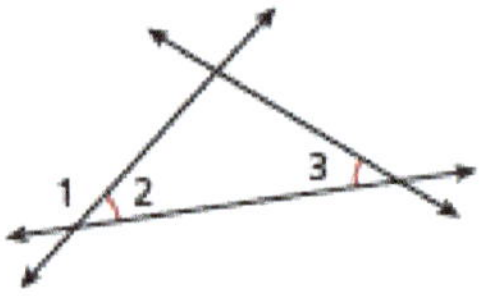

Example 3

3. Use the given plan to write a two-column proof.

Given: X is the midpoint of $\overline{AY}$, and Y is the midpoint of $\overline{XB}$.

Prove: $\overline{AX} \cong \overline{YB}$

Plan: By the definition of midpoint, $\overline{AX} \cong \overline{XY}$, and $\overline{XY} \cong \overline{YB}$. Use the Transitive Property to conclude that $\overline{AX} \cong \overline{YB}$.

Reasoning and Writing Geometric Proofs

Mathematical Proof

Flowchart and Paragraph Proofs

Example 1

1. Use the given flowchart proof to write a two-column proof.

Given: $\angle 1 \cong \angle 2$

Prove: $\angle 1$ and $\angle 2$ are right angles.

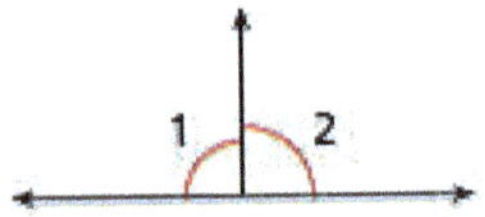

Flowchart proof:

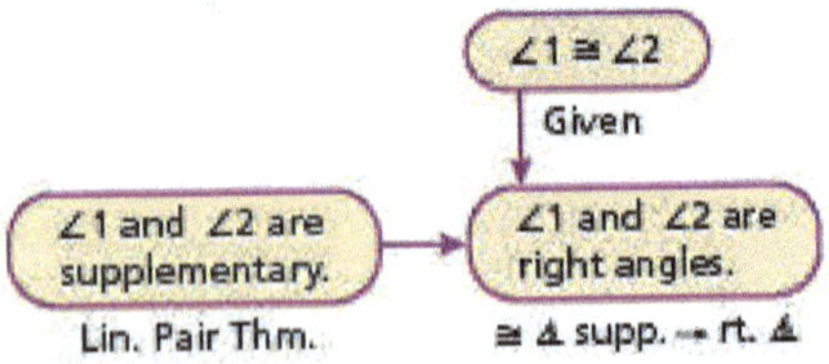

2. Use the given two-column proof to write a flowchart proof.

Given: $\angle 2$ and $\angle 4$ are supplementary.

Prove: $m\angle 2 = m\angle 3$

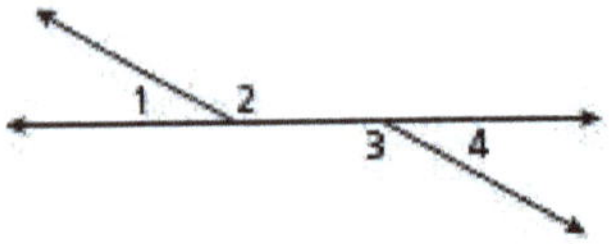

Two-column proof:

Statements	Reasons
1. $\angle 2$ and $\angle 4$ are supplementary.	1. Given
2. $\angle 3$ and $\angle 4$ are supplementary.	2. Lin. Pair Thm.
3. $\angle 2 \cong \angle 3$	3. $\cong$ Supps. Thm. *Steps 1, 2*
4. $m\angle 2 = m\angle 3$	4. Def. of $\cong$ ∡

Example 2

3. Use the given paragraph proof to write a two-column proof.
Given: $\angle 2 \cong \angle 4$
Prove: $\angle 1 \cong \angle 3$
Paragraph proof:
By the Vertical Angles Theorem, $\angle 1 \cong \angle 2$, and $\angle 3 \cong \angle 4$. It is given that $\angle 2 \cong \angle 4$. By the Transitive Property of Congruence, $\angle 1 \cong \angle 4$, and thus $\angle 1 \cong \angle 3$.

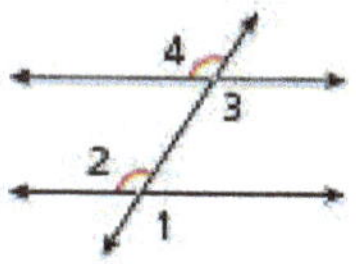

Example 3

4. Use the given two-column proof to write a paragraph proof.
Given: $\overrightarrow{BD}$ bisects $\angle ABC$.
Prove: $\overrightarrow{BG}$ bisects $\angle FBH$.
Two-column proof:

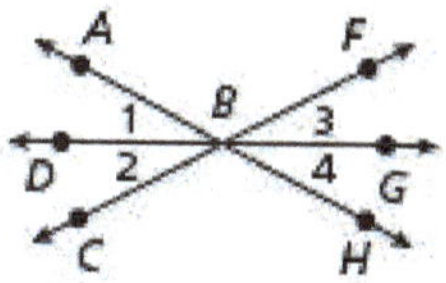

Statements	Reasons
1. $\overrightarrow{BD}$ bisects $\angle ABC$.	1. Given
2. $\angle 1 \cong \angle 2$	2. Def. of $\angle$ bisector
3. $\angle 1 \cong \angle 4$, $\angle 2 \cong \angle 3$	3. Vert. $\angle$s Thm.
4. $\angle 4 \cong \angle 2$	4. Trans. Prop. of $\cong$ *Steps 2, 3*
5. $\angle 4 \cong \angle 3$	5. Trans. Prop. of $\cong$ *Steps 3, 4*
6. $\overrightarrow{BG}$ bisects $\angle FBH$.	6. Def. of $\angle$ bisector

2.2.1 Algebraic Proof

Multi-Step Solve each equation. Write a justification for each step.

1. $5x - 3 = 4(x + 2)$

2. $1.6 = 3.2n$

3. $\frac{z}{3} - 2 = -10$

4. $-(h + 3) = 72$

5. $9y + 17 = -19$

6. $\frac{1}{2}(p - 16) = 13$

7. **Ecology** The equation $T = 0.03c + 0.05b$ relates the numbers of cans c and bottles b collected in a recycling rally to the total dollars T raised. How many cans were collected if \$147 was raised and 150 bottles were collected? Solve the equation for c and justify each step.

Write a justification for each step.

8. $m\angle XYZ = m\angle 2 + m\angle 3$
$4n - 6 = 58 + (2n - 12)$
$4n - 6 = 2n + 46$
$2n - 6 = 46$
$2n = 52$
$n = 26$

9. $m\angle WYV = m\angle 1 + m\angle 2$
$5n = 3(n - 2) + 58$
$5n = 3n - 6 + 58$
$5n = 3n + 52$
$2n = 52$
$n = 26$

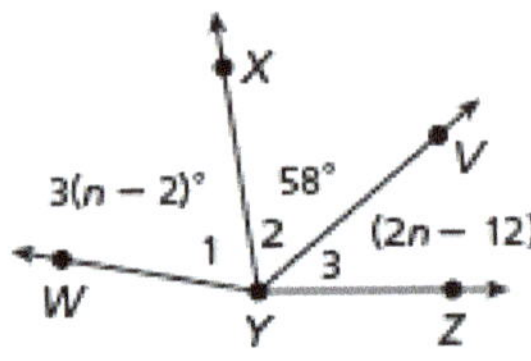

$m\angle WYV = 5n°$
$m\angle XYZ = (4n - 6)°$

Identify the property that justifies each statement.

Reasoning and Writing Geometric Proofs

Review 2.2 / Mathematical Proof

10. $\overline{KL} \cong \overline{PR}$, so $\overline{PR} \cong \overline{KL}$.

11. $412 = 412$

12. If $a = b$ and $b = 0$, then $a = 0$.

13. figure $A \cong$ figure A

2.2.2 Geometric Proof

14. Write a justification for each step, given that $\overrightarrow{BX}$ bisects $\angle ABC$ and $m\angle XBC = 45°$.

1. $\overrightarrow{BX}$ bisects $\angle ABC$.
2. $\angle ABX \cong \angle XBC$
3. $m\angle ABX = m\angle XBC$
4. $m\angle XBC = 45°$
5. $m\angle ABX = 45°$
6. $m\angle ABX + m\angle XBC = m\angle ABC$
7. $45° + 45° = m\angle ABC$
8. $90° = m\angle ABC$
9. $\angle ABC$ is a right angle.

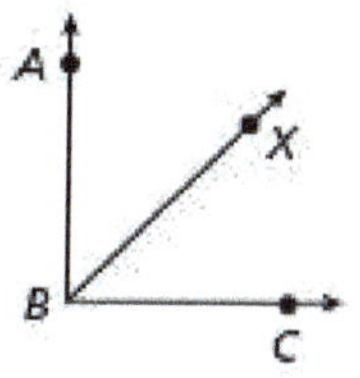

Fill in the blanks to complete each two-column proof.

Reasoning and Writing Geometric Proofs

Review 2.2 / Mathematical Proof

15. **Given:** ∠1 and∠2 are supplementary, and
∠3 and ∠4 are supplementary.
∠2 ≅ ∠3
Prove: ∠1 ≅ ∠4

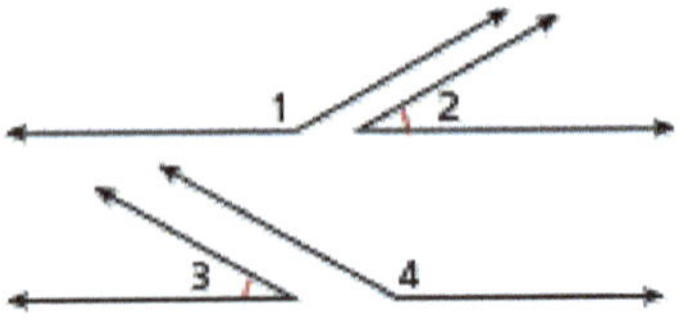

Proof:

Statements	Reasons
1. ∠1 and ∠2 are supplementary. ∠3 and ∠4 are supplementary.	1. Given
2. a. ___?___	2. Def. of supp. ∠s
3. m∠1 + m∠2 = m∠3 + m∠4	3. b. ___?___
4. ∠2 ≅ ∠3	4. Given
5. m∠2 = m∠3	5. Def. of ≅ ∠s
6. c. ___?___	6. Subtr. Prop. of = *Steps 3, 5*
7. ∠1 ≅ ∠4	7. d. ___?___

16. **Given:** ∠*BAC* is a right angle. ∠2 ≅ ∠3
Prove: ∠1 and ∠3 are complementary.

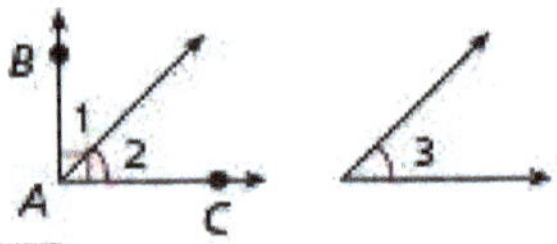

Proof:

Statements	Reasons
1. ∠*BAC* is a right angle.	1. Given
2. m∠*BAC* = 90°	2. a. ___?___
3. b. ___?___	3. ∠ Add. Post.
4. m∠1 + m∠2 = 90°	4. Subst. *Steps 2, 3*
5. ∠2 ≅ ∠3	5. Given
6. c. ___?___	6. Def. of ≅ ∠s
7. m∠1 + m∠3 = 90°	7. d. ___?___ *Steps 4, 6*
8. e. ___?___	8. Def. of comp. ∠s

Use the given plan to write a two-column proof.

17. **Given:** $\overline{BE} \cong \overline{CE}$, $\overline{DE} \cong \overline{AE}$
Prove: $\overline{AB} \cong \overline{CD}$
Plan: Use the definition of congruent segments to write the given information in terms of lengths. Then use the Segment Addition Postulate to show that $AB = CD$ and thus $\overline{AB} \cong \overline{CD}$.

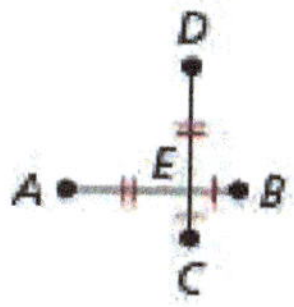

18. **Given:** $\angle 1$ and $\angle 3$ are complementary, and $\angle 2$ and $\angle 4$ are complementary. $\angle 3 \cong \angle 4$
Prove: $\angle 1 \cong \angle 2$
Plan: Since $\angle 1$ and $\angle 3$ are complementary and $\angle 2$ and $\angle 4$ are complementary, both pairs of angle measures add to 90°. Use substitution to show that the sums of both pairs are equal. Since $\angle 3 \cong \angle 4$, their measures are equal. Use the Subtraction Property of Equality and the definition of congruent angles to conclude that $\angle 1 \cong \angle 2$.

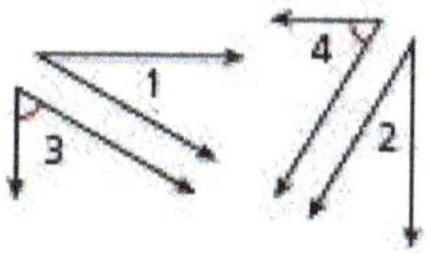

2.2.3 Flowchart and Paragraph Proofs

Reasoning and Writing Geometric Proofs

Review 2.2 / Mathematical Proof

19. Use the given flowchart proof to write a two-column proof.

Given: B is the midpoint of $\overline{AC}$.

$AD = EC$

Prove: $DB = BE$

Flowchart proof:

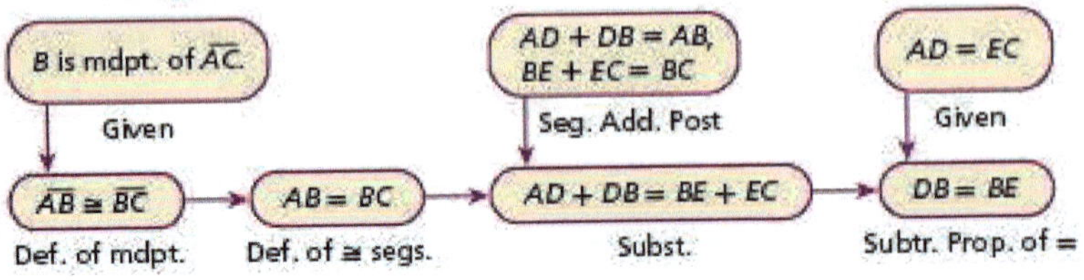

20. Use the given two-column proof to write a flowchart proof.

Given: ∠3 is a right angle.

Prove: ∠4 is a right angle.

Two-column proof:

Statements	Reasons
1. ∠3 is a right angle.	1. Given
2. m∠3 = 90°	2. Def. of rt. ∠
3. ∠3 and ∠4 are supplementary.	3. Lin. Pair Thm.
4. m∠3 + m∠4 = 180°	4. Def. of supp. ∠s
5. 90° + m∠4 = 180°	5. Subst. *Steps 2, 4*
6. m∠4 = 90°	6. Subtr. Prop. of =
7. ∠4 is a right angle.	7. Def. of rt. ∠

Reasoning and Writing Geometric Proofs

Review 2.2 / Mathematical Proof

21. Use the given paragraph proof to write a two-column proof.

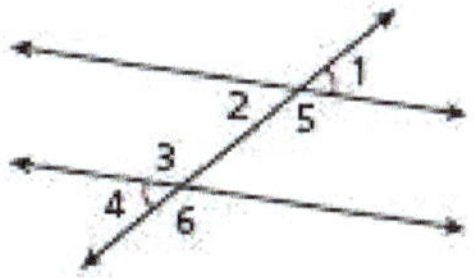

Given: ∠1 ≅ ∠4

Prove: ∠2 and ∠3 are supplementary.

Paragraph proof:

∠4 and ∠3 form a linear pair, so they are supplementary by the Linear Pair Theorem. Therefore, m∠4 + m∠3 = 180°. Also, ∠1 and ∠2 are vertical angles, so ∠1 ≅ ∠2 by the Vertical Angles Theorem. It is given that ∠1 ≅ ∠4. So by the Transitive Property of Congruence, ∠4 ≅ ∠2, and by the definition of congruent angles, m∠4 = m∠2. By substitution, m∠2 + m∠3 = 180°, so ∠2 and ∠3 are supplementary by the definition of supplementary angles.

22. Use the given two-column proof to write a paragraph proof.

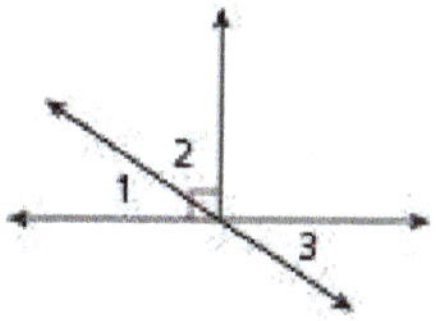

Given: ∠1 and ∠2 are complementary.

Prove: ∠2 and ∠3 are complementary.

Two-column proof:

Statements	Reasons
1. ∠1 and ∠2 are complementary.	1. Given
2. m∠1 + m∠2 = 90°	2. Def. of comp. ∠s
3. ∠1 ≅ ∠3	3. Vert. ∠s Thm.
4. m∠1 = m∠3	4. Def. of ≅ ∠s
5. m∠3 + m∠2 = 90°	5. Subst. *Steps 2, 4*
6. ∠2 and ∠3 are complementary.	6. Def. of comp. ∠s

Parallel and Perpendicular Lines
Lines with Transversals
Planes, Lines, and Angles

Example 1

Identify each of the following.

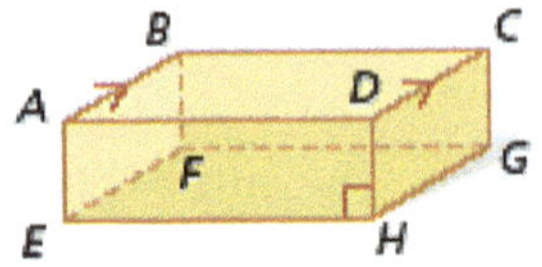

1. one pair of perpendicular segments

2. one pair of skew segments

3. one pair of parallel segments

4. one pair of parallel planes

Example 2

Give an example of each angle pair.

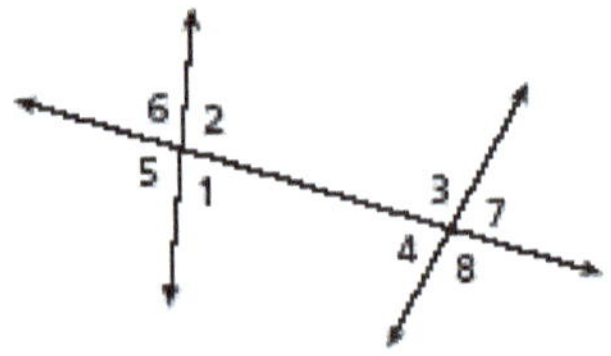

5. alternate interior angles

6. alternate exterior angles

7. corresponding angles

8. same-side interior angles

Planes, Lines, and Angles

Identify the transversal and classify each angle pair.

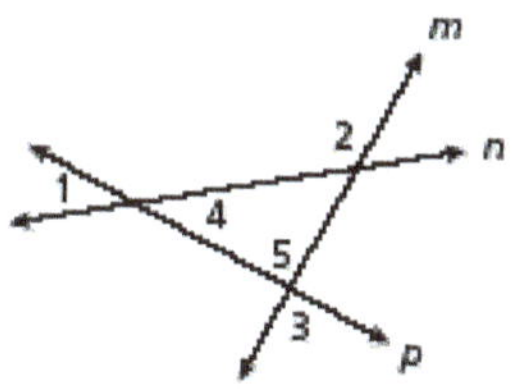

9. ∠1 and ∠2

10. ∠2 and ∠3

11. ∠2 and ∠4

12. ∠4 and ∠5

Parallel and Perpendicular Lines

Lines with Transversals

Angles, Parallel Lines, and Transversals

Example 1

Find each angle measure.

1. m∠*JKL*

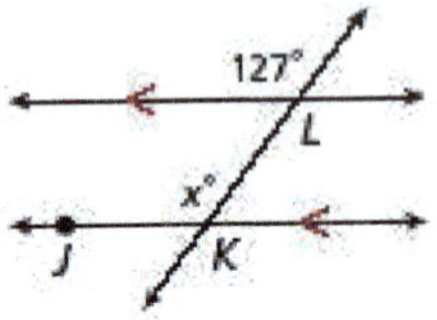

2. m∠*BEF*

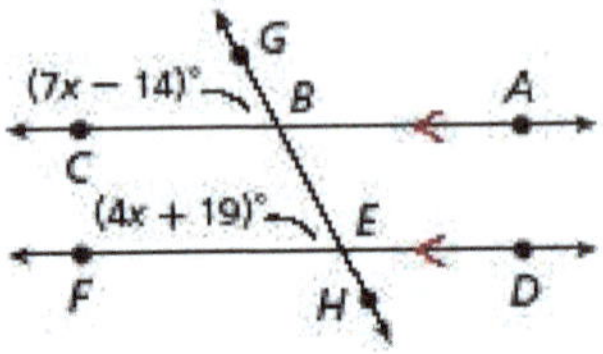

3. m∠1

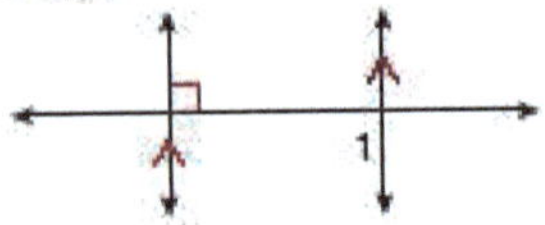

Example 2

4. **Safety** The railing of a wheelchair ramp is parallel to the ramp. Find *x* and *y* in the diagram.

Hint: Use the Alternate Interior Angles Theorem to write a system of equations and then solve.

Example 3

Find each angle measure.

5. m∠*CBY*

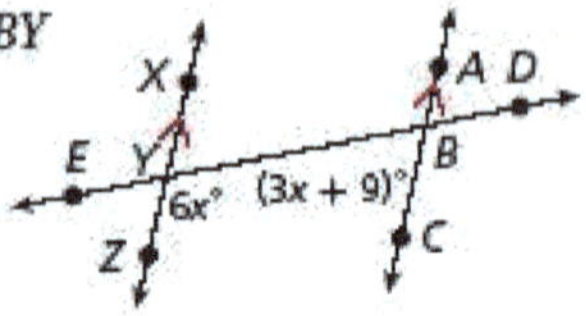

Parallel and Perpendicular Lines

Lines with Transversals

Proving that Lines are Parallel

Example 1

Use the Converse of the Corresponding Angles Postulate and the given information to show that $p \parallel q$.

1. $\angle 4 \cong \angle 5$

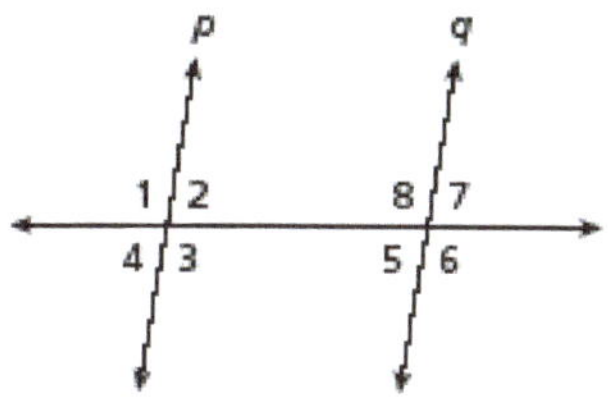

2. $m\angle 1 = (4x + 16)^\circ$, $m\angle 8 = (5x - 12)^\circ$, $x = 28$

3. $m\angle 4 = (6x - 19)^\circ$, $m\angle 5 = (3x + 14)^\circ$, $x = 11$

Use the theorems and given information to show that $r \parallel s$.

4. $\angle 1 \cong \angle 5$

5. $\angle 3 \cong \angle 7$

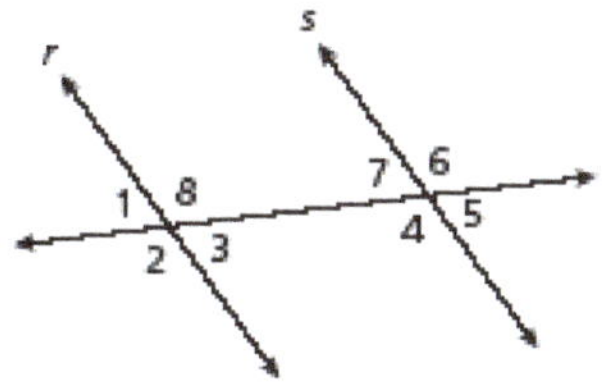

6. $m\angle 4 = (13x - 4)^\circ$, $m\angle 8 = (9x + 16)^\circ$, $x = 5$

7. $m\angle 2 = (25x + 7)^\circ$, $m\angle 6 = (24x + 12)^\circ$, $x = 5$

Example 2

Use the theorems and given information to show that $r \parallel s$.

8. $m\angle 3 + m\angle 4 = 180^\circ$

9. $m\angle 8 = (17x + 37)^\circ$, $m\angle 7 = (9x - 13)^\circ$, $x = 6$

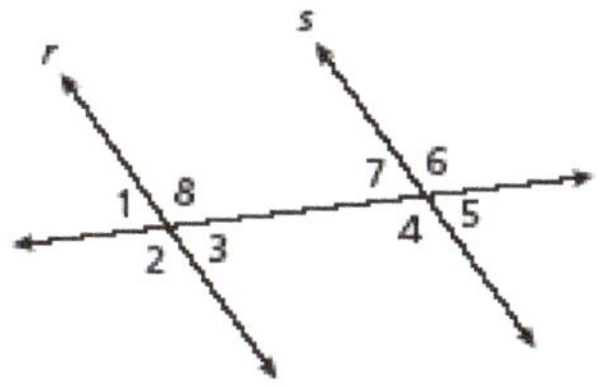

Example 3

10. Complete the following two-column proof.

Given: $\angle 1 \cong \angle 2$, $\angle 3 \cong \angle 1$
Prove: $XY \parallel WV$

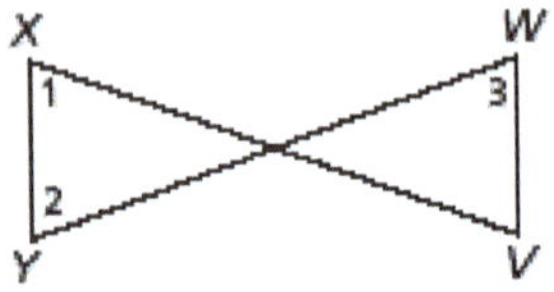

Proof:

Statements	Reasons
1. $\angle 1 \cong \angle 2$, $\angle 3 \cong \angle 1$	1. Given
2. $\angle 2 \cong \angle 3$	2. a. ___?___
3. b. ___?___	3. c. ___?___

11. **Architecture** In the fire escape, $m\angle 1 = (17x + 9)^\circ$, $m\angle 2 = (14x + 18)^\circ$, and $x = 3$. Show that the two landings are parallel.

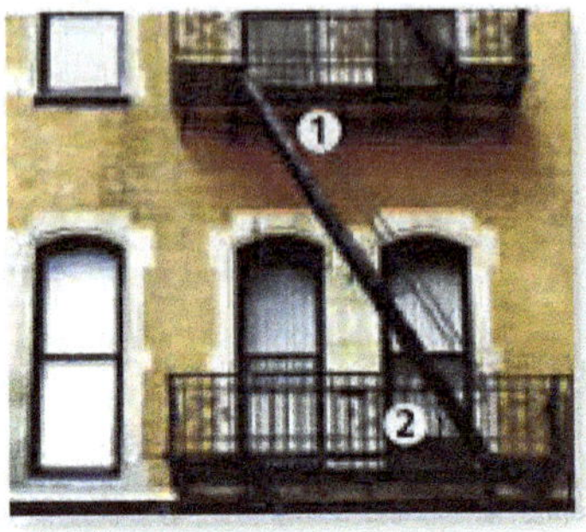

Properties of Perpendicular Lines

Example 1

1. Name the shortest segment from point E to $\overleftrightarrow{AD}$.

2. Write and solve an inequality for x.

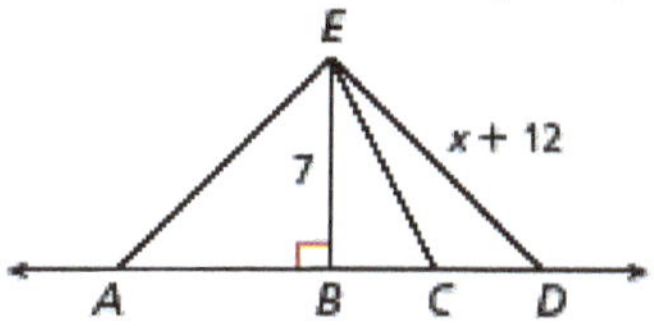

Example 2

3. Complete the two-column proof.
Given: $\angle ABC \cong \angle CBE$, $\overleftrightarrow{DE} \perp \overleftrightarrow{AF}$
Prove: $\overleftrightarrow{CB} \parallel \overleftrightarrow{DE}$
Proof:

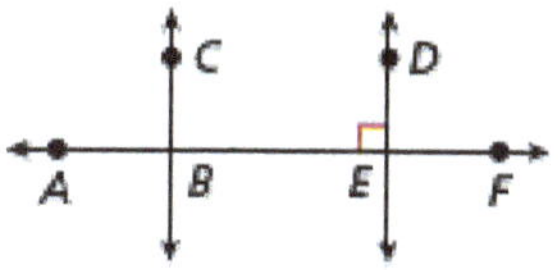

Statements	Reasons
1. $\angle ABC \cong \angle CBE$	1. Given
2. $\overleftrightarrow{CB} \perp \overleftrightarrow{AF}$	2. a. ___?___
3. b. ___?___	3. Given
4. $\overleftrightarrow{CB} \parallel \overleftrightarrow{DE}$	4. c. ___?___

Properties of Perpendicular Lines

4. **Sports** The center line in a tennis court is perpendicular to both service lines. Explain why the service lines must be parallel to each other.

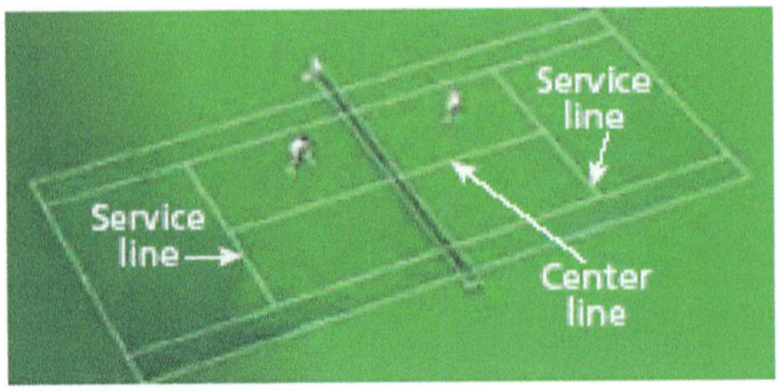

Parallel and Perpendicular Lines

Review 3.1 / Lines with Transversals

3.1.1 Planes, Lines, and Angles

Identify each of the following.

1. one pair of parallel segments

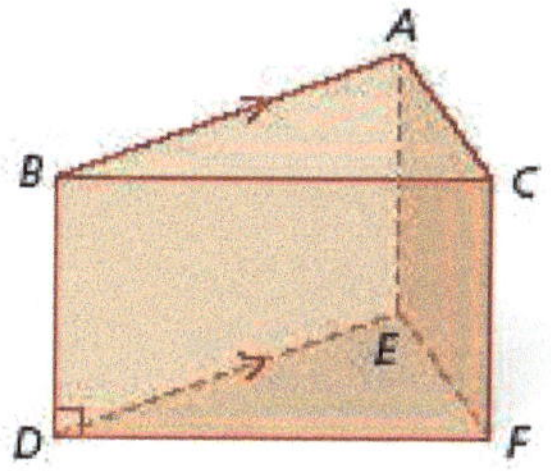

2. one pair of skew segments

3. one pair of perpendicular segments

4. one pair of parallel planes

Give an example of each angle pair.

5. same-side interior angles

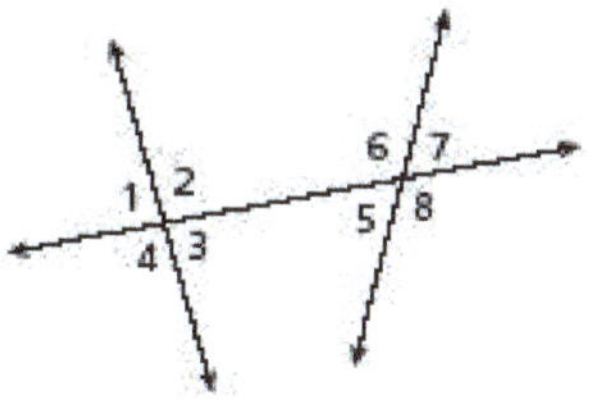

6. alternate exterior angles

7. corresponding angles

8. alternate interior angles

Identify the transversal and classify each angle pair.

9. $\angle 2$ and $\angle 3$

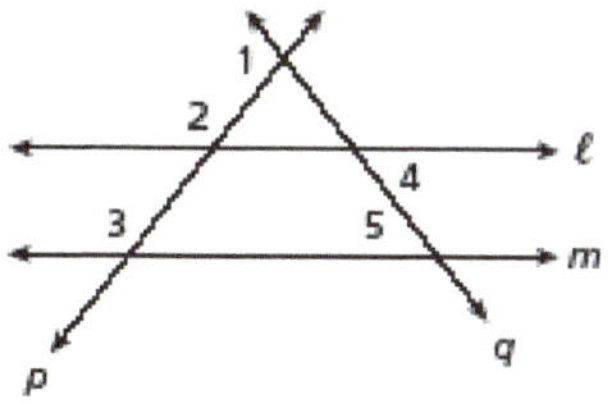

10. $\angle 4$ and $\angle 5$

11. $\angle 2$ and $\angle 4$

12. $\angle 1$ and $\angle 2$

3.1.2 Angles, Parallel Lines, and Transversals

Find each angle measure.

13. $m\angle KLM$

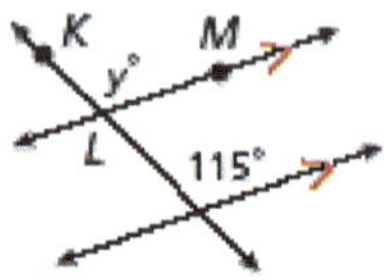

14. $m\angle VYX$

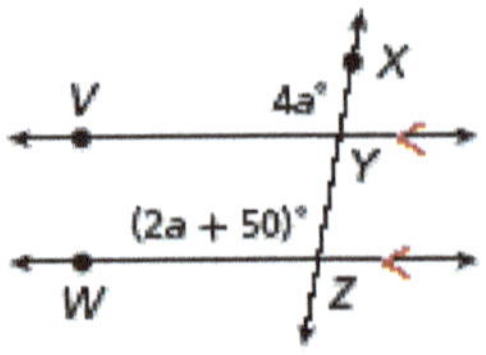

15. $m\angle ABC$

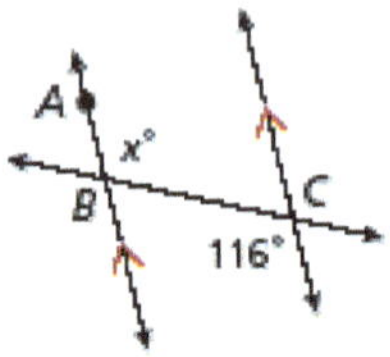

16. $m\angle EFG$

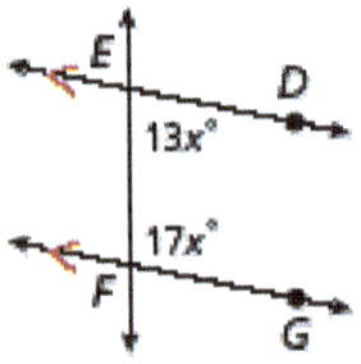

17. $m\angle PQR$

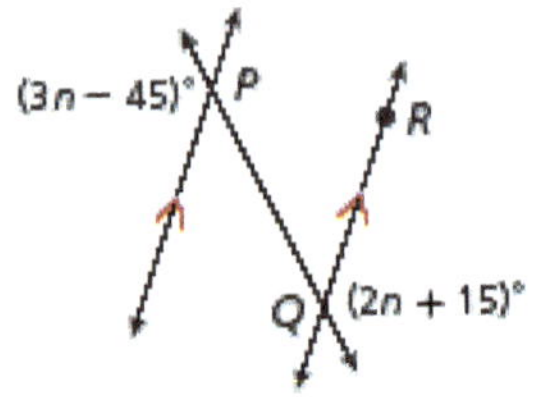

18. $m\angle STU$

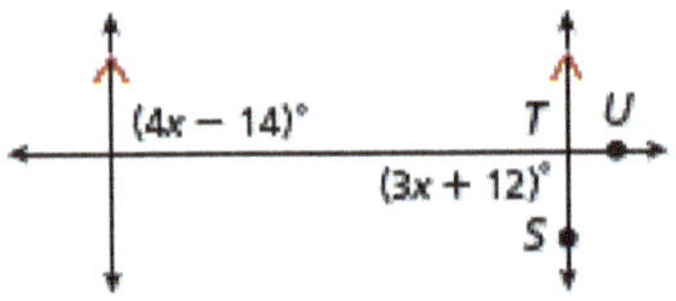

19. **Parking** In the parking lot shown, the lines that mark the width of each space are parallel.
$m\angle 1 = (2x - 3y)°$
$m\angle 2 = (x + 3y)°$
Find x and y.

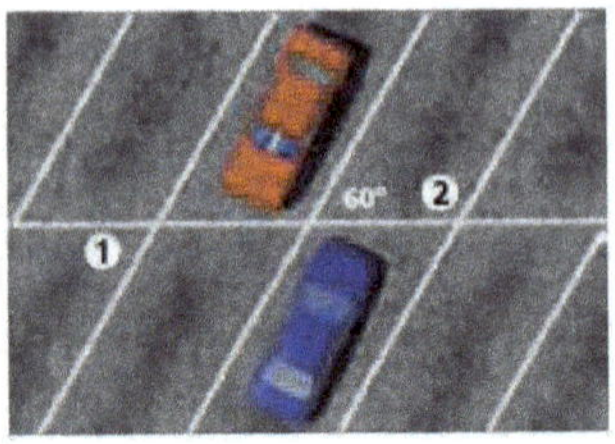

3.1.3 Proving that Lines are Parallel

Use the Converse of the Corresponding Angles Postulate and the given information to show that $\ell \parallel m$.

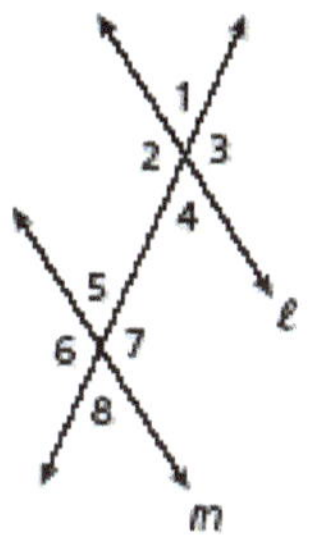

20. $\angle 3 \cong 7$

21. $m\angle 4 = 54°$, $m\angle 8 = (7x + 5)°$, $x = 7$

22. $m\angle 2 = (8x + 4)°$, $m\angle 6 = (11x - 41)°$, $x = 15$

23. $m\angle 1 = (3x + 19)°$, $m\angle 5 = (4x + 7)°$, $x = 12$

Use the theorems and given information to show that $n \parallel p$.

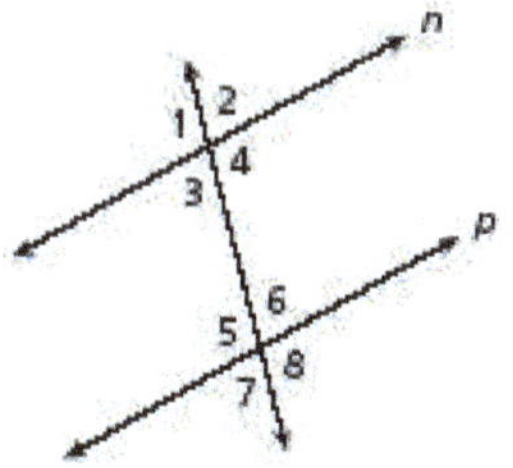

24. $\angle 3 \cong \angle 6$

25. $\angle 2 \cong \angle 7$

26. $m\angle 4 + m\angle 6 = 180°$

27. $m\angle 1 = (8x - 7)°$, $m\angle 8 = (6x + 21)°$, $x = 14$

28. $m\angle 4 = (4x + 3)°$, $m\angle 5 = (5x - 22)°$, $x = 25$

29. $m\angle 3 = (2x + 15)°$, $m\angle 5 = (3x + 15)°$, $x = 30$

30. Complete the following two-column proof.

Given: $\overline{AB} \parallel \overline{CD}$, $\angle 1 \cong \angle 2$, $\angle 3 \cong \angle 4$
Prove: $\overline{BC} \parallel \overline{DE}$

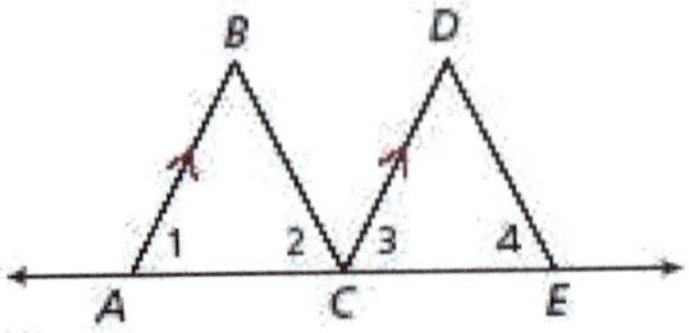

Proof:

Statements	Reasons
1. $\overline{AB} \parallel \overline{CD}$	1. Given
2. $\angle 1 \cong \angle 3$	2. a. ___?___
3. $\angle 1 \cong \angle 2$, $\angle 3 \cong \angle 4$	3. b. ___?___
4. $\angle 2 \cong \angle 4$	4. c. ___?___
5. d. ___?___	5. e. ___?___

31. **Art** Edmund Dulac used perspective when drawing the floor titles in this illustration for *The Wind's Tale* by Hans Christian Andersen. Show that $DJ \parallel EK$ if $m\angle 1 = (3x + 2)^\circ$, $m\angle 2 = (5x - 10)^\circ$, and $x = 6$.

3.1.4 Properties of Perpendicular Lines

32. Name the shortest segment from point W to $\overline{XZ}$.

33. Write and solve an inequality for x.

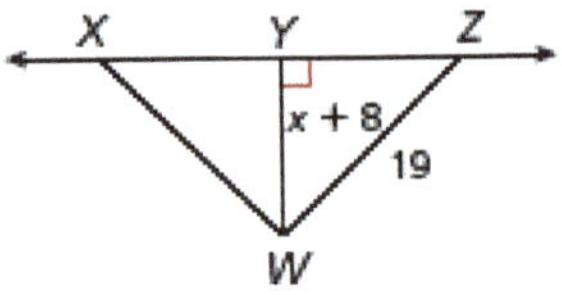

34. Complete the two-column proof below.

Given: $\overleftrightarrow{AB} \perp \overleftrightarrow{BC}$, $m\angle 1 + m\angle 2 = 180^\circ$
Prove: $\overleftrightarrow{BC} \perp \overleftrightarrow{CD}$

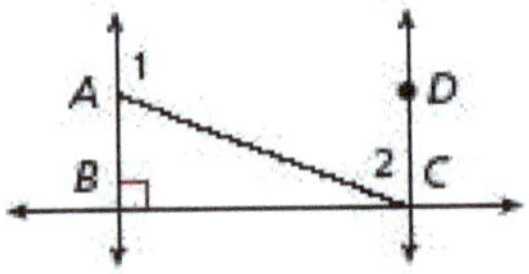

Proof:

Statements	Reasons
1. $\overleftrightarrow{AB} \perp \overleftrightarrow{BC}$	1. Given
2. $m\angle 1 + m\angle 2 = 180^\circ$	2. a. ___?___
3. $\angle 1$ and $\angle 2$ are supplementary.	3. Def. of supplementary
4. b. ___?___	4. Converse of the Same-Side Interior Angles Theorem
5. $\overleftrightarrow{BC} \perp \overleftrightarrow{CD}$	5. c. ___?___

35. **Music** The *frets* on a guitar are all perpendicular to one of the strings. Explain why the frets must be parallel to each other.

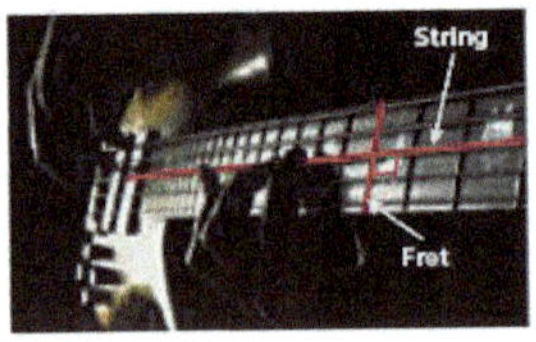

Parallel and Perpendicular Lines

Slope and the Equation of a Line

Finding the Slope Given Two Points

Example 1

Use the slope formula to determine the slope of each line.

1. $\overleftrightarrow{MN}$

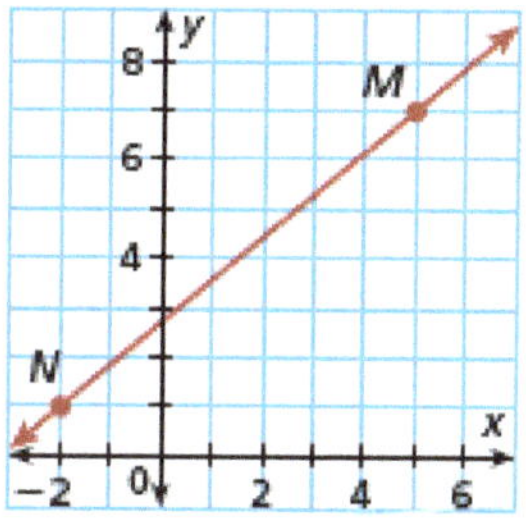

2. $\overleftrightarrow{CD}$

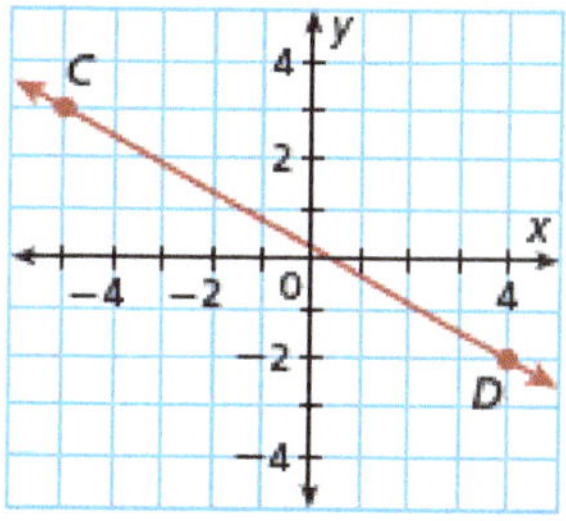

3. $\overleftrightarrow{AB}$

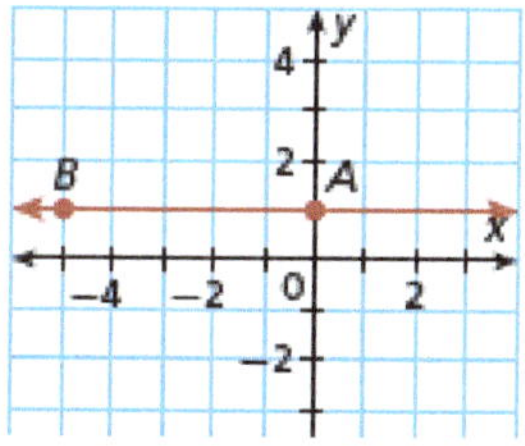

4. $\overleftrightarrow{ST}$

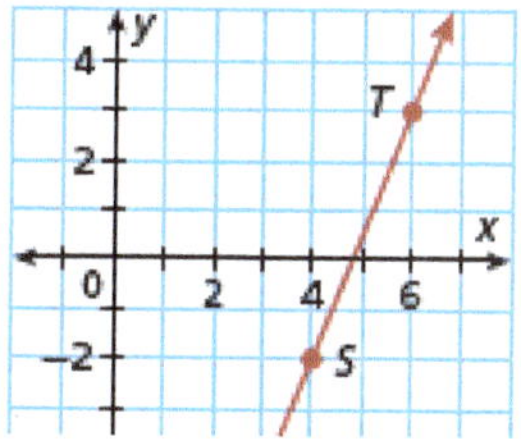

Example 2

5. **Biology** A migrating bird flying at a constant speed travels 80 miles by 8:00 A.M. and 200 miles by 11:00 A.M. Graph the line that represents the bird's distance traveled. Find and interpret the slope of the line.

Example 3

Graph each pair of lines. Use slopes to determine whether the lines are parallel, perpendicular, or neither.

6. $\overleftrightarrow{HJ}$ and $\overleftrightarrow{KM}$ for $H(3, 2)$, $J(4, 1)$, $K(-2, -4)$, and $M(-1, -5)$

7. $\overleftrightarrow{LM}$ and $\overleftrightarrow{NP}$ for $L(-2, 2)$, $M(2, 5)$, $N(0, 2)$, and $P(3, -2)$

8. $\overleftrightarrow{QR}$ and $\overleftrightarrow{ST}$ for $Q(6, 1)$, $R(-2, 4)$, $S(5, 3)$, and $T(-3, -1)$

Example 1

Graph each line given the slope and *y*-intercept.

1. slope $= \frac{1}{3}$, y-intercept $= -3$
2. slope $= 0.5$, y-intercept $= 3.5$
3. slope $= 5$, y-intercept $= -1$
4. slope $= -2$, y-intercept $= 2$

Example 2

Write the equation that describes each line in slope-intercept form.

5. slope $= 8$, y-intercept $= 2$
6. slope $= \frac{1}{2}$, y-intercept $= -6$
7. slope $= 0$, y-intercept $= -3$
8. slope $= 5$, the point $(2, 7)$ is on the line

Example 3

Write each equation in slope-intercept form. Then graph the line described by the equation.

9. $y = \frac{2}{5}x - 6$
10. $3x - y = -1$
11. $2x + y = 4$

Example 4

12. Helen is in a bicycle race. She has already biked 10 miles and is now biking at a rate of 18 miles per hour. Her distance as a function of time is shown in the graph.

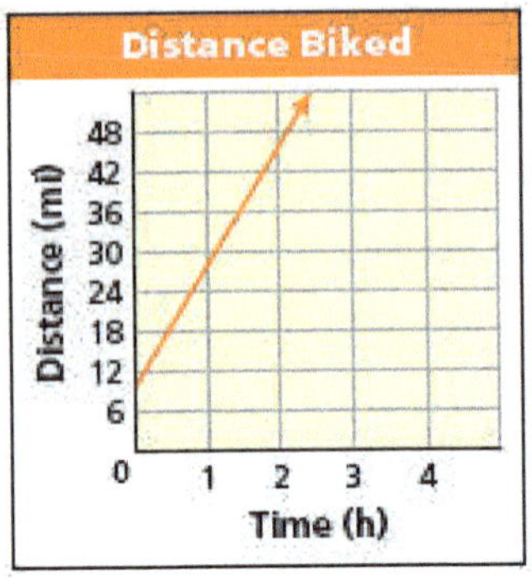

a. Write an equation that represents the distance Helen has biked as a function of time.

b. Identify the slope and y-intercept and describe their meanings.

c. How far will Helen have biked after 2 hours?

Point-Slope Form

Example 1

Graph the line with the given slope that contains the given point.

1. slope = 1; $(1, 0)$
2. slope = -1; $(3, 1)$
3. slope = -2; $(-4, -2)$

Example 2

Graph the line described by each equation.

4. $y - 1 = -(x - 3)$
5. $y + 2 = -2(x + 4)$
6. $y + 1 = -\frac{1}{2}(x + 4)$

Example 3

Write the equation that describes each line in slope-intercept form.

7. slope = $-\frac{1}{3}$, $(-3, 8)$ is on the line.
8. slope = 2, $(1, 1)$ is on the line.
9. $(-2, 2)$ and $(2, -2)$ are on the line.
10. $(1, 1)$ and $(-5, 3)$ are on the line.
11. x-intercept = 8, y-intercept = 4
12. x-intercept = -2, y-intercept = 3

Example 4

Each pair of points is on a line. Find the intercepts.

13. $(5, 2)$ and $(7, 4)$ **14.** $(-1, 5)$ and $(-3, -5)$ **15.** $(2, 9)$ and $(-4, -9)$

Example 5

16. **Measurement** An oil tank is being filled at a constant rate. The depth of the oil is a function of the number of minutes the tank has been filling, as shown in the table. Write an equation in slope-intercept form that represents this linear function. Then find the depth of the oil after one-half hour.

Time (min)	Depth (ft)
0	3
10	5
15	6

Parallel and Perpendicular Lines

Slope and the Equation of a Line

Slopes of Parallel and Perpendicular Lines

Example 1

Identify which lines are parallel.

1. $y = 6$; $y = 6x + 5$; $y = 6x - 7$; $y = -8$

2. $y = \frac{3}{4}x - 1$; $y = -2x$; $y - 3 = \frac{3}{4}(x - 5)$; $y - 4 = -2(x + 2)$

Example 2

3. **Geometry** Show that *ABCD* is a trapezoid. (*Hint:* In a trapezoid, exactly one pair of opposite sides is parallel.)

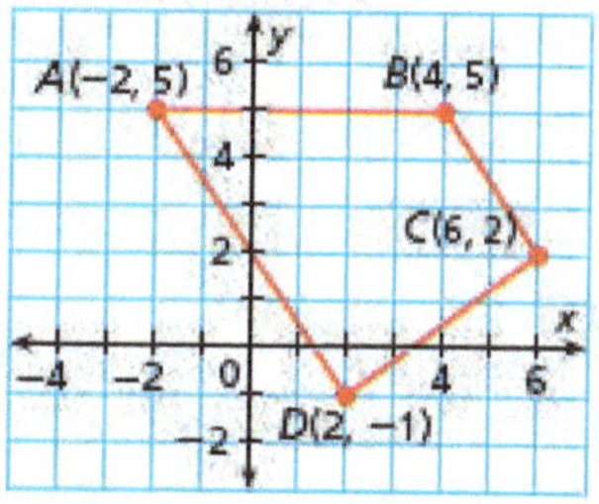

Example 3

Identify which lines are perpendicular.

4. $y = \frac{2}{3}x - 4$; $y = -\frac{3}{2}x + 2$; $y = -1$; $x = 3$

5. $y = -\frac{3}{7}x - 4$; $y - 4 = -7(x + 2)$;
$y - 1 = \frac{1}{7}(x - 4)$; $y - 7 = \frac{7}{3}(x - 3)$

Example 4

6. **Geometry** Show that *PQRS* is a rectangle. (*Hint:* In a rectangle, all four angles are right angles.)

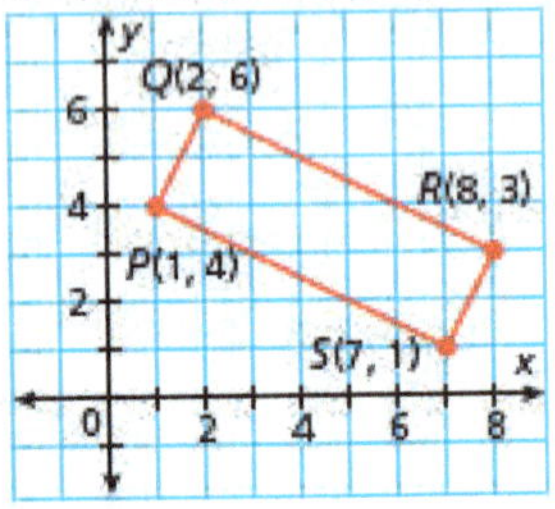

Example 5

7. Write an equation in slope-intercept form for the line that passes through (5, 0) and is perpendicular to the line described by $y = -\frac{5}{2}x + 6$.

Parallel and Perpendicular Lines

Review 3.2 / Slope and the Equation of a Line

3.2.1 Finding the Slope Given Two Points

Use the slope formula to determine the slope of each line.

1. $\overleftrightarrow{AB}$

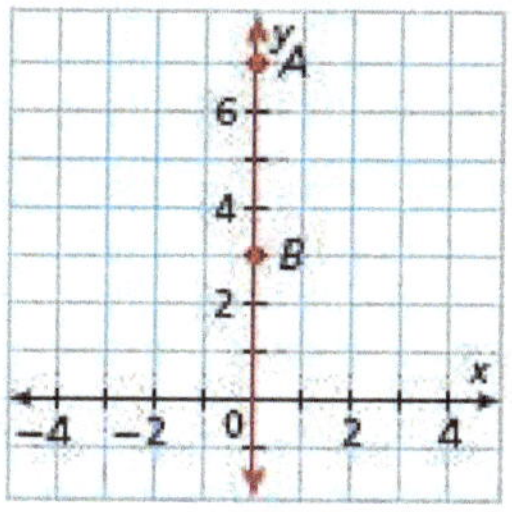

2. $\overleftrightarrow{CD}$

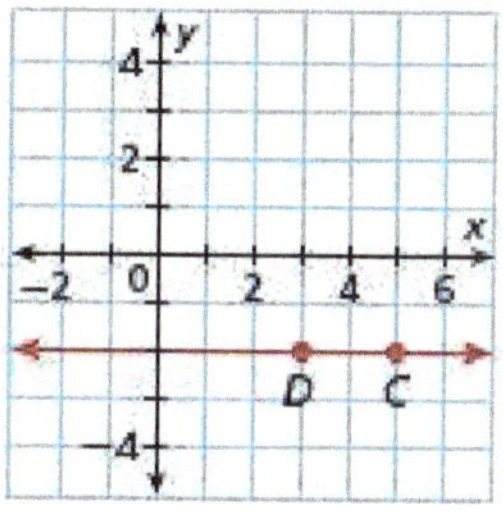

3. $\overleftrightarrow{EF}$

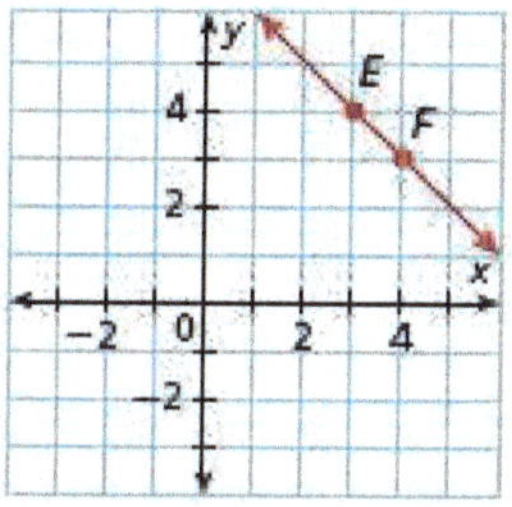

4. $\overleftrightarrow{GH}$

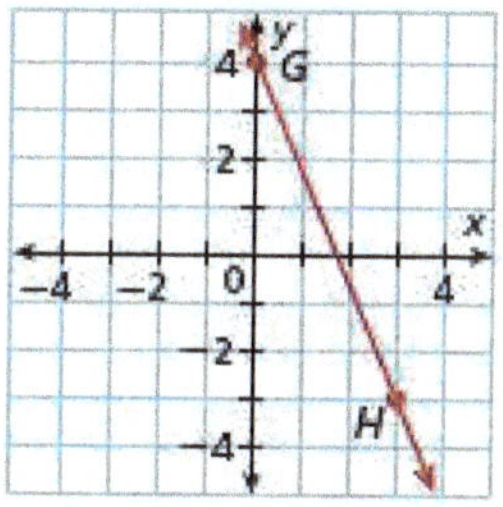

5. **Aviation** A pilot traveling at a constant speed flies 100 miles by 2:30 P.M. and 475 miles by 5:00 P.M. Graph the line that represents the pilot's distance flown. Find and interpret the slope of the line.

3.2.2 Slope-Intercept Form

Write the equation of each line in slope-intercept form.

6. slope $= -\frac{2}{3}$, y-intercept $= 2$

7. slope $= 0$, $(6, -8)$ is on the line

8. line through $(5, 2)$ and $(-2, 2)$

Parallel and Perpendicular Lines

Review 3.2 / Slope and the Equation of a Line

Graph each line.

9. $y = \frac{1}{2}x - 2$

10. $y = 2$

3.2.3 Point-Slope Form

Write the equation of each line in point-slope form.

11. the line through $(0, -2)$ and $(4, 6)$

12. the line through $(6, -4)$ with slope $\frac{2}{3}$

13. **Graph.** $y - 7 = x + 4$

3.2.4 Slopes of Parallel and Perpendicular Lines

Graph each pair of lines. Use slopes to determine whether the lines are parallel, perpendicular, or neither.

14. $\overleftrightarrow{AB}$ and $\overleftrightarrow{CD}$ for $A(2, -1)$, $B(7, 2)$, $C(2, -3)$, and $D(-3, -6)$

Parallel and Perpendicular Lines

Review 3.2 / Slope and the Equation of a Line

15. $\overleftrightarrow{XY}$ and $\overleftrightarrow{ZW}$ for $X(-2, 5)$, $Y(6, -2)$, $Z(-3, 6)$, and $W(4, 0)$

16. $\overleftrightarrow{JK}$ and $\overleftrightarrow{JL}$ for $J(-4, -2)$, $K(4, -2)$, and $L(-4, 6)$

17. **Identify which lines are parallel.**
$y = -3x + 2$; $y = \frac{1}{2}x - 1$; $-x + 2y = 17$; $3x + y = 27$

18. **Identify which lines are perpendicular.**
$y - 9 = 3(x + 1)$; $y = -\frac{1}{3}x + 5$; $y = 0$; $x = 6$

19. **Geometry** Show that $LMNP$ is a parallelogram.

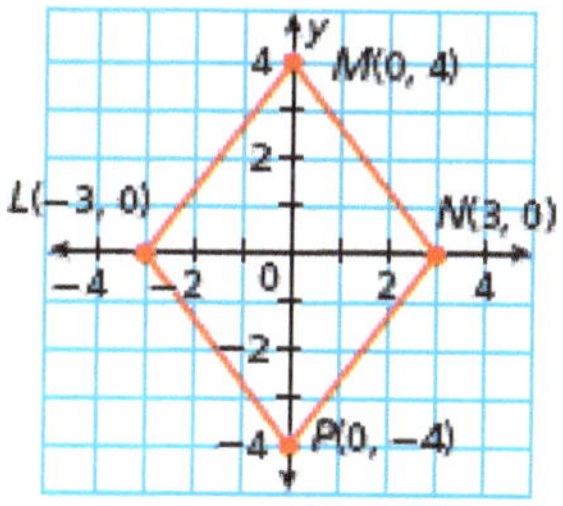

20. **Geometry** Show that ABC is a right triangle.

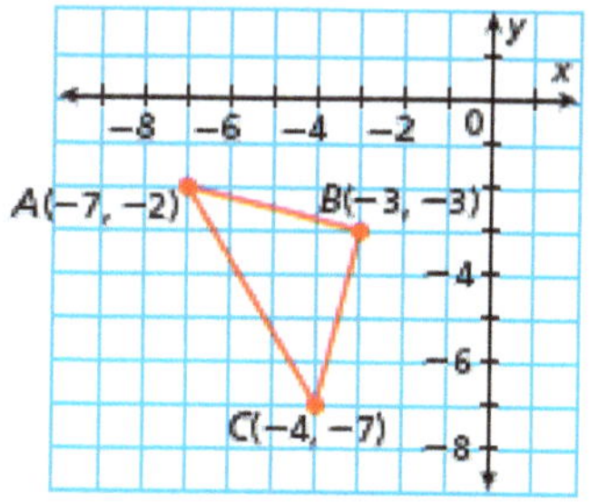

Example 1

Classify each triangle by its angle measures.

1. $\triangle DBC$
2. $\triangle ABD$
3. $\triangle ADC$

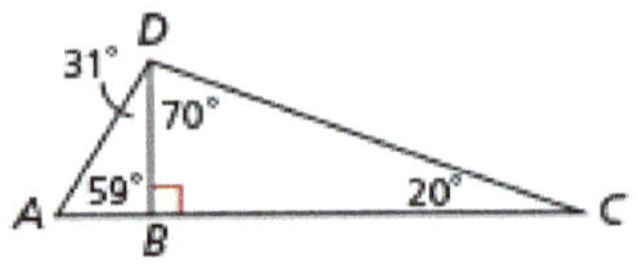

Example 2

Classify each triangle by its side lengths.

4. $\triangle EGH$
5. $\triangle EFH$
6. $\triangle HFG$

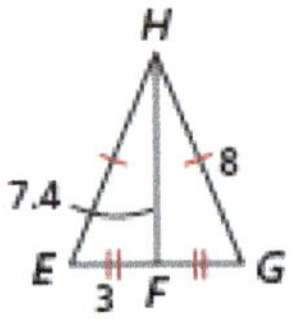

Example 3

Multi-Step Find the side lengths of each triangle.

7.

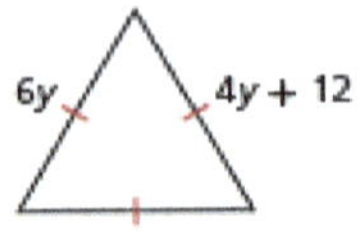

8.

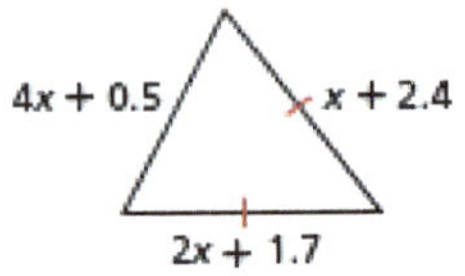

Example 4

9. **Crafts** A jeweler creates triangular earrings by bending pieces of silver wire. Each earring is an isosceles triangle with the dimensions shown. How many earrings can be made from a piece of wire that is 50 cm long?

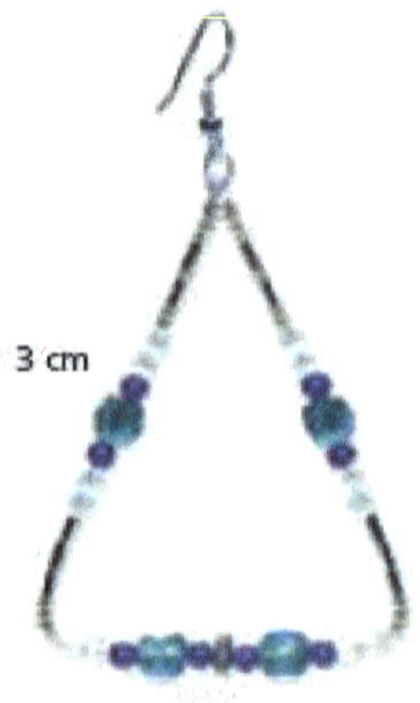

Example 1

Astronomy **Use the following information for Exercises 1 and 2.**

An *asterism* is a group of stars that is easier to recognize than a constellation. One popular asterism is the Summer Triangle, which is composed of the stars Deneb, Altair, and Vega.

1. What is the value of y?

2. What is the measure of each angle in the Summer Triangle?

Example 2

The measure of one of the acute angles in a right triangle is given. What is the measure of the other acute angle?

3. 20.8°

4. $y°$

5. $24\frac{2}{3}°$

Example 3

Find each angle measure.

6. $m\angle M$

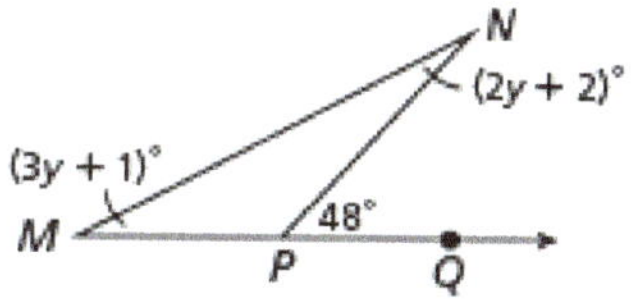

7. $m\angle L$

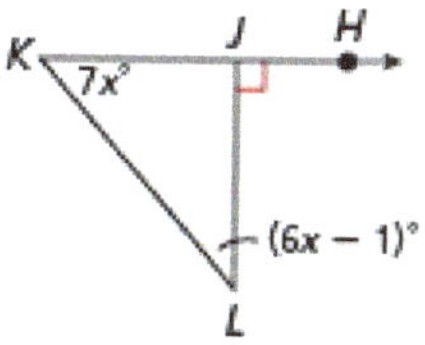

8. In $\triangle ABC$, $m\angle A = 65°$, and the measure of an exterior angle at C is $117°$.
Find $m\angle B$ and the $m\angle BCA$.

Example 4

Find each angle measure.

9. $m\angle C$ and $m\angle F$

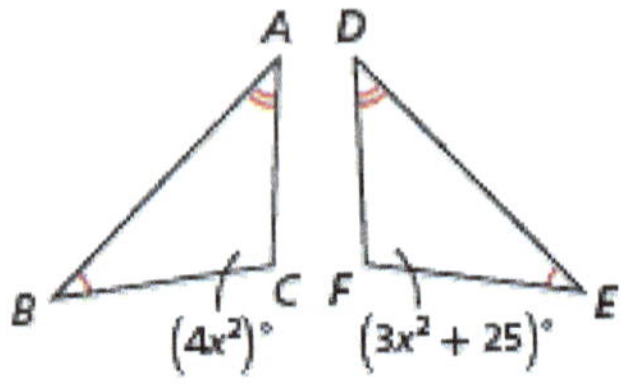

10. $m\angle S$ and $m\angle U$

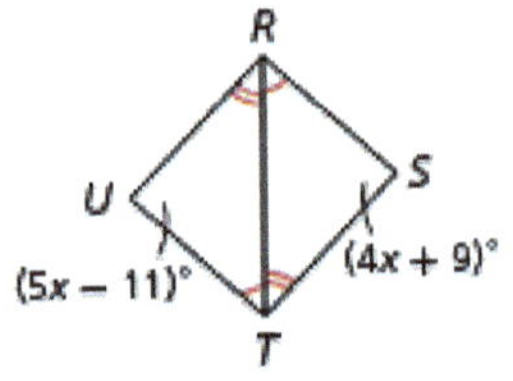

11. For $\triangle ABC$ and $\triangle XYZ$, $m\angle A = m\angle X$ and $m\angle B = m\angle Y$.
Find the measures of $\angle C$ and $\angle Z$ if $m\angle C = 4x + 7$ and $m\angle Z = 3(x + 5)$.

Triangle Congruence

Triangles and Congruence

Congruent Triangles

Example 1

Given: $\triangle RST \cong \triangle LMN$. **Identify the congruent corresponding parts.**

1. $\overline{RS} \cong$ ___?___
2. $\overline{LN} \cong$ ___?___
3. $\angle S \cong$ ___?___
4. $\overline{TS} \cong$ ___?___
5. $\angle L \cong$ ___?___
6. $\angle N \cong$ ___?___

Example 2

Given: $\triangle FGH \cong \triangle JKL$. **Find each value.**

7. KL
8. x

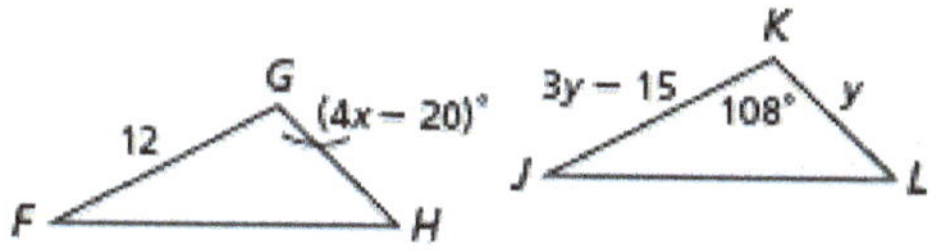

Example 3

9. **Given:** E is the midpoint of $\overline{AC}$ and $\overline{BD}$. $\overline{AB} \cong \overline{CD}$, $\overline{AB} \parallel \overline{CD}$

Prove: $\triangle ABE \cong \triangle CDE$

Proof:

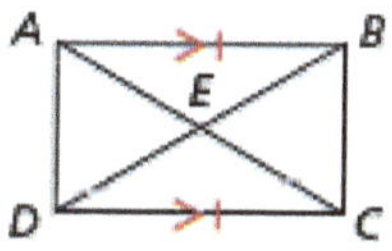

Statements	Reasons
1. $\overline{AB} \parallel \overline{CD}$	1. a. ___?___
2. $\angle ABE \cong \angle CDE$, $\angle BAE \cong \angle DCE$	2. b. ___?___
3. $\overline{AB} \cong \overline{CD}$	3. c. ___?___
4. E is the mdpt. of $\overline{AC}$ and $\overline{BD}$.	4. d. ___?___
5. e. ___?___	5. Def. of mdpt.
6. $\angle AEB \cong \angle CED$	6. f. ___?___
7. $\triangle ABE \cong \triangle CDE$	7. g. ___?___

Example 4

10. **Engineering** The McDonald Observatory has four research telescopes and is a leading center for astronomical study. Prove that the triangles that make up the observatory dome are congruent.

Given: $\overline{SU} \cong \overline{ST} \cong \overline{SR}$, $\overline{TU} \cong \overline{TR}$, $\angle UST \cong \angle RST$, and $\angle U \cong \angle R$

Prove: $\triangle RTS \cong \triangle UTS$

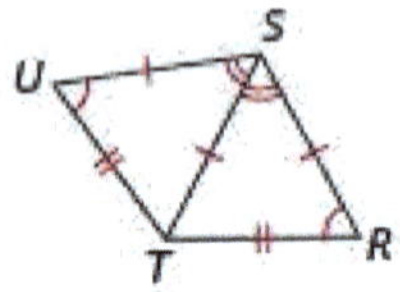

4.1.1 Classifying Triangles

Classify each triangle by its angle measures.

1. $\triangle BEA$

2. $\triangle DBC$

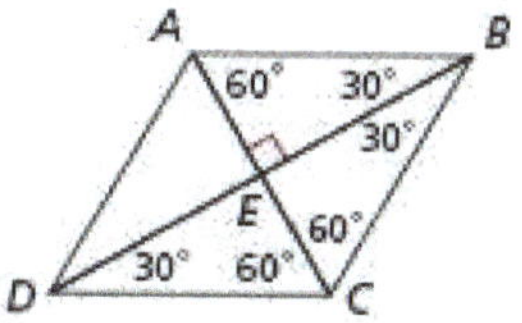

3. $\triangle ABC$

Classify each triangle by its side lengths.

4. $\triangle PST$

5. $\triangle RSP$

6. $\triangle RPT$

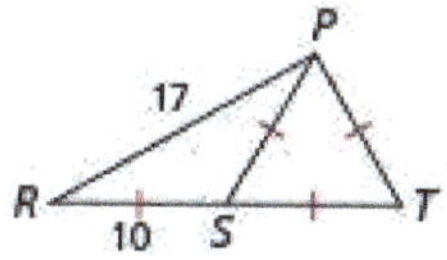

Multi-Step **Find the side lengths of each triangle.**

7. 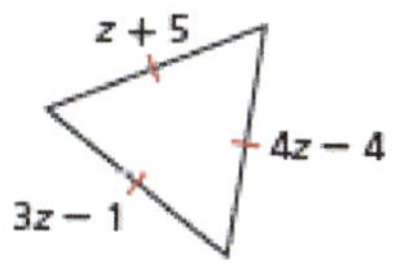

8. 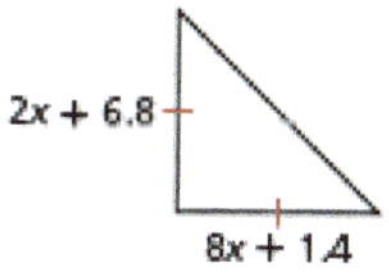

9. Draw a triangle large enough to measure. Label the vertices X, Y, and Z.
   a. Name the three sides and three angles of the triangle.

b. Use a ruler and protractor to classify the triangle by its side lengths and angle measures.

Carpentry **Use the following information for Exercises 10 and 11.** A manufacturer makes trusses, or triangular supports, for the roofs of houses. Each truss is the shape of an isosceles triangle in which $\overline{PQ} \cong \overline{PR}$. The length of the base $\overline{QR}$ is $\frac{4}{3}$ the length of each of the congruent sides.

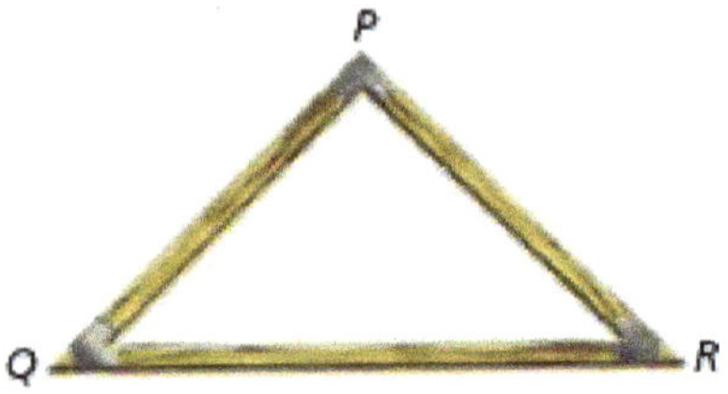

10. The perimeter of each truss is 60 ft. Find each side length.

11. How many trusses can the manufacturer make from 150 feet of lumber?

4.1.2 Angle Relationships in Triangles

12. **Navigation** A sailor on ship A measures the angle between ship B and the pier and finds that it is 39°. A sailor on ship B measures the angle between ship A and the pier and finds that it is 57°. What is the measure of the angle between ships A and B?

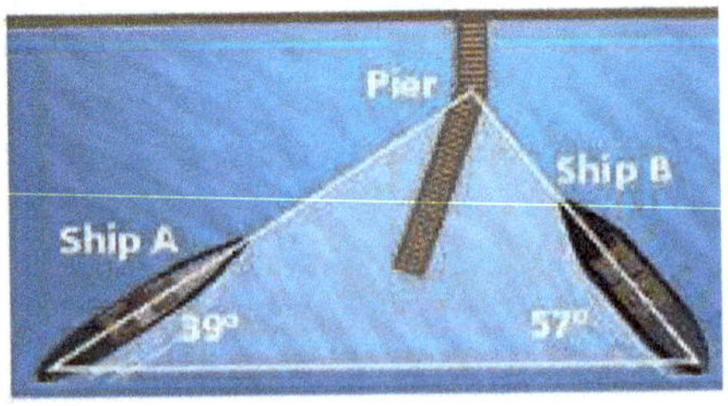

The measure of one of the acute angles in a right triangle is given. What is the measure of the other acute angle?

13. $76\frac{1}{4}^\circ$

14. $2x^\circ$

15. 56.8°

Find each angle measure.

16. $m\angle XYZ$

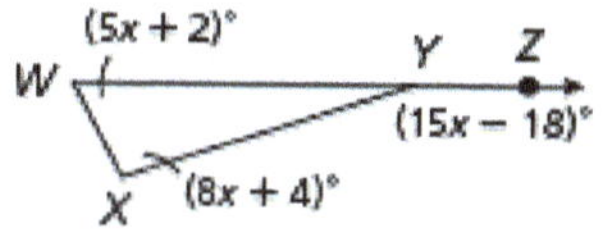

17. $m\angle C$

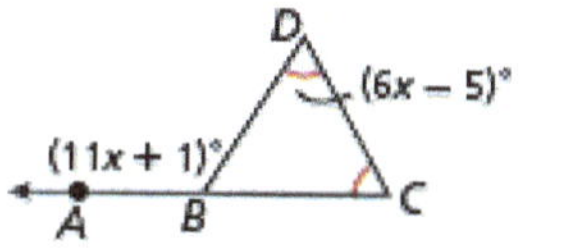

18. $m\angle N$ and $m\angle P$

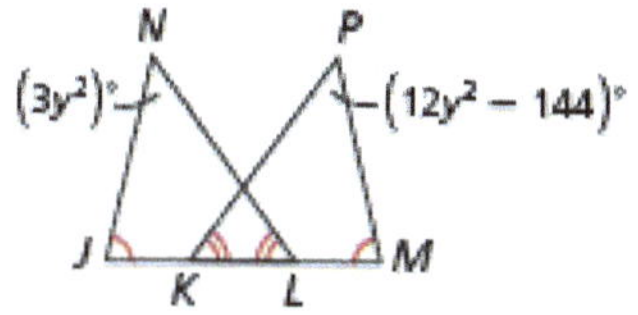

19. $m\angle Q$ and $m\angle S$

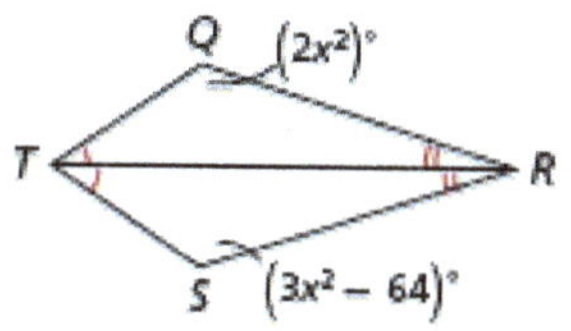

4.1.3 Congruent Triangles

Given: Polygon $CDEF \cong$ polygon $KLMN$. Identify the congruent corresponding parts.

20. $\overline{DE} \cong$ ___?___

21. $\overline{KN} \cong$ ___?___

Triangle Congruence

Review 4.1 / Triangles and Congruence

22. $\angle F \cong$ __?__

23. $\angle L \cong$ __?__

Given: $\triangle ABD \cong \triangle CBD$. **Find each value.**

24. $m\angle C$

25. y

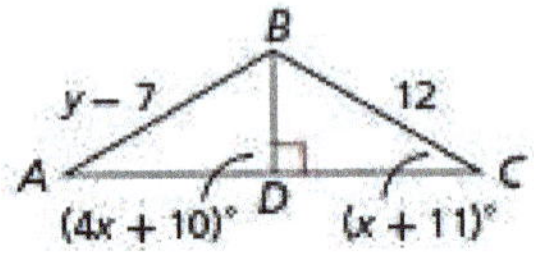

26. **Given:** $\overline{MP}$ bisects $\angle NMR$. P is the midpoint of $\overline{NR}$. $\overline{MN} \cong \overline{MR}$, $\angle N \cong \angle R$

Prove: $\triangle MNP \cong \triangle MRP$

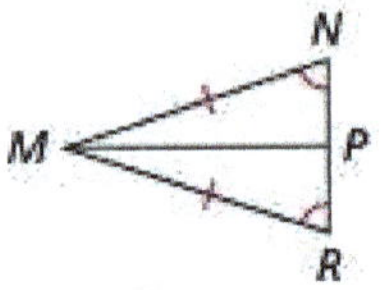

Proof:

Statements	Reasons
1. $\angle N \cong \angle R$	1. a. __?__
2. $\overline{MP}$ bisects $\angle NMR$.	2. b. __?__
3. c. __?__	3. Def. of $\angle$ bisector
4. d. __?__	4. Third $\triangle$ Thm.
5. P is the mdpt. of $\overline{NR}$.	5. e. __?__
6. f. __?__	6. Def. of mdpt.
7. $\overline{MN} \cong \overline{MR}$	7. g. __?__
8. $\overline{MP} \cong \overline{MP}$	8. h. __?__
9. $\triangle MNP \cong \triangle MRP$	9. Def. of $\cong$ $\triangle$s

27. **Hobbies** In a garden, triangular flower beds are separated by straight rows of grass as shown.

Given: $\angle ADC$ and $\angle BCD$ are right angles.
$\overline{AC} \cong \overline{BD}$, $\overline{AD} \cong \overline{BC}$
$\angle DAC \cong \angle CBD$

Prove: $\triangle ADC \cong \triangle BCD$

Triangle Congruence

Proving Triangle Congruence

Triangle Congruence: SSS and SAS

Example 1

Use SSS to explain why the triangles in each pair are congruent.

1. $\triangle ABD \cong \triangle CDB$

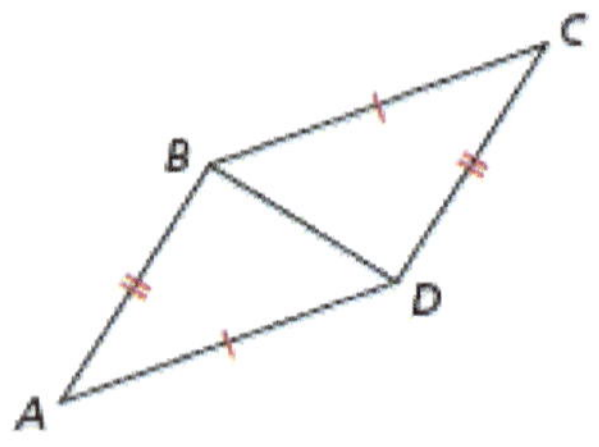

2. $\triangle MNP \cong \triangle MQP$

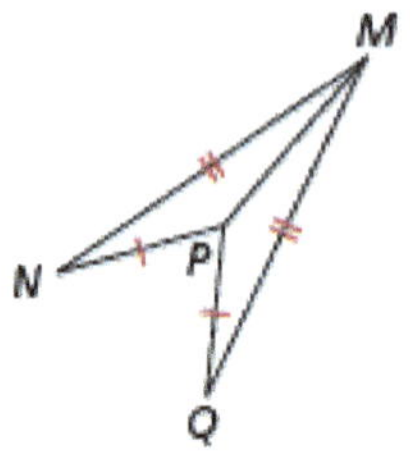

Example 2

3. **Design** This Texas flag consists of a blue, perpendicular stripe with a white star in the center. The star consists of five triangles. $GJ = LG = 20$ in., and $GK = GH = 13$ in. Use SAS to explain why $\triangle JGK \cong \triangle LGH$.

Example 3

Show that the triangles are congruent for the given value of the variable.

4. $\triangle GHJ \cong \triangle IHJ$, $x = 4$

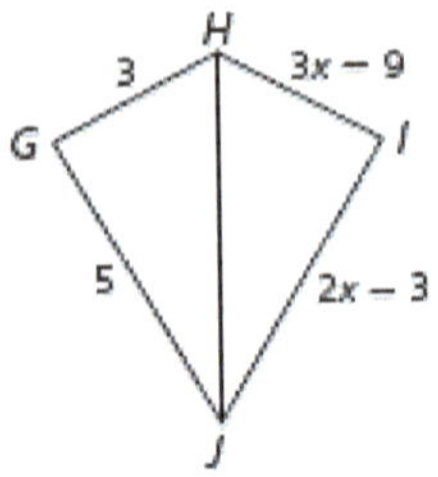

5. $\triangle RST \cong \triangle TUR$, $x = 18$

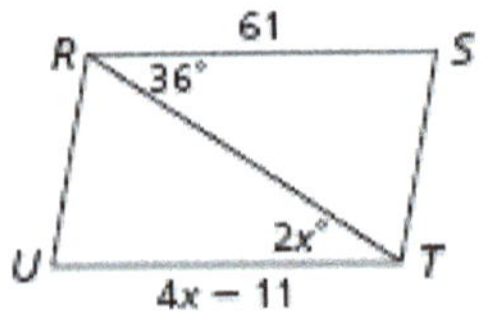

Example 4

6. **Given:** $\overline{JK} \cong \overline{ML}$, $\angle JKL \cong \angle MLK$
Prove: $\triangle JKL \cong \triangle MLK$

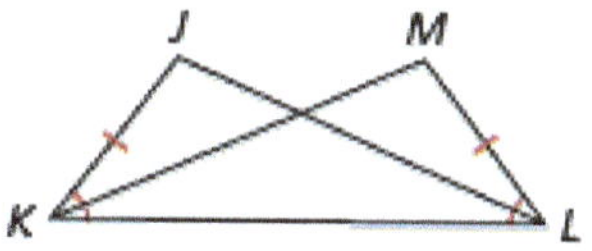

Proof:

Statements	Reasons
1. $\overline{JK} \cong \overline{ML}$	1. a. ___?___
2. b. ___?___	2. Given
3. $\overline{KL} \cong \overline{LK}$	3. c. ___?___
4. $\triangle JKL \cong \triangle MLK$	4. d. ___?___

Triangle Congruence: ASA, AAS, and HL

Example 1 & 2

Determine if you can use ASA to prove the triangles congruent. Explain.

1. $\triangle VRS$ and $\triangle VTS$, given that $\overline{VS}$ bisects $\angle RST$ and $\angle RVT$

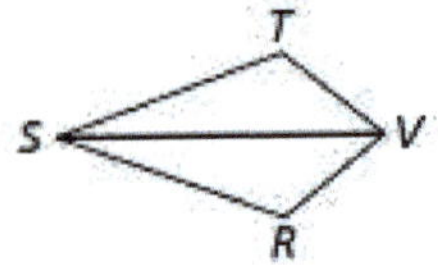

2. $\triangle DEH$ and $\triangle FGH$

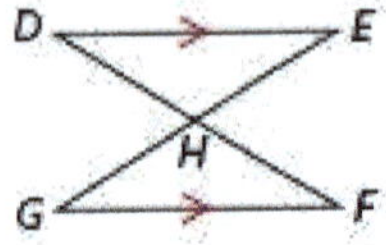

Example 3

3. Use AAS to prove the triangles congruent.
   Given: $\angle R$ and $\angle P$ are right angles.
   $\overline{QR} \parallel \overline{SP}$
   Prove: $\triangle QPS \cong \triangle SRQ$
   Proof:

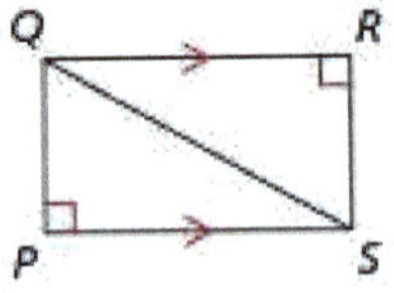

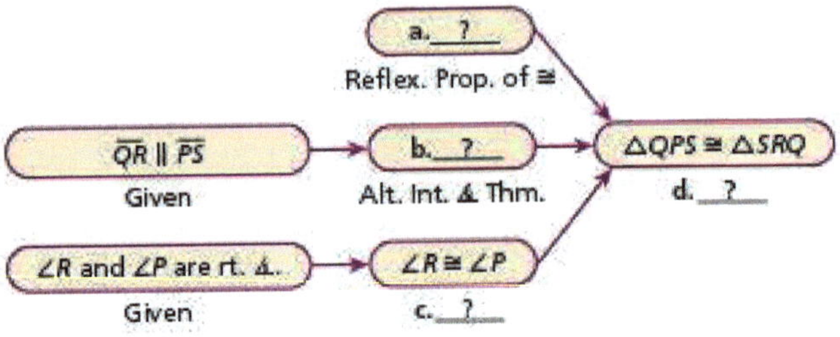

Triangle Congruence

Proving Triangle Congruence

Triangle Congruence: ASA, AAS, and HL

Example 4

Determine if you can use the HL Congruence Theorem to prove the triangles congruent. If not, tell what else you need to know.

4. $\triangle ABC$ and $\triangle CDA$

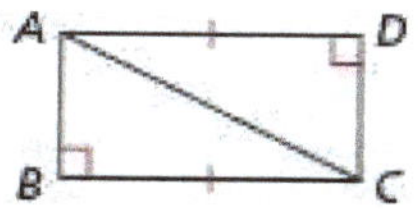

5. $\triangle XYV$ and $\triangle ZYV$

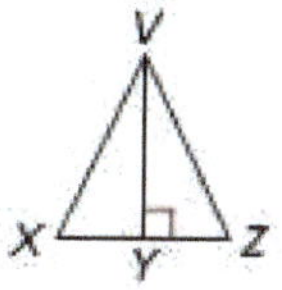

Example 1

1. **Engineering** To find the height of a windmill, a rancher places a marker at C and steps off the distance from C to B. Then the rancher walks the same distance from C in the opposite direction and places a marker at D. If $DE = 6.3$ m, what is AB?

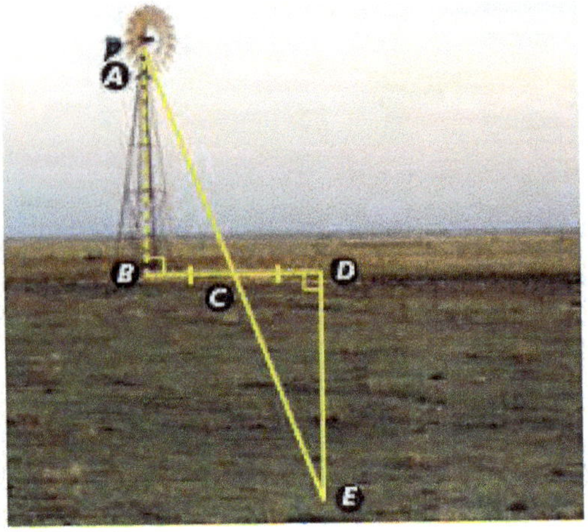

Example 2

2. **Given:** X is the midpoint of $\overline{ST}$. $\overline{RX} \perp \overline{ST}$
Prove: $\overline{RS} \cong \overline{RT}$
Proof:

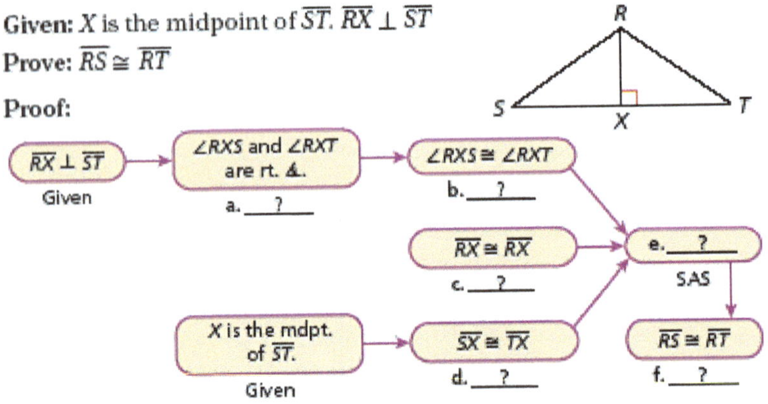

Triangle Congruence: CPCTC

Example 3

3. Given: $\overline{AC} \cong \overline{AD}$, $\overline{CB} \cong \overline{DB}$
Prove: $\overline{AB}$ bisects $\angle CAD$.

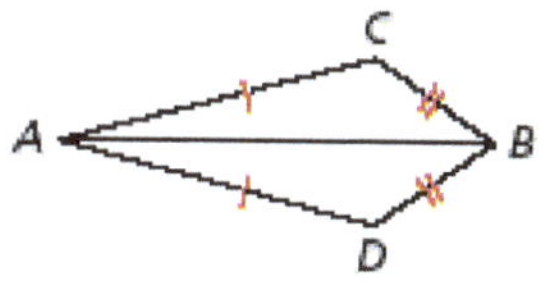

Proof:

Statements	Reasons
1. $\overline{AC} \cong \overline{AD}$, $\overline{CB} \cong \overline{DB}$	1. a. ?
2. b. ?	2. Reflex. Prop. of $\cong$
3. $\triangle ACB \cong \triangle ADB$	3. c. ?
4. $\angle CAB \cong \angle DAB$	4. d. ?
5. $\overline{AB}$ bisects $\angle CAD$	5. e. ?

Example 4

Multi-Step Use the given set of points to prove each congruence statement.

4. $E(-3, 3)$, $F(-1, 3)$, $G(-2, 0)$, $J(0, -1)$, $K(2, -1)$, $L(1, 2)$; $\angle EFG \cong \angle JKL$

5. $A(2, 3)$, $B(4, 1)$, $C(1, -1)$, $R(-1, 0)$, $S(-3, -2)$, $T(0, -4)$; $\angle ACB \cong \angle RTS$

Example 1

Position each figure in the coordinate plane.

1. a rectangle with a length of 4 units and width of 1 unit

2. a right triangle with leg lengths of 1 unit and 3 units

Example 2

Write a proof using coordinate geometry.

3. **Given:** Right $\triangle PQR$ has coordinates $P(0, 6)$, $Q(8, 0)$, and $R(0, 0)$. A is the midpoint of $\overline{PR}$. B is the midpoint of $\overline{QR}$.
   Prove: $AB = \frac{1}{2}PQ$

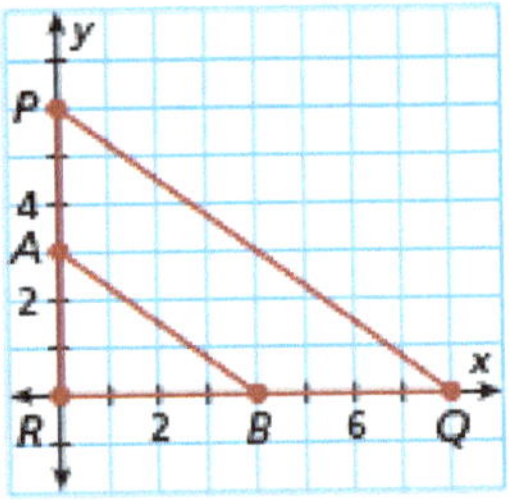

Example 3

Position each figure in the coordinate plane and give the coordinates of each vertex.

4. a right triangle with leg lengths m and n

5. a rectangle with length a and width b

Example 4

Multi-Step **Assign coordinates to each vertex and write a coordinate proof.**

6. **Given:** $\angle R$ is a right angle in $\triangle PQR$. A is the midpoint of $\overline{PR}$. B is the midpoint of $\overline{QR}$.

   Prove: $AB = \frac{1}{2}PQ$

Isosceles and Equilateral Triangles

Example 1

1. **Surveying** To find the distance *QR* across a river, a surveyor locates three points *Q*, *R*, and *S*. $QS = 41$ m, and $m\angle S = 35°$. The measure of exterior $\angle PQS = 70°$. Draw a diagram and explain how you can find *QR*.

Example 2

Find each angle measure.

2. $m\angle ECD$

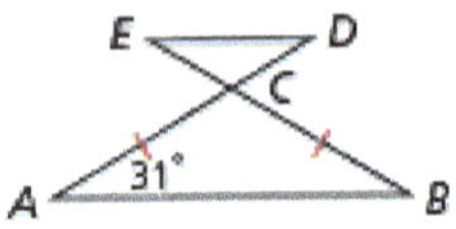

3. $m\angle K$

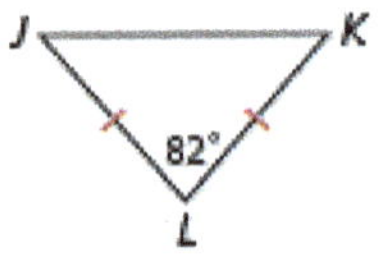

4. $m\angle X$

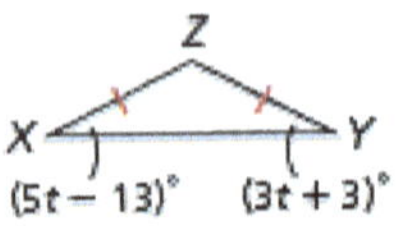

5. $m\angle A$

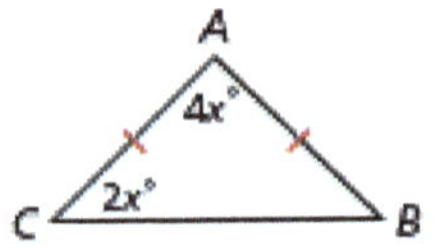

Example 3

Find each value.

6. y

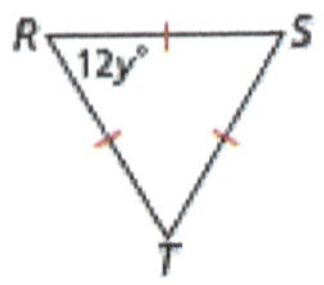

7. x

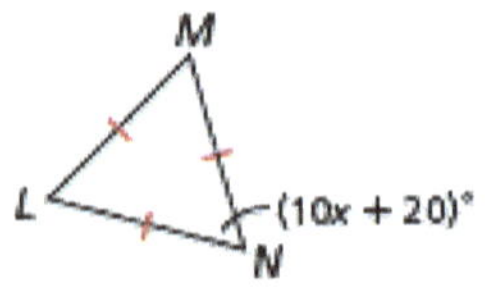

Isosceles and Equilateral Triangles

8. BC

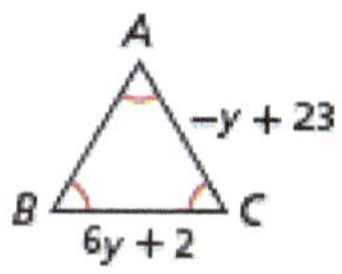

9. JK

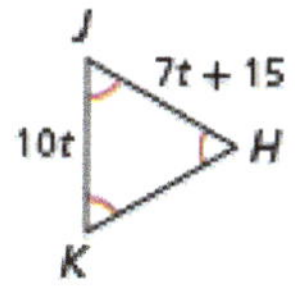

Example 4

10. **Given:** $\triangle ABC$ is right isosceles. X is the midpoint of $\overline{AC}$. $\overline{AB} \cong \overline{BC}$

Prove: $\triangle AXB$ is isosceles.

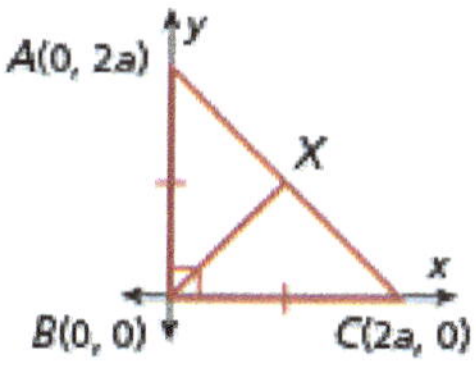

4.2.1 Triangle Congruence: SSS and SAS

Use SSS to explain why the triangles in each pair are congruent.

1. $\triangle BCD \cong \triangle EDC$

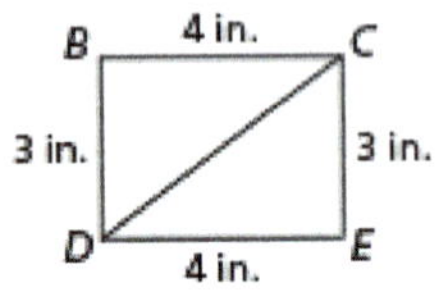

2. $\triangle GJK \cong \triangle GJL$

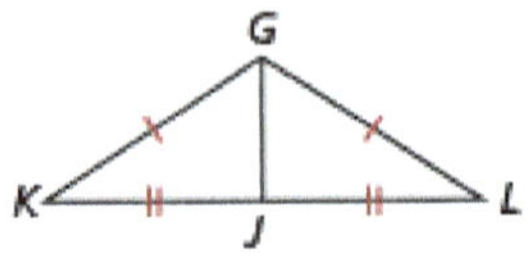

3. **Theater** The lights shining on a stage appear to form two congruent right triangles. Given $\overline{EC} \cong \overline{DB}$, use SAS to explain why $\triangle ECB \cong \triangle DBC$.

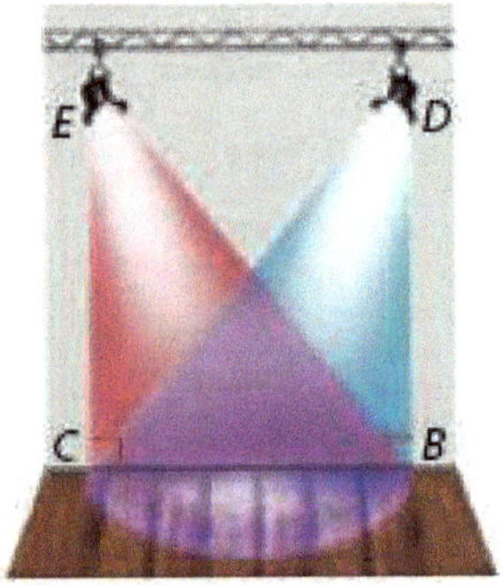

Show that the triangles are congruent for the given value of the variable.

4. $\triangle MNP \cong \triangle QNP$, $y = 3$

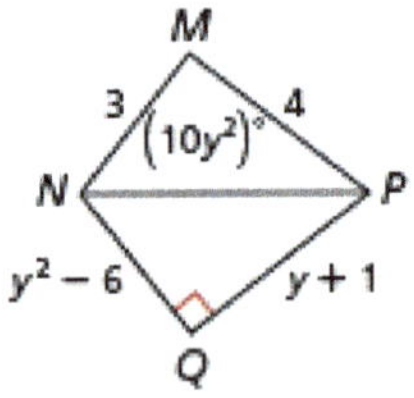

5. $\triangle XYZ \cong \triangle STU$, $t = 5$

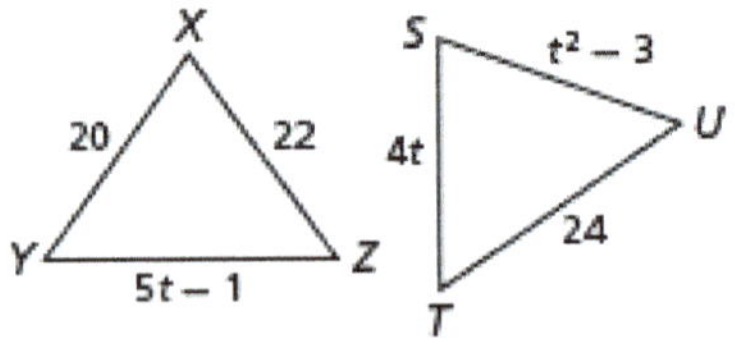

6. **Given:** B is the midpoint of $\overline{DC}$. $\overline{AB} \perp \overline{DC}$
Prove: $\triangle ABD \cong \triangle ABC$

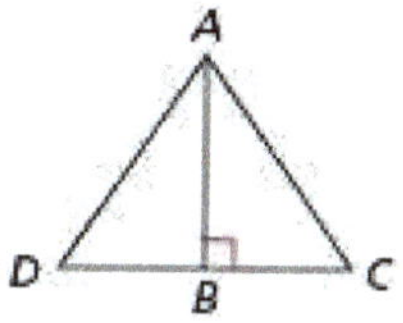

Proof:

Statements	Reasons
1. B is the mdpt. of $\overline{DC}$.	1. a. ?
2. b. ?	2. Def. of mdpt.
3. c. ?	3. Given
4. $\angle ABD$ and $\angle ABC$ are rt. ∠s.	4. d. ?
5. $\angle ABD \cong \angle ABC$	5. e. ?
6. f. ?	6. Reflex. Prop. of ≅
7. $\triangle ABD \cong \triangle ABC$	7. g. ?

4.2.2 Triangle Congruence: ASA, AAS, and HL

Determine if you can use ASA to prove the triangles congruent. Explain.

7. $\triangle MKJ$ and $\triangle MKL$

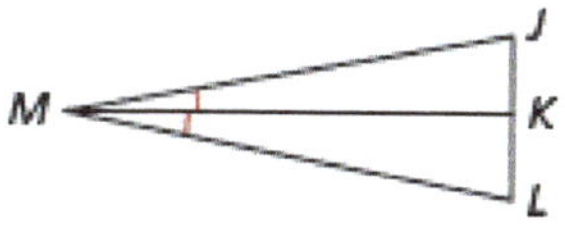

8. $\triangle RST$ and $\triangle TUR$

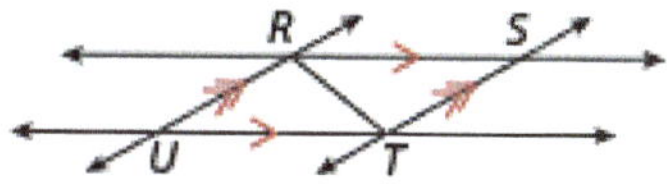

Triangle Congruence

Review 4.2 / Proving Triangle Congruence

9. Given: $\overline{AB} \cong \overline{DE}$, $\angle C \cong \angle F$
Prove: $\triangle ABC \cong \triangle DEF$

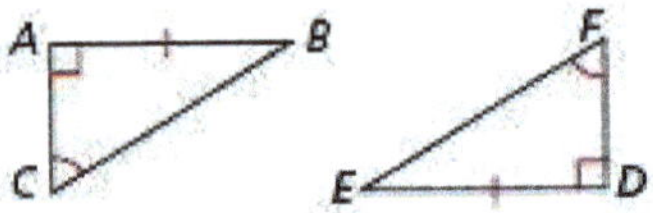

Proof:

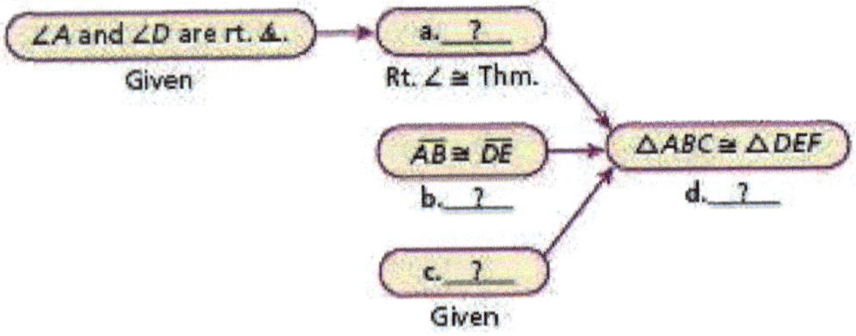

Determine if you can use the HL Congruence Theorem to prove the triangles congruent. If not, tell what else you need to know.

10. $\triangle GHJ$ and $\triangle JKG$

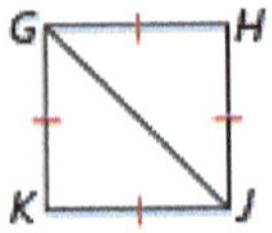

11. $\triangle ABE$ and $\triangle DCE$, given that E is the midpoint of $\overline{AD}$ and $\overline{BC}$

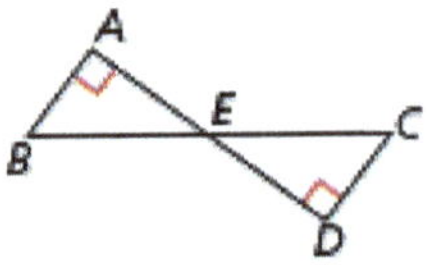

4.2.3 Triangle Congruence: CPCTC

12. **Surveying** To find the distance AB across a river, a surveyor first locates point C. He measures the distance from C to B. Then he locates point D the same distance east of C. If $DE = 420$ ft, what is AB?

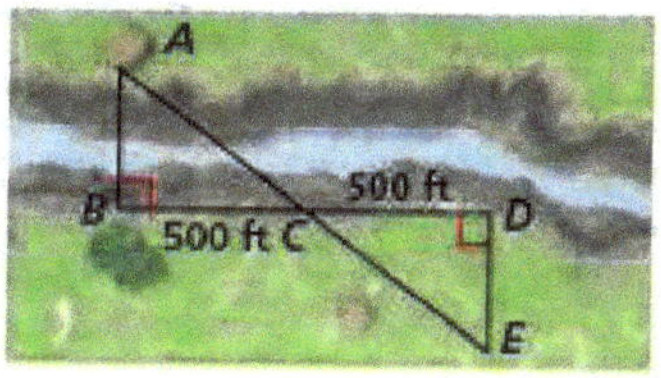

13. **Given:** M is the midpoint of $\overline{PQ}$ and $\overline{RS}$.
Prove: $\overline{QR} \cong \overline{PS}$

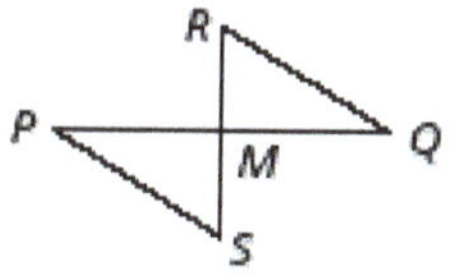

14. **Given:** $\overline{WX} \cong \overline{XY} \cong \overline{YZ} \cong \overline{ZW}$
Prove: $\angle W \cong \angle Y$

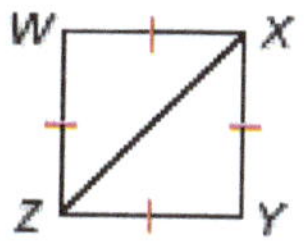

15. **Given:** G is the midpoint of $\overline{FH}$.
$\overline{EF} \cong \overline{EH}$
Prove: $\angle 1 \cong \angle 2$

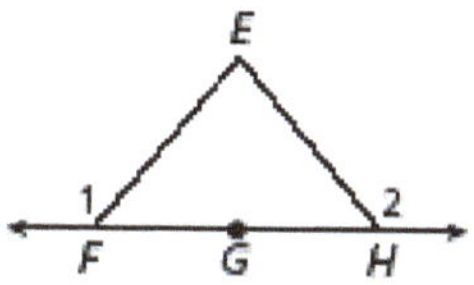

16. **Given:** $\overline{LM}$ bisects $\angle JLK$. $\overline{JL} \cong \overline{KL}$
Prove: M is the midpoint of $\overline{JK}$.

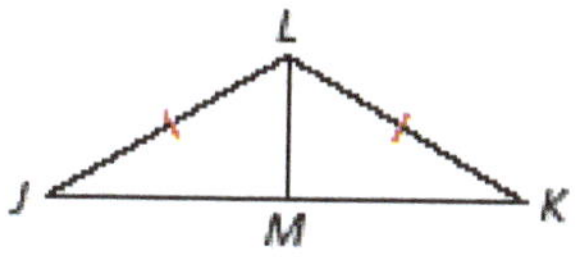

Multi-Step Use the given set of points to prove each congruence statement.

17. $R(0, 0)$, $S(2, 4)$, $T(-1, 3)$, $U(-1, 0)$, $V(-3, -4)$, $W(-4, -1)$; $\angle RST \cong \angle UVW$

18. $A(-1, 1)$, $B(2, 3)$, $C(2, -2)$, $D(2, -3)$, $E(-1, -5)$, $F(-1, 0)$; $\angle BAC \cong \angle EDF$

4.2.4 Introduction to Coordinate Proof

Position each figure in the coordinate plane.

19. a square with side lengths of 2 units

20. a right triangle with leg lengths of 1 unit and 5 units

Write a proof using coordinate geometry.

21. **Given:** Rectangle $ABCD$ has coordinates $A(0, 0)$, $B(0, 10)$, $C(6, 10)$, and $D(6, 0)$. E is the midpoint of $\overline{AB}$, and F is the midpoint of $\overline{CD}$.
Prove: $EF = BC$

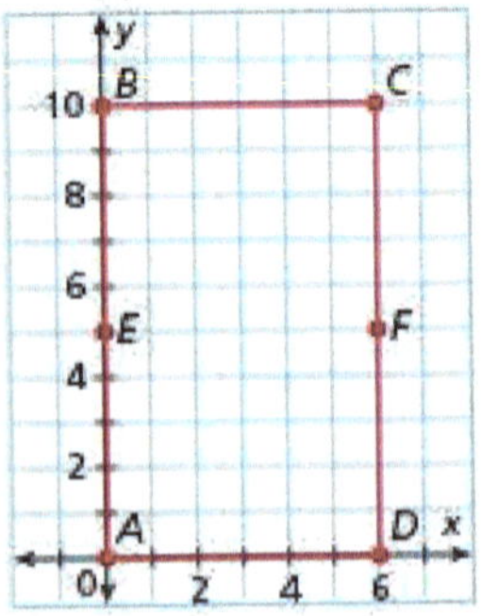

Position each figure in the coordinate plane and give the coordinates of each vertex.

22. a square with side length $2m$

23. a rectangle with dimensions x and $3x$

Multi-Step **Assign coordinates to each vertex and write a coordinate proof.**

24. **Given:** E is the midpoint of $\overline{AB}$ in rectangle $ABCD$. F is the midpoint of $\overline{CD}$.
Prove: $EF = AD$

4.2.5 Isosceles and Equilateral Triangles

25. **Aviation** A plane is flying parallel to the ground along $\overrightarrow{AC}$. When the plane is at A, an air-traffic controller in tower T measures the angle to the plane as 40°. After the plane has traveled 2.4 mi to B, the angle to the plane is 80°. How can you find BT?

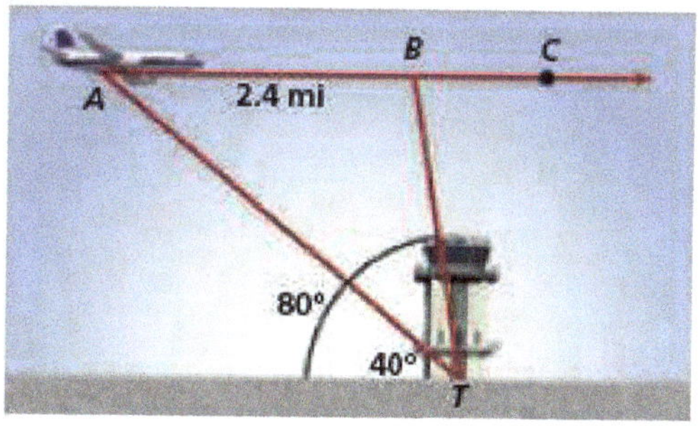

Find each angle measure.

26. m∠*E*

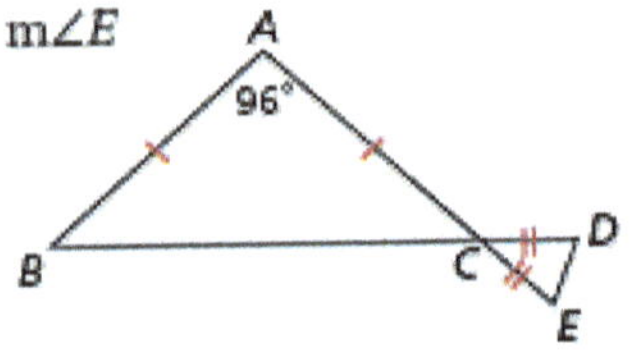

27. m∠*TRU*

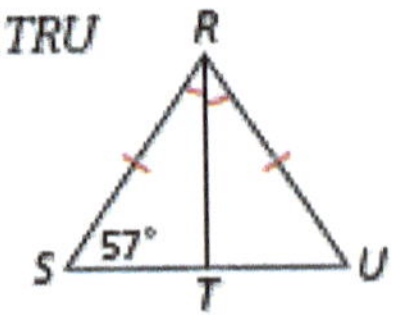

28. m∠*F*

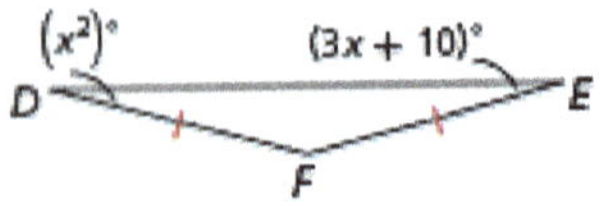

29. m∠*A*

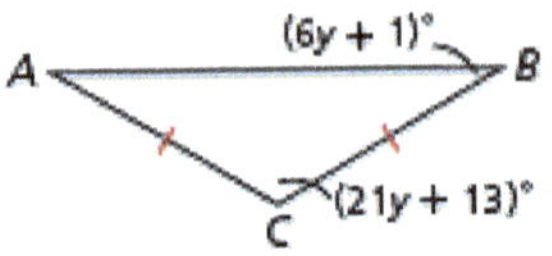

Find each value.

30. *z*

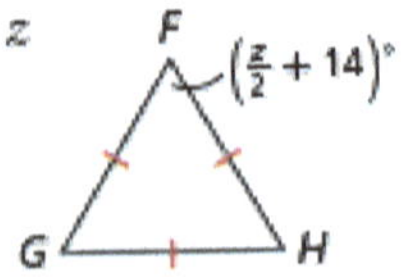

31. *y*

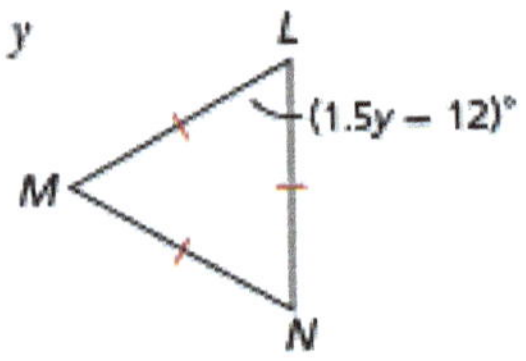

32. *BC*

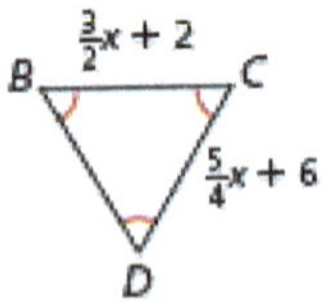

33. *XZ*

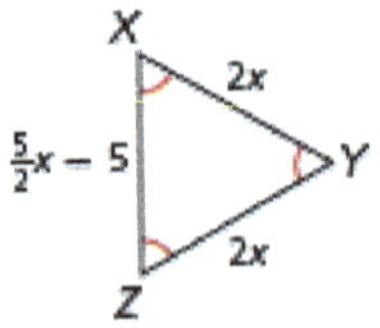

34. **Given:** $\triangle ABC$ is isosceles. P is the midpoint of $\overline{AB}$. Q is the midpoint of $\overline{AC}$. $\overline{AB} \cong \overline{AC}$
Prove: $\overline{PC} \cong \overline{QB}$

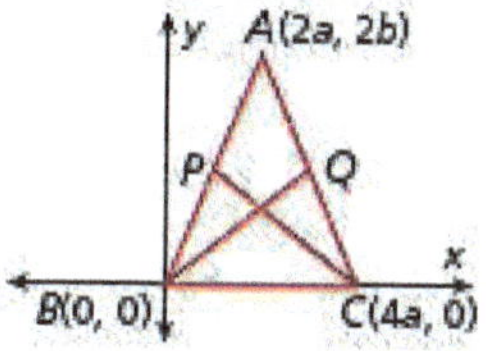

Perpendicular and Angle Bisector Theorems

Example 1

Use the diagram for Exercises 1-3.

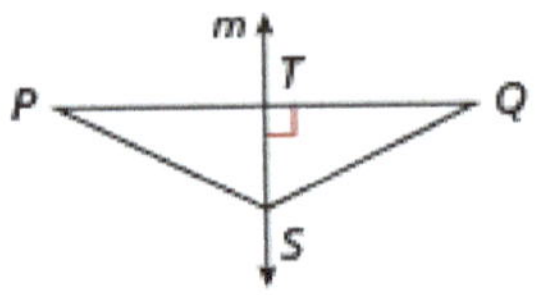

1. Given that $PS = 53.4$, $QT = 47.7$, and $QS = 53.4$, find PQ.

2. Given that m is the perpendicular bisector of $\overline{PQ}$ and $SQ = 25.9$, find SP.

3. Given that m is the perpendicular bisector of $\overline{PQ}$, $PS = 4a$, and $QS = 2a + 26$, find QS.

Example 2

Use the diagram for Exercises 4-6.

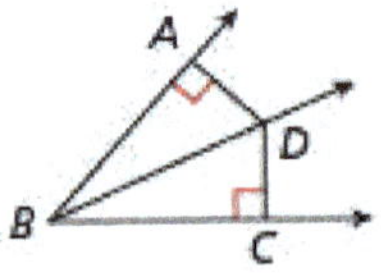

4. Given that $\overrightarrow{BD}$ bisects $\angle ABC$ and $CD = 21.9$, find AD.

5. Given that $AD = 61$, $CD = 61$, and $m\angle ABC = 48°$, find $m\angle CBD$.

6. Given that $DA = DC$, $m\angle DBC = (10y + 3)°$, and $m\angle DBA = (8y + 10)°$, find $m\angle DBC$.

7. **Carpentry** For a king post truss to be constructed correctly, P must lie on the bisector of $\angle JLN$. How can braces $\overline{PK}$ and $\overline{PM}$ be used to ensure that P is in the proper location?

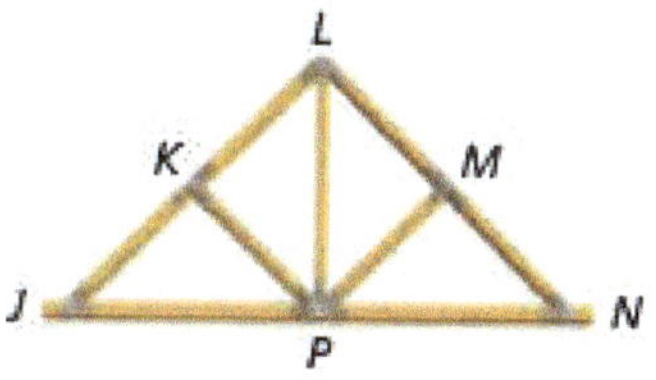

Example 3

Write an equation in point-slope form for the perpendicular bisector of the segment with the given endpoints.

8. $M(-5, 4), N(1, -2)$

9. $U(2, -6), V(4, 0)$

10. $J(-7, 5), K(1, -1)$

Example 1

Use ΔACD for Exercises 1-3.

1. If $\overline{BD}$ is a median of ΔACD and $AC = 10$, what is AB?

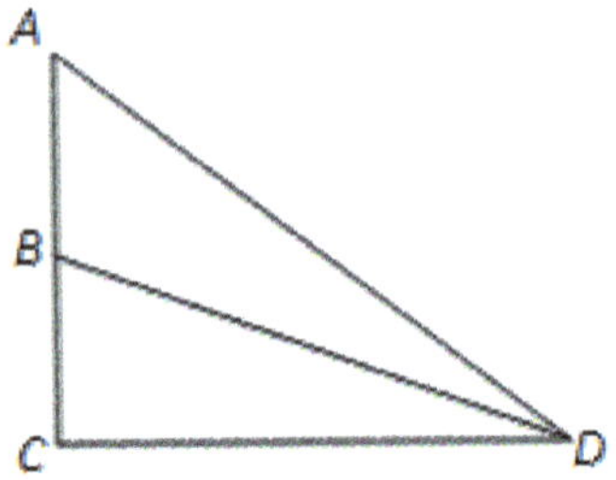

2. If $AB = 2x$ and $BC = 3x - 6$ and $x = 6$, explain why $\overline{BD}$ is a median of ΔACD.

3. If $AB = x + 2$ and $AC = 3x - 1$ and $\overline{BD}$ is a median of ΔACD, what is x?

Example 2

4. ΔDEF has vertices $D(-4, -3)$, $E(3, -4)$, and $F(2, 1)$. Find the equation of the median through E in slope-intercept form.

Example 3

5. ΔDEF has vertices $D(-4, -3)$, $E(3, -4)$, and $F(2, 1)$. Find the equation of the altitude through D in slope-intercept form.

Example 4

Find each measure.

6. *NM* **7.** *XZ*

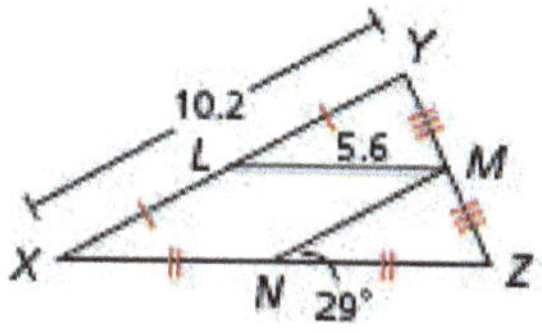

8. *NZ* **9.** m∠*LMN* **10.** m∠*YXZ* **11.** m∠*XLM*

Properties and Attributes of Triangles
Review 5.1 / Segments in Triangles

5.1.1 Perpendicular and Angle Bisector Theorems

Use the diagram for Exercises 1-3.

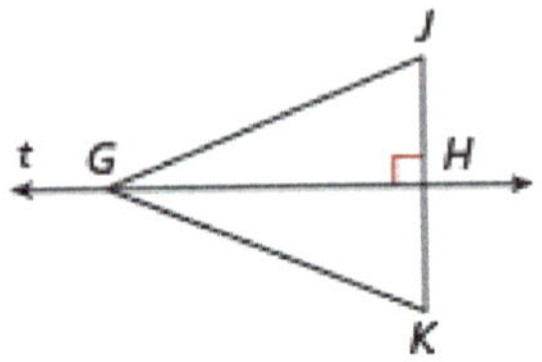

1. Given that line t is the perpendicular bisector of $\overline{JK}$ and $GK = 8.25$, find GJ.

2. Given that line t is the perpendicular bisector of $\overline{JK}$, $JG = x + 12$, and $KG = 3x - 17$, find KG.

3. Given that $GJ = 70.2$, $JH = 26.5$, and $GK = 70.2$, find JK.

Use the diagram for Exercises 4-6.

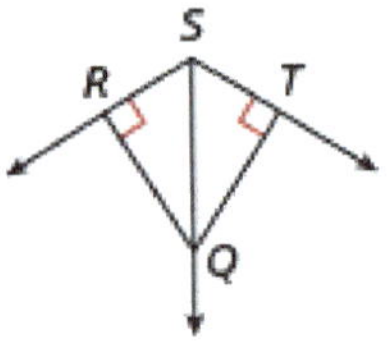

4. Given that $m\angle RSQ = m\angle TSQ$ and $TQ = 1.3$, find RQ.

5. Given that $m\angle RSQ = 58°$, $RQ = 49$, and $TQ = 49$, find $m\angle RST$.

6. Given that $RQ = TQ$, $m\angle QSR = (9a + 48)°$, and $m\angle QST = (6a + 50)°$, find $m\angle QST$.

7. **City Planning** The planners for a new section of the city want every location on Main Street to be equidistant from Elm Street and Grove Street. How can the planners ensure that this is the case?

Write an equation in point-slope form for the perpendicular bisector of the segment with the given endpoints.

8. $E(-4, -7), F(0, 1)$

9. $X(-7, 5), Y(-1, -1)$

10. $M(-3, -1), N(7, -5)$

5.1.2 Medians, Altitudes, and Midsegments in Triangles

Find each measure.

11. GJ

12. RQ

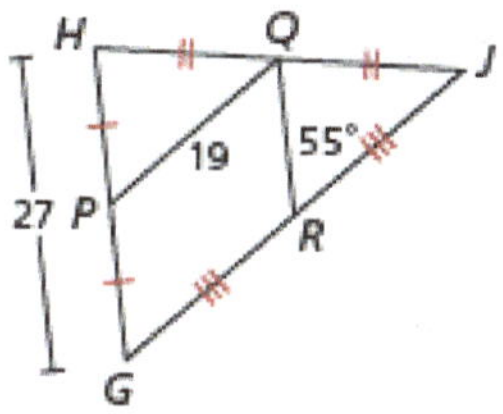

13. RJ

14. $m\angle PQR$

15. $m\angle HGJ$

16. $m\angle GPQ$

17. **Carpentry** In each support for the garden swing, the crossbar $\overline{DE}$ is attached at the midpoints of legs $\overline{BA}$ and $\overline{BC}$. The distance AC is $4\frac{1}{2}$ feet. The carpenter has a timber that is 30 inches long. Is this timber long enough to be used as one of the crossbars? Explain.

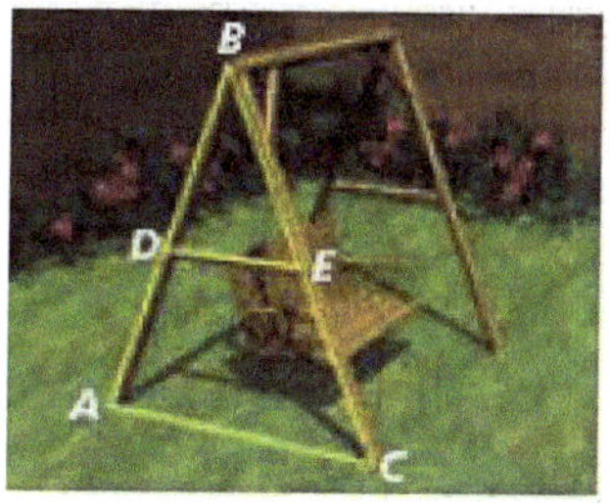

Indirect Proof and Inequalities in One Triangle

Example 1

Write an indirect proof of each statement.

1. A scalene triangle cannot have two congruent angles.

2. An isosceles triangle cannot have a base angle that is a right angle.

Example 2

3. Write the angles in order from smallest to largest.

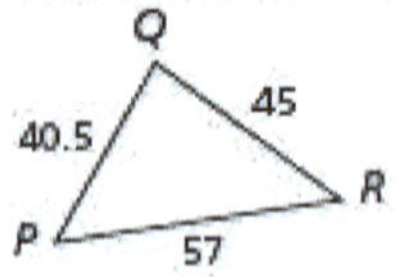

4. Write the sides in order from shortest to longest.

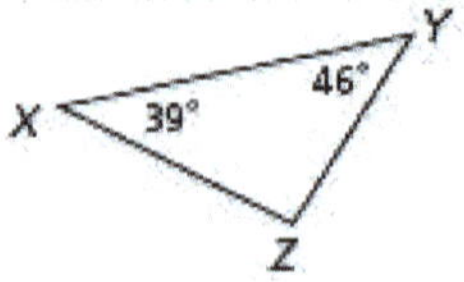

Example 3

Tell whether a triangle can have sides with the given lengths. Explain.

5. 4, 7, 10

6. 2, 9, 12

7. $3\frac{1}{2}, 3\frac{1}{2}, 6$

8. 3, 1.1, 1.7

9. $3x, 2x - 1, x^2$, when $x = 5$

10. $7c + 6, 10c - 7, 3c^2$, when $c = 2$

Example 4

The lengths of two sides of a triangle are given. Find the range of possible lengths for the third side.

11. 8 mm, 12 mm

12. 16 ft, 16 ft

13. 11.4 cm, 12 cm

Example 5

14. **Design** The refrigerator, stove, and sink in a kitchen are at the vertices of a path called the work triangle.

a. If the angle at the sink is the largest, which side of the work triangle will be the longest?

b. The designer wants the longest side of this triangle to be 9 feet long. Can the lengths of the other sides be 5 feet and 4 feet? Explain.

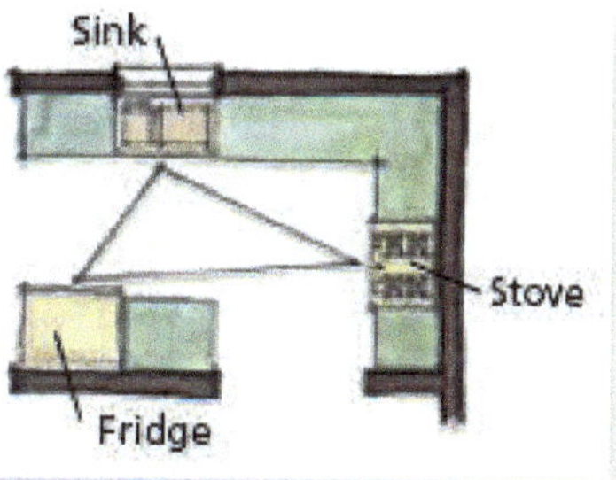

Example 1

Compare the given measures.

1. AC and XZ

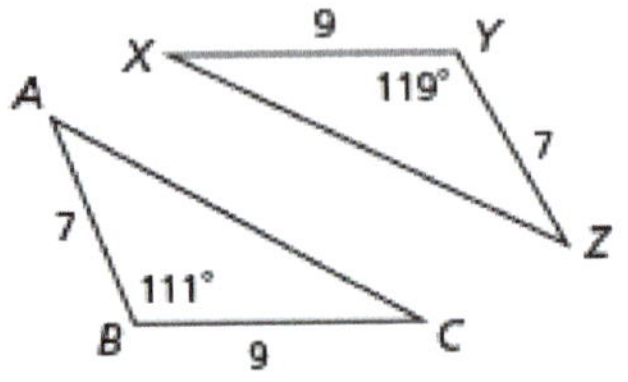

2. m∠SRT and m∠QRT

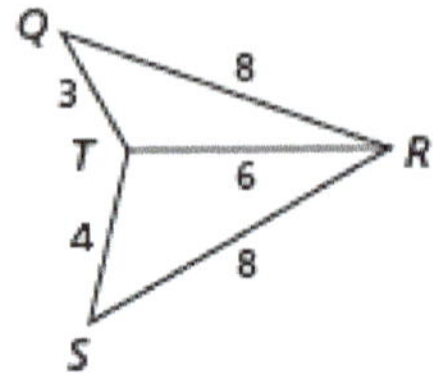

3. KL and KN

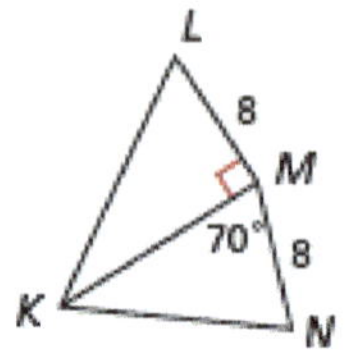

Find the range of values for x.

4.

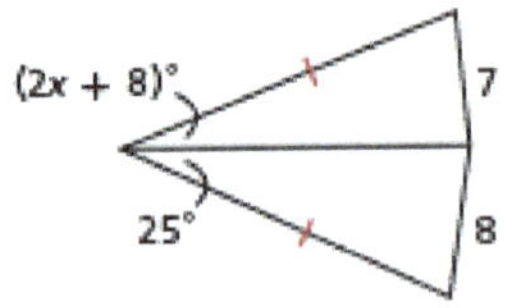

5.

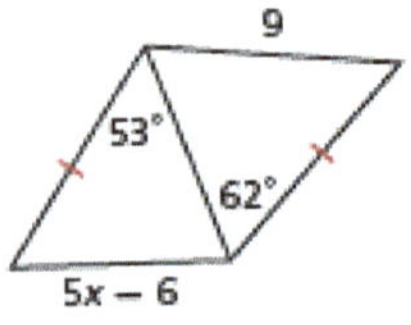

6.

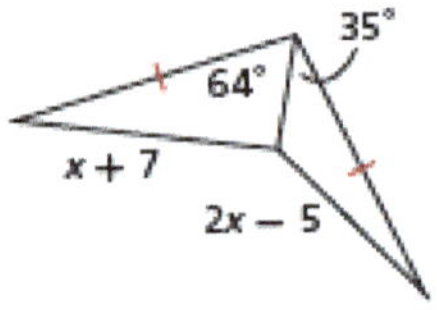

Example 2

7. Health A therapist can take measurements to gauge the flexibility of a patient's elbow joint. In which position is the angle measure at the elbow joint greater? Explain.

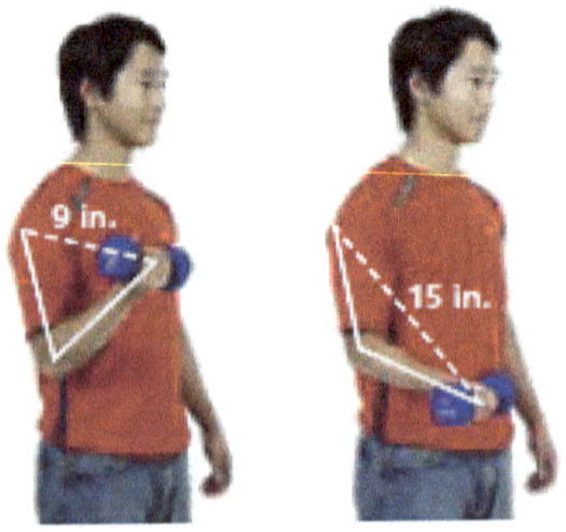

Example 3

8. Write a two-column proof.
Given: $\overline{FH}$ is a median of $\triangle DFG$.
$m\angle DHF > m\angle GHF$
Prove: $DF > GF$

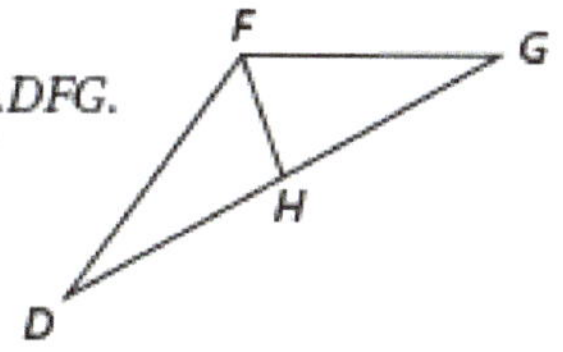

Properties and Attributes of Triangles

Relationships in Triangles

The Pythagorean Theorem

Example 1

Find the value of x. Give your answer in simplest radical form.

1.

2.

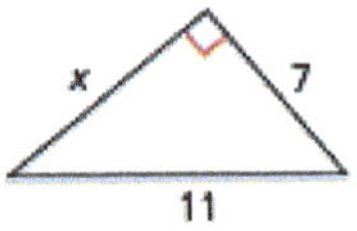

3.

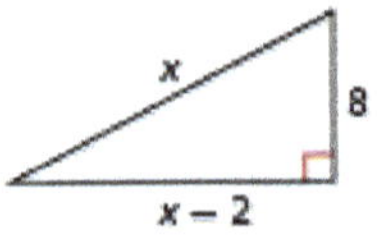

Example 2

4. **Computers** The size of a computer monitor is usually given by the length of its diagonal. A monitor's aspect ratio is the ratio of its width to its height. This monitor has a diagonal length of 19 inches and an aspect ratio of 5:4. What are the width and height of the monitor? Round to the nearest tenth of an inch.

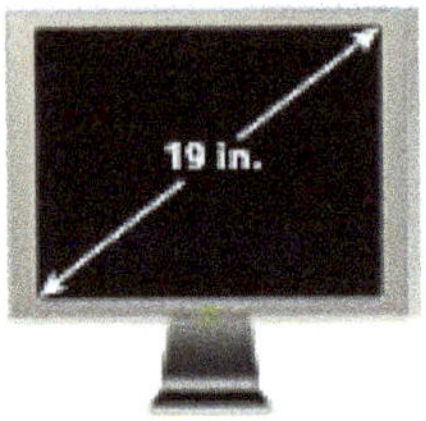

Example 3

Find the missing side length. Tell if the side lengths form a Pythagorean triple. Explain.

5.

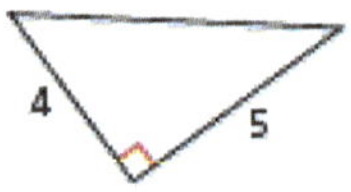

6.

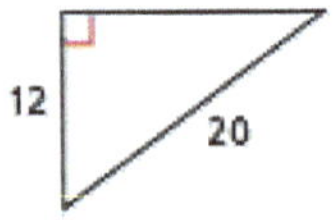

7.

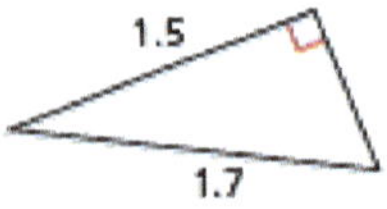

The Pythagorean Theorem

Example 4

Multi-Step **Tell if the measures can be the side lengths of a triangle. If so, classify the triangle as acute, obtuse, or right.**

8. 7, 10, 12

9. 9, 11, 15

10. 9, 40, 41

11. $1\frac{1}{2}, 1\frac{3}{4}, 3\frac{1}{4}$

12. 5.9, 6, 8.4

13. 11, 13, $7\sqrt{6}$

Applying Special Right Triangles

Example 1

Find the value of x. Give your answer in simplest radical form.

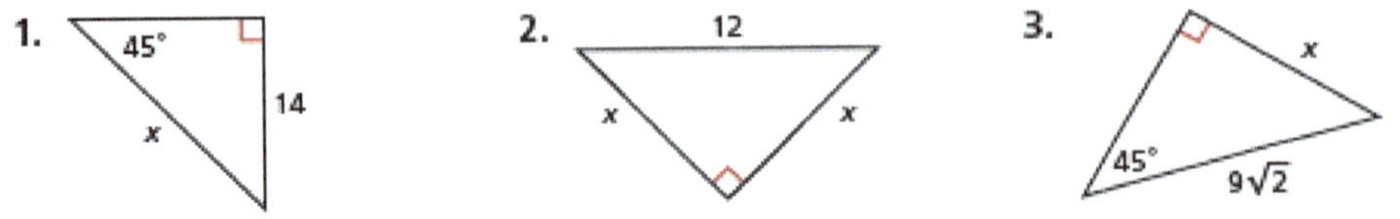

Example 2

4. **Transportation** The two arms of the railroad sign are perpendicular bisectors of each other. In Pennsylvania, the lengths marked in red must be 19.5 inches. What is the distance labeled d? Round to the nearest tenth of an inch.

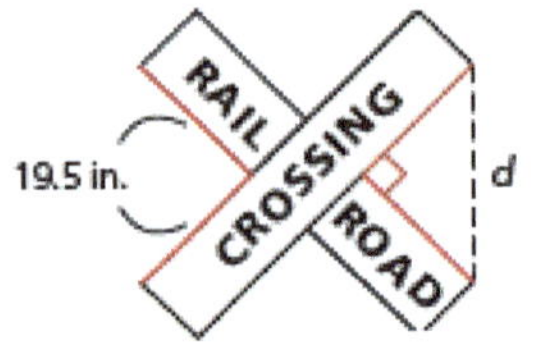

Example 3

Find the values of x and y. Give your answers in simplest radical form.

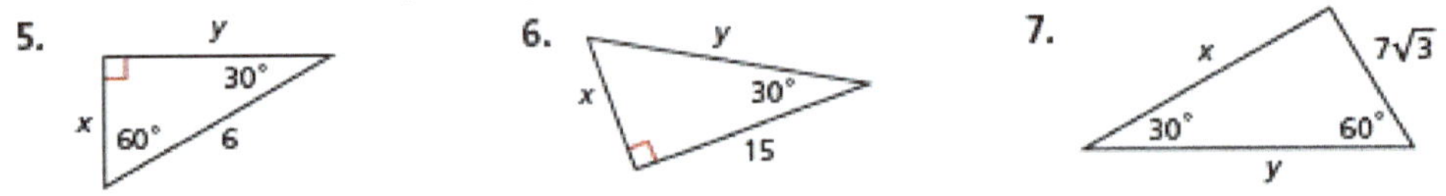

Example 4

8. **Entertainment** Regulation billiard balls are $2\frac{1}{4}$ inches in diameter. The rack used to group 15 billiard balls is in the shape of an equilateral triangle. What is the approximate height of the triangle formed by the rack? Round to the nearest quarter of an inch.

5.2.1 Indirect Proof and Inequalities in One Triangle

Write an indirect proof of each statement.

1. A scalene triangle cannot have two congruent midsegments.

2. Two supplementary angles cannot both be obtuse angles.

3. Write the angles in order from smallest to largest.

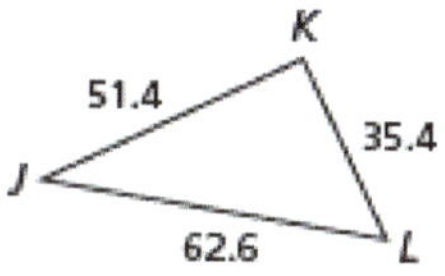

4. Write the sides in order from shortest to longest.

Tell whether a triangle can have sides with the given lengths. Explain.

5. 6, 10, 15

6. 14, 18, 32

7. 11.9, 5.8, 5.8

8. 103, 41.9, 62.5

9. $z + 8, 3z + 5, 4z - 11$, when $z = 6$

10. $m + 11, 8m, m^2 + 1$, when $m = 3$

The lengths of two sides of a triangle are given. Find the range of possible lengths for the third side.

11. 4 yd, 19 yd

12. 28 km, 23 km

13. 9.2 cm, 3.8 cm

14. 3.07 m, 1.89 m

15. $2\frac{1}{8}$ in., $3\frac{5}{8}$ in.

16. $3\frac{5}{6}$ ft, $6\frac{1}{2}$ ft

17. **Bicycles** The five steel tubes of this mountain bike frame form two triangles. List the five tubes in order from shortest to longest. Explain your answer.

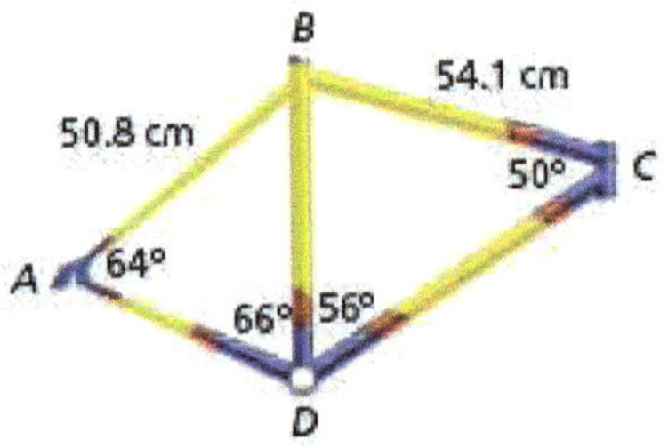

5.2.2 Inequalities in Two Triangles

Compare the given measures.

18. m∠*DCA* and m∠*BCA*

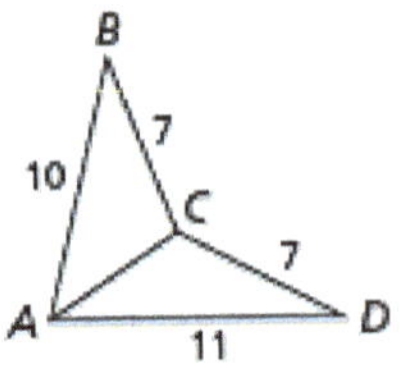

19. m∠*GHJ* and m∠*KLM*

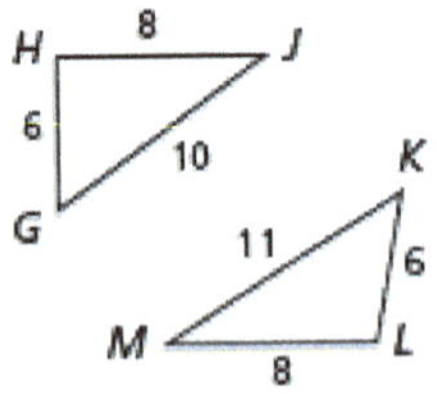

20. *TU* and *SV*

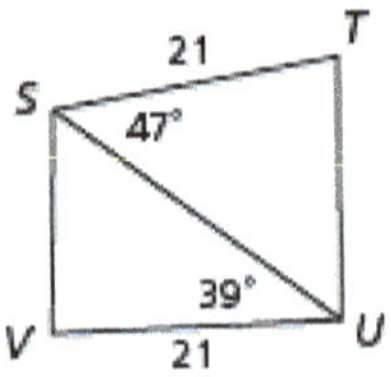

Find the range of values for z.

21.

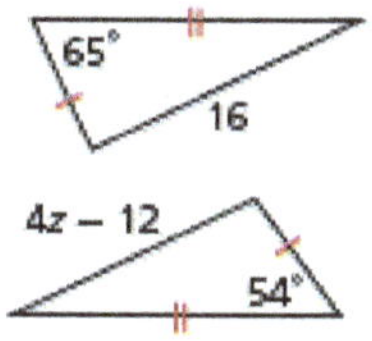

22.

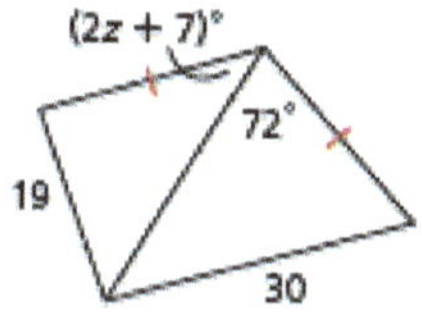

23.

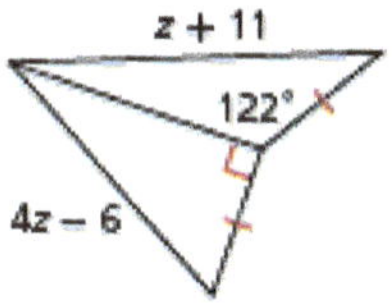

24. **Industry** The operator of a backhoe changes the distance between the cab and the bucket by changing the angle formed by the arms. In which position is the distance from the cab to the bucket greater? Explain.

25. Write a two-column proof.
Given: $\overline{JK} \cong \overline{NM}$, $\overline{KP} \cong \overline{MQ}$, $JQ > NP$
Prove: $m\angle K > m\angle M$

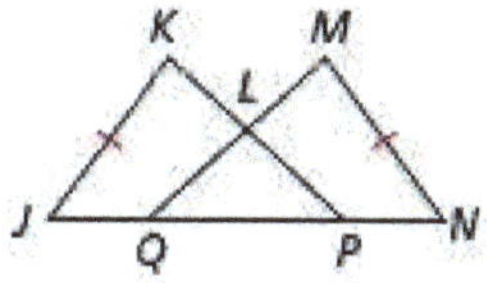

5.2.3 The Pythagorean Theorem

Find the value of x. Give your answer in simplest radical form.

26.

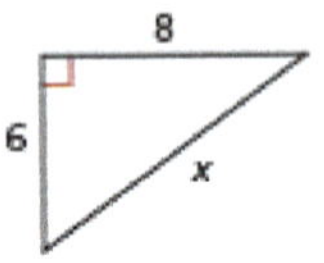

27.

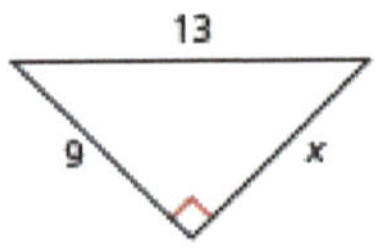

28.

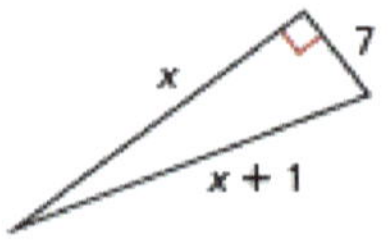

29. **Safety** The safety rules for a playground state that the height of the slide and the distance from the base of the ladder to the front of the slide must be in a ratio of 3:5. If a slide is about 8 feet long, what are the height of the slide and the distance from the base of the ladder to the front of the slide? Round to the nearest inch.

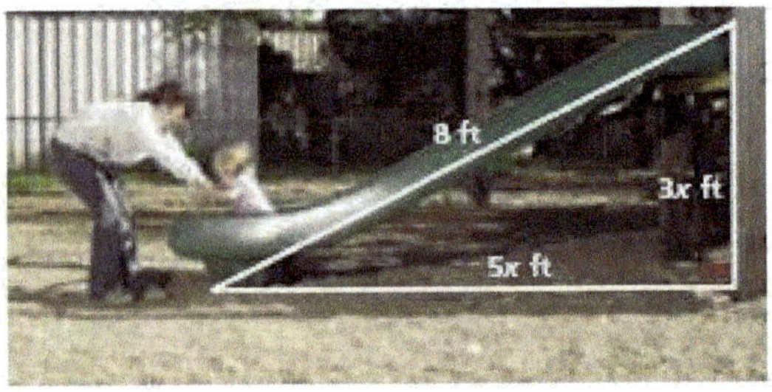

Find the missing side length. Tell if the side lengths form a Pythagorean triple. Explain.

30.

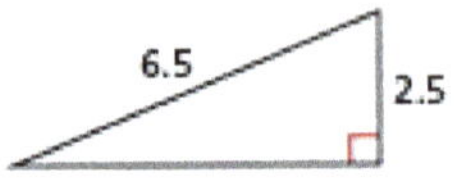

31.

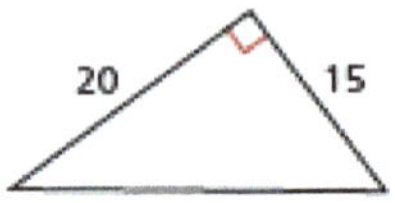

32.

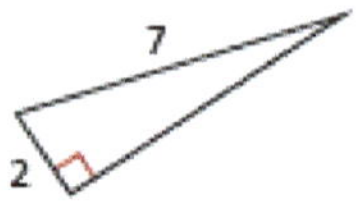

Multi-Step **Tell if the measures can be the side lengths of a triangle. If so, classify the triangle as acute, obtuse, or right.**

33. 10, 12, 15

34. 8, 13, 23

35. 9, 14, 17

36. $1\frac{1}{2}, 2, 2\frac{1}{2}$ **37.** 0.7, 1.1, 1.7 **38.** 7, 12, $6\sqrt{5}$

5.2.4 Applying Special Right Triangles

Find the value of x. Give your answer in simplest radical form.

39.

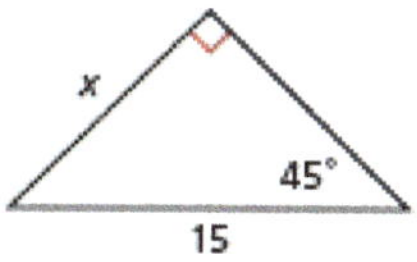

40.

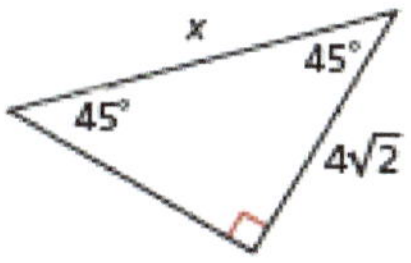

41.

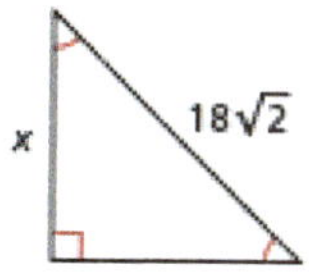

42. **Design** This tabletop is an isosceles right triangle. The length of the front edge of the table is 48 inches. What is the length w of each side edge? Round to the nearest tenth of an inch.

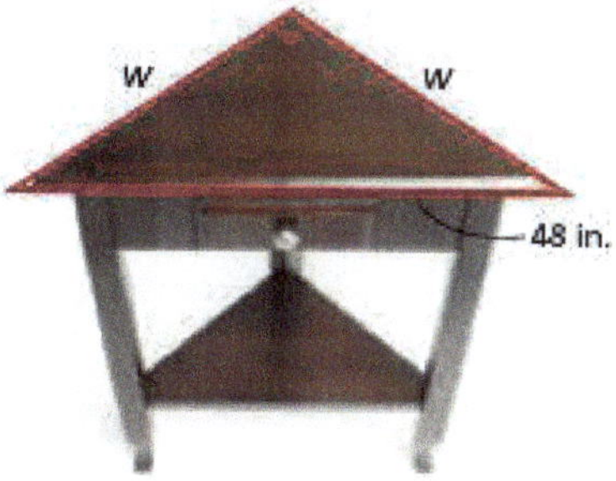

Find the value of x and y. Give your answers in simplest radical form.

43.

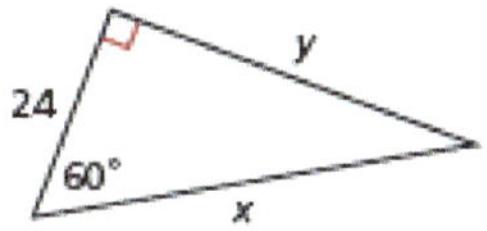

44.

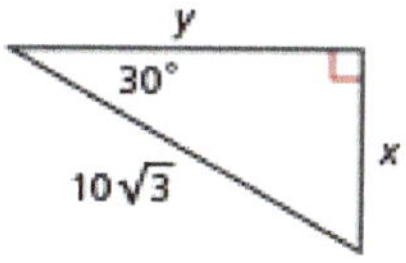

45.

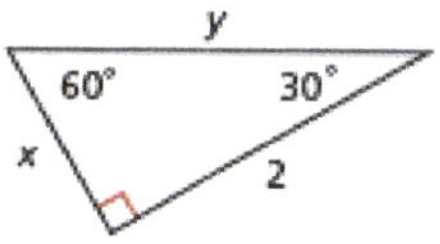

Properties and Attributes of Triangles

Review 5.2 / Relationships in Triangles

46. **Pets** A dog walk is used in dog agility competitions. In this dog walk, each ramp makes an angle of 30° with the ground.

a. How long is one ramp?

b. How long is the entire dog walk, including both ramps?

Properties and Attributes of Polygons

Example 1

Tell whether each outlined shape is a polygon. If it is a polygon, name it by the number of its sides.

1. 

2. 

3. 

4. 

Example 2

Tell whether each polygon is regular or irregular. Tell whether it is concave or convex.

5. 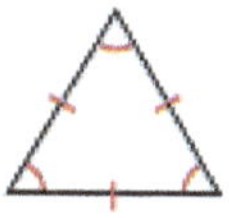

6. 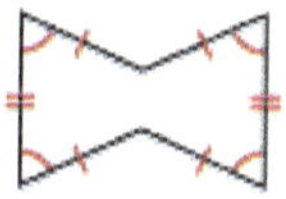

7. 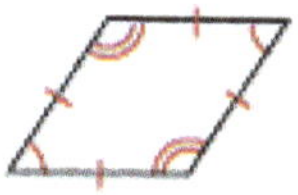

Example 3

8. Find the measure of each interior angle of pentagon *ABCDE*.

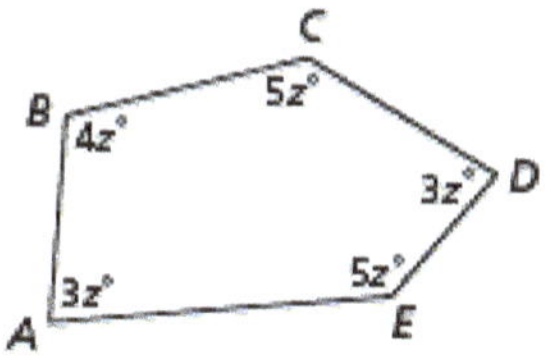

9. Find the measure of each interior angle of a regular dodecagon.

10. Find the sum of the interior angle measures of a convex 20-gon.

Example 4

11. Find the value of y in polygon *JKLM*.

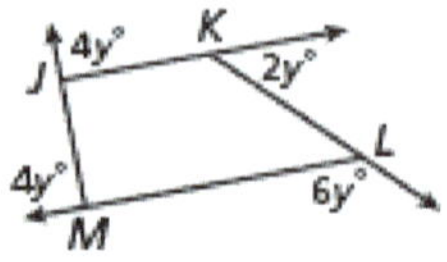

12. Find the measure of each exterior angle of a regular pentagon.

Example 5

Safety Use the photograph of the traffic sign for Exercises 13 and 14.

13. Name the polygon by the number of its sides.

14. In the polygon, $\angle P$, $\angle R$, and $\angle T$ are right angles, and $\angle Q \cong \angle S$. What are $m\angle Q$ and $m\angle S$?

Example 1

Safety The handrail is made from congruent parallelograms. In ▱$ABCD$, $AB = 17.5$, $DE = 18$, and $m\angle BCD = 110°$. Find each measure.

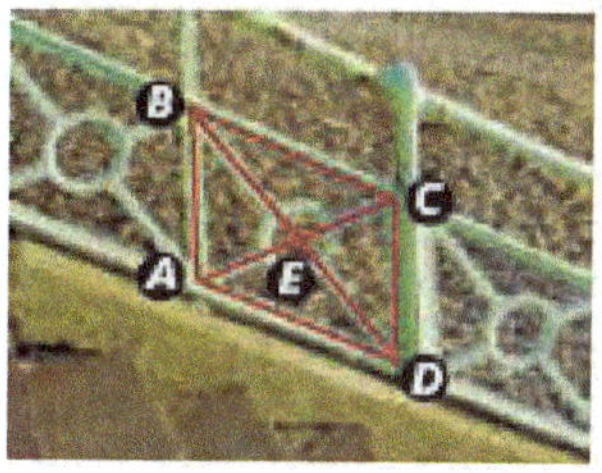

1. BD

2. CD

3. BE

4. $m\angle ABC$

5. $m\angle ADC$

6. $m\angle DAB$

Example 2

$JKLM$ is a parallelogram. Find each measure.

7. JK

8. LM

9. $m\angle L$

10. $m\angle M$

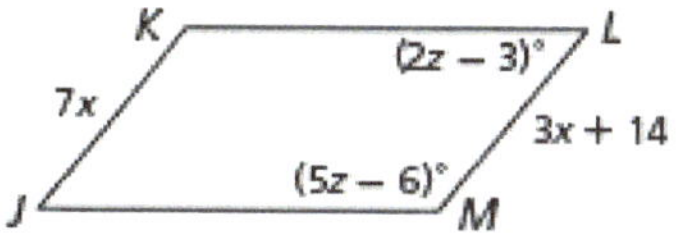

Example 3

11. **Multi-Step** Three vertices of ▱$DFGH$ are $D(-9, 4)$, $F(-1, 5)$, and $G(2, 0)$. Find the coordinates of vertex H.

Example 4

12. Write a two-column proof.
Given: $PSTV$ is a parallelogram. $\overline{PQ} \cong \overline{RQ}$
Prove: $\angle STV \cong \angle R$

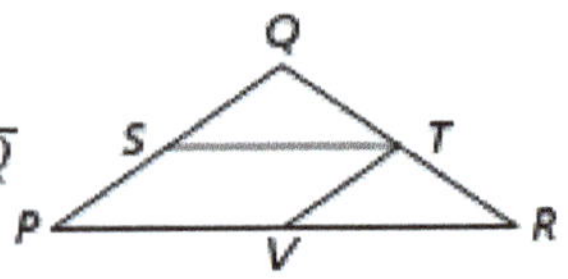

Conditions for Parallelograms

Example 1

1. Show that *EFGH* is a parallelogram for $s = 5$ and $t = 6$.

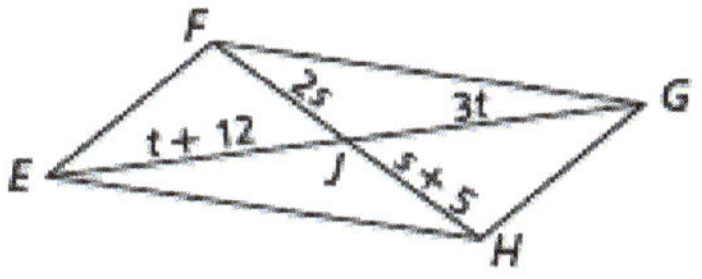

2. Show that *KLPQ* is a parallelogram for $m = 14$ and $n = 12.5$.

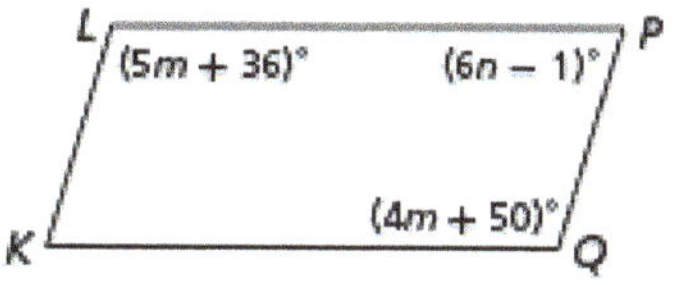

Example 2

Determine if each quadrilateral must be a parallelogram. Justify your answer.

3. 

4. 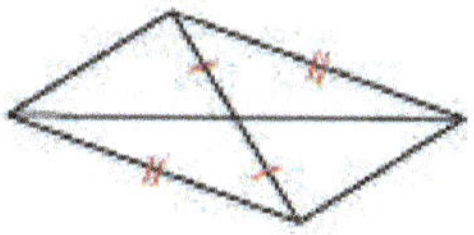

5. 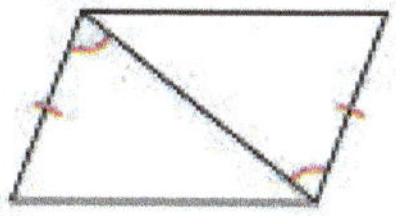

Example 3

Show that the quadrilateral with the given vertices is a parallelogram.

6. $W(-5, -2)$, $X(-3, 3)$, $Y(3, 5)$, $Z(1, 0)$

7. $R(-1, -5)$, $S(-2, -1)$, $T(4, -1)$, $U(5, -5)$

Example 4

8. **Navigation** A parallel rule can be used to plot a course on a navigation chart. The tool is made of two rulers connected at hinges to two congruent crossbars $\overline{AD}$ and $\overline{BC}$. You place the edge of one ruler on your desired course and then move the second ruler over the compass rose on the chart to read the bearing for your course. If $\overline{AD} \parallel \overline{BC}$, why is $\overline{AB}$ always parallel to $\overline{CD}$?

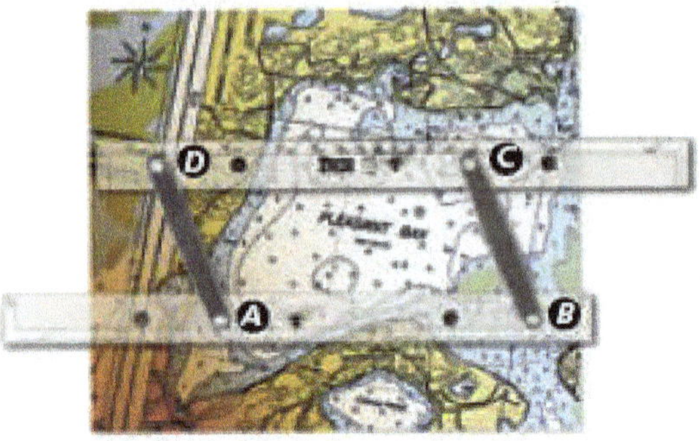

6.1.1 Properties and Attributes of Polygons

Tell whether each figure is a polygon. If it is a polygon, name it by the number of its sides.

1. 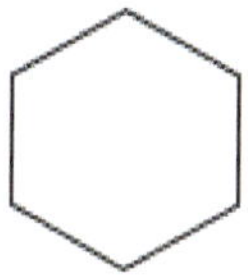

2. 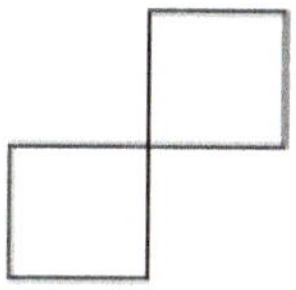

3. 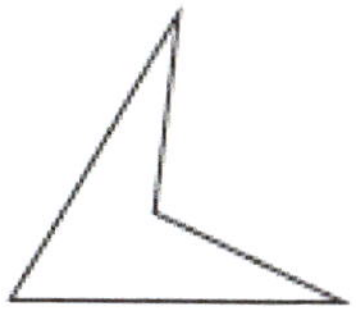

Tell whether each polygon is regular or irregular. Tell whether it is concave or convex.

4. 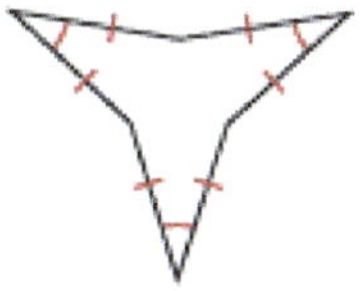

5. 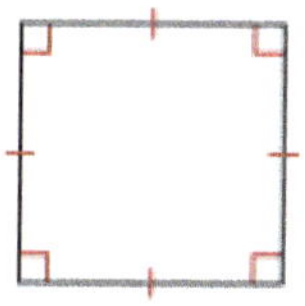

6. 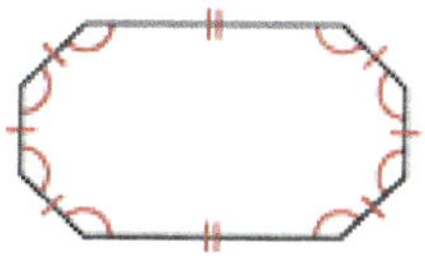

7. Find the measure of each interior angle of quadrilateral *RSTV*.

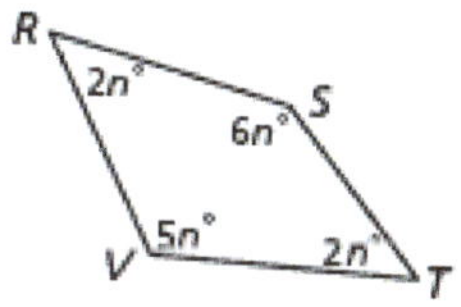

8. Find the measure of each interior angle of a regular 18-gon.

9. Find the sum of the interior angle measures of a convex heptagon.

10. Find the measure of each exterior angle of a regular nonagon.

11. A pentagon has exterior angle measures of $5a°$, $4a°$, $10a°$, $3a°$, and $8a°$. Find the value of a.

Crafts **The folds on the lid of the gift box form a regular hexagon. Find each measure.**

12. $m\angle JKM$

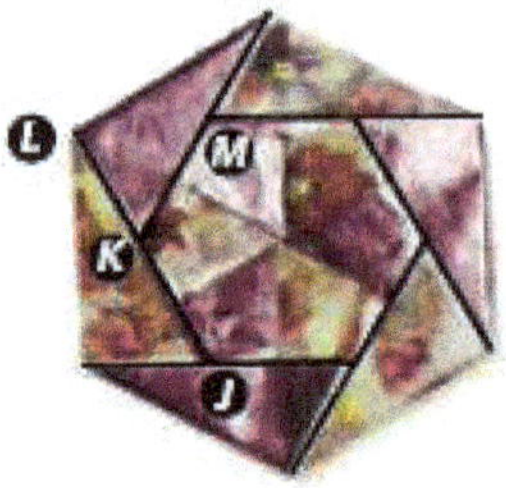

13. $m\angle MKL$

6.1.2 Properties of Parallelograms

Shipping Cranes can be used to load cargo onto ships. In ▱$JKLM$, $JL = 165.8$, $JK = 110$, and $m\angle JML = 50°$. Find the measure of each part of the crane.

14. JN

15. LM

16. LN

17. $m\angle JKL$

18. $m\angle KLM$

19. $m\angle MJK$

WXYZ is a parallelogram. Find each measure.

20. *WV*

21. *YW*

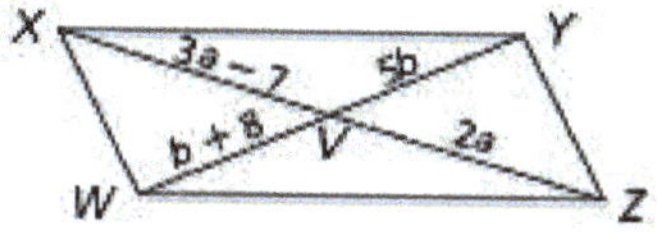

22. *XZ*

23. *ZV*

24. **Multi-Step** Three vertices of ▱*PRTV* are $P(-4, -4)$, $R(-10, 0)$, and $V(5, -1)$. Find the coordinates of vertex *T*.

25. Write a two-column proof.
Given: *ABCD* and *AFGH* are parallelograms.
Prove: $\angle C \cong \angle G$

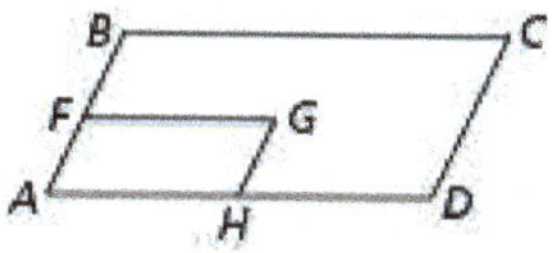

6.1.3 Conditions for Parallelograms

26. Show that *BCGH* is a parallelogram for $x = 3.2$ and $y = 7$.

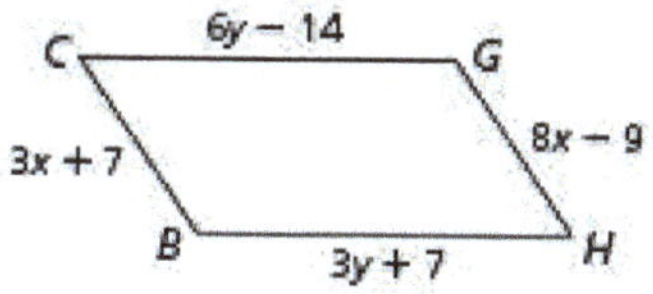

27. Show that *TUVW* is a parallelogram for for $a = 19.5$ and $b = 22$.

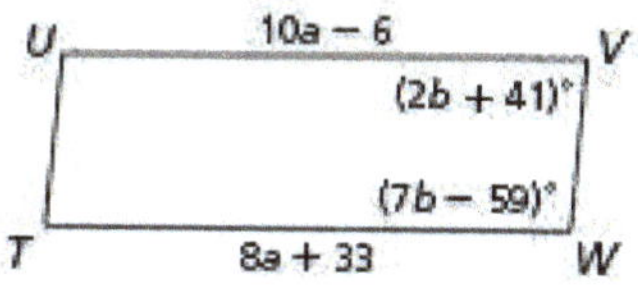

Determine if each quadrilateral must be a parallelogram. Justify your answer.

28. 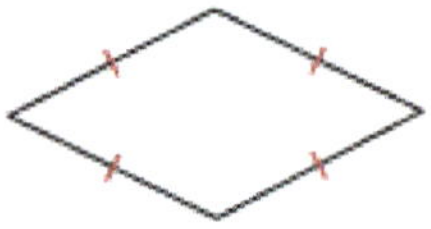

29. 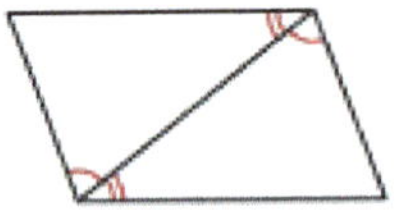

30. 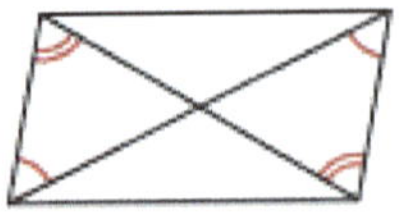

Show that the quadrilateral with the given vertices is a parallelogram.

31. $J(-1, 0)$, $K(-3, 7)$, $L(2, 6)$, $M(4, -1)$

32. $P(-8, -4)$, $Q(-5, 1)$, $R(1, -5)$, $S(-2, -10)$

33. **Design** The toolbox has cantilever trays that pull away from the box so that you can reach the items beneath them. Two congruent brackets connect each tray to the box. Given that $AD = BC$, how do the brackets $\overline{AB}$ and $\overline{CD}$ keep the tray horizontal?

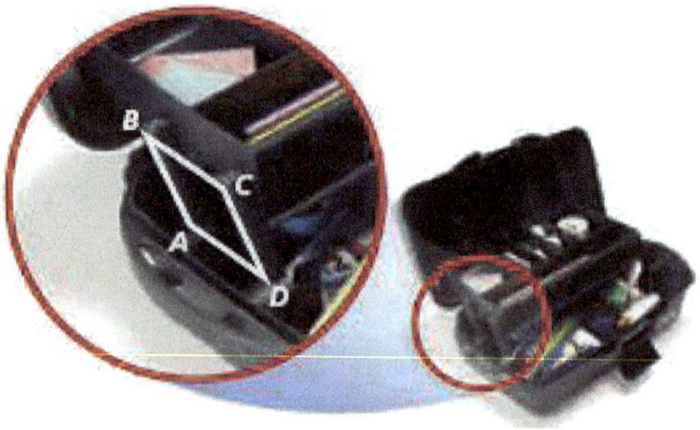

Example 1

Engineering **The braces of the bridge support lie along the diagonals of rectangle *PQRS*. *RS* = 160 ft, and *QS* = 380 ft. Find each length.**

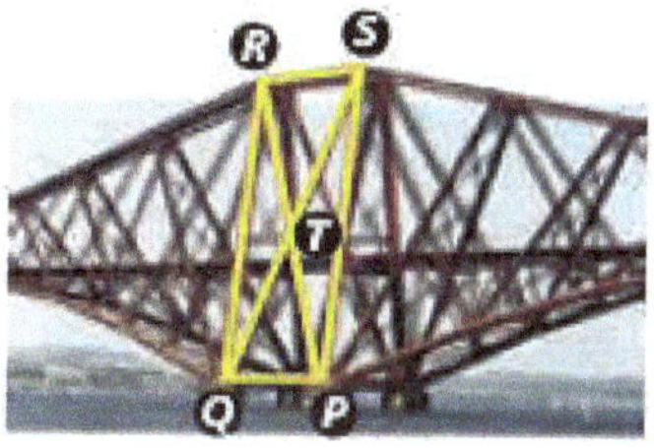

1. *TQ*
2. *PQ*
3. *ST*
4. *PR*

Example 2

***ABCD* is a rhombus. Find each measure.**

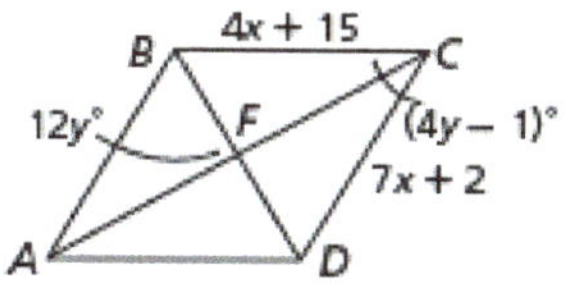

5. *AB*
6. m∠*ABC*

Example 3

7. **Multi-Step** The vertices of square *JKLM* are *J*(−3, −5), *K*(−4, 1), *L*(2, 2), and *M*(3, −4). Show that the diagonals of square *JKLM* are congruent perpendicular bisectors of each other.

Example 4

8. **Given:** $RECT$ is a rectangle. $\overline{RX} \cong \overline{TY}$
   Prove: $\triangle REY \cong \triangle TCX$

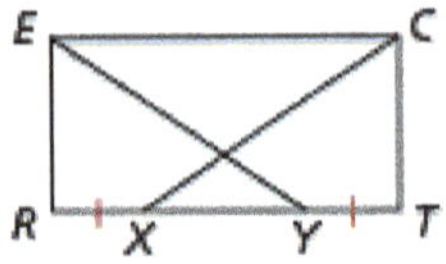

Example 1

1. **Gardening** A city garden club is planting a square garden. They drive pegs into the ground at each corner and tie strings between each pair. The pegs are spaced so that $\overline{WX} \cong \overline{XY} \cong \overline{YZ} \cong \overline{ZW}$. How can the garden club use the diagonal strings to verify that the garden is a square?

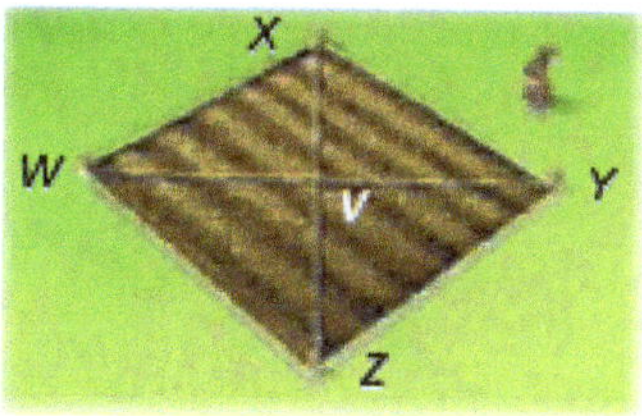

Example 2

Determine if the conclusion is valid. If not, tell what additional information is needed to make it valid.

2. Given: $\overline{AC} \cong \overline{BD}$
   Conclusion: $ABCD$ is a rectangle.

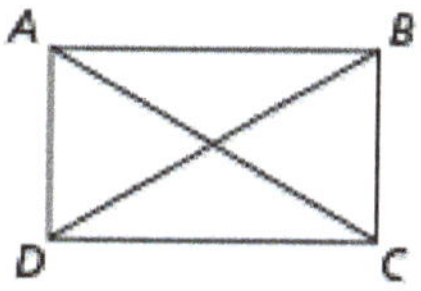

3. Given: $\overline{AB} \parallel \overline{CD}$, $\overline{AB} \cong \overline{CD}$, $\overline{AB} \perp \overline{BC}$
   Conclusion: $ABCD$ is a rectangle.

Example 3

Multi-Step **Use the diagonals to determine whether a parallelogram with the given vertices is a rectangle, rhombus, or square. Give all the names that apply.**

4. $P(-5, 2)$, $Q(4, 5)$, $R(6, -1)$, $S(-3, -4)$

5. $W(-6, 0)$, $X(1, 4)$, $Y(2, -4)$, $Z(-5, -8)$

Example 1

Vocabulary **Apply the vocabulary from this lesson to answer each question.**

1. **Crafts** The edges of the kite-shaped glass in the sun catcher are sealed with lead strips. *JH*, *KH*, and *LH* are 2.75 inches, and *MH* is 5.5 inches. How much lead is needed to seal the edges of the sun catcher? If the craftsperson has two 3-foot lengths of lead, how many sun catchers can be sealed?

Example 2

In kite $WXYZ$, **m**$\angle WXY = 104°$, **and m**$\angle VYZ = 49°$. **Find each measure.**

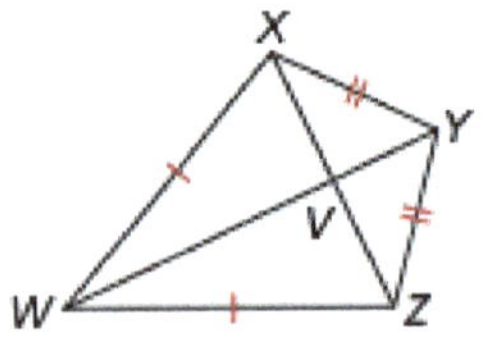

2. m$\angle VZY$

3. m$\angle VXW$

4. m$\angle XWZ$

Example 3

5. Find m$\angle A$.

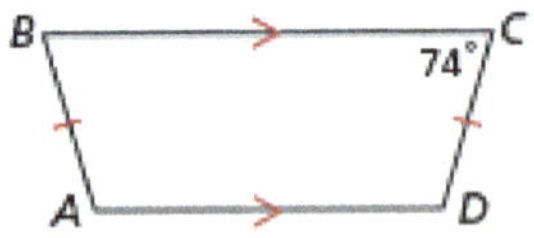

6. $RW = 17.7$, and $SV = 23.3$. Find TW.

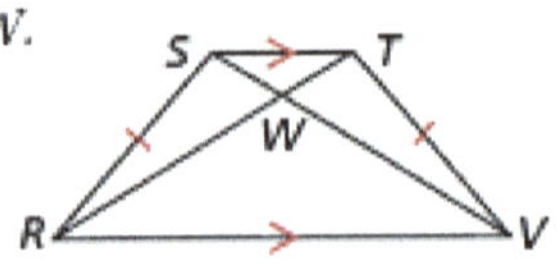

Example 4

7. Find the value of z so that $EFGH$ is isosceles.

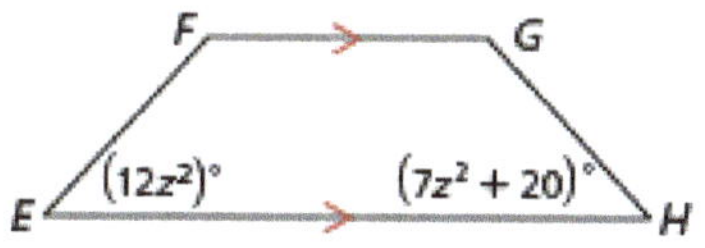

8. $MQ = 7y - 6$, and $LP = 4y + 11$. Find the value of y so that $LMPQ$ is isosceles.

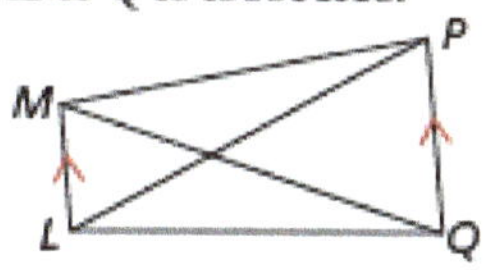

Example 5

9. Find QR.

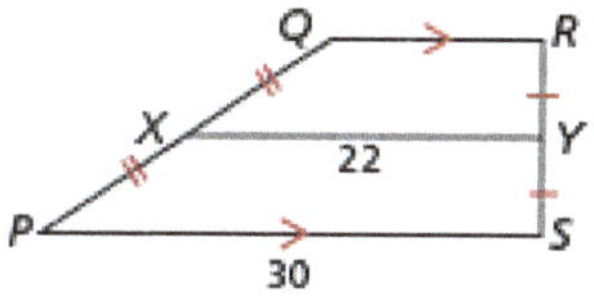

10. Find AZ.

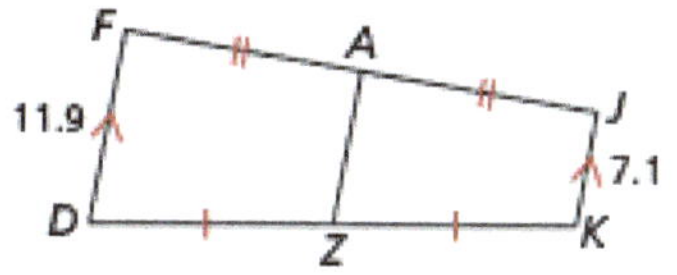

6.2.1 Properties of Special Parallelograms

Carpentry A carpenter measures the diagonals of a piece of wood. In rectangle $JKLM$, $JM = 25$ in., and $JP = 14\frac{1}{2}$ in. Find each length.

1. JL

2. KL

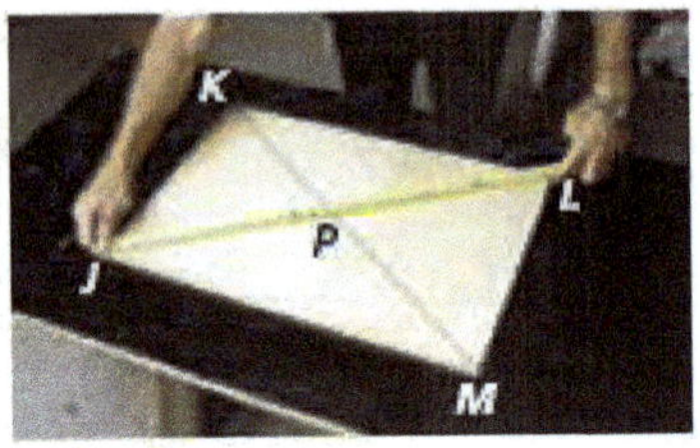

3. KM

4. MP

$VWXY$ is a rhombus. Find each measure.

5. VW

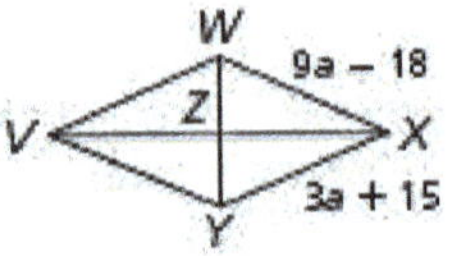

6. $m\angle VWX$ and $m\angle WYX$ if $m\angle WVY = (4b + 10)^\circ$ and $m\angle XZW = (10b - 5)^\circ$

7. **Multi-Step** The vertices of square $PQRS$ are $P(-4, 0)$, $Q(4, 3)$, $R(7, -5)$, and $S(-1, -8)$. Show that the diagonals of square $PQRS$ are congruent perpendicular bisectors of each other.

8. **Given:** *RHMB* is a rhombus with diagonal $\overline{HB}$.
Prove: $\angle HMX \cong \angle HRX$

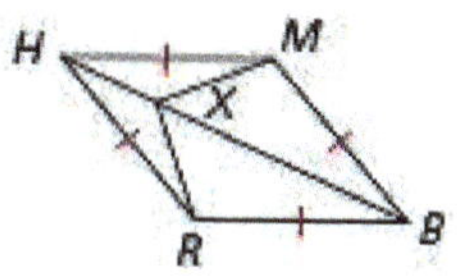

6.2.2 Conditions for Special Parallelograms

9. **Crafts** A framer uses a clamp to hold together the pieces of a picture frame. The pieces are cut so that $\overline{PQ} \cong \overline{RS}$ and $\overline{QR} \cong \overline{SP}$. The clamp is adjusted so that *PZ*, *QZ*, *RZ*, and *SZ* are all equal. Why must the frame be a rectangle?

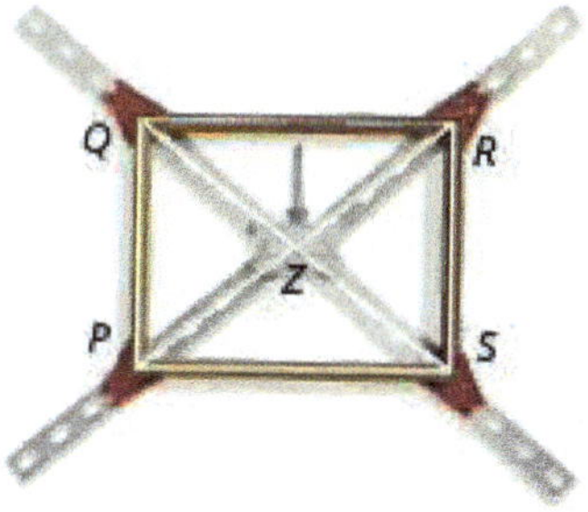

Determine if the conclusion is valid. If not, tell what additional information is needed to make it valid.

10. Given: $\overline{EG}$ and $\overline{FH}$ bisect each other. $\overline{EG} \perp \overline{FH}$
Conclusion: *EFGH* is a rhombus.

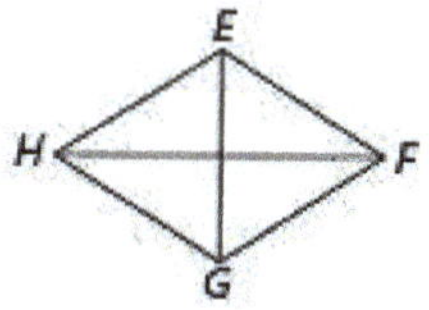

11. Given: $\overline{FH}$ bisects $\angle EFG$ and $\angle EHG$.
Conclusion: *EFGH* is a rhombus.

Multi-Step **Use the diagonals to determine whether a parallelogram with the given vertices is a rectangle, rhombus, or square. Give all the names that apply.**

12. $A(-10, 4), B(-2, 10), C(4, 2), D(-4, -4)$

13. $J(-9, -7), K(-4, -2), L(3, -3), M(-2, -8)$

6.2.3 Properties of Kites and Trapezoids

14. **Design** Each square section in the iron railing contains four small kites. The figure shows the dimensions of one kite. What length of iron is needed to outline one small kite? How much iron is needed to outline one complete section, including the square?

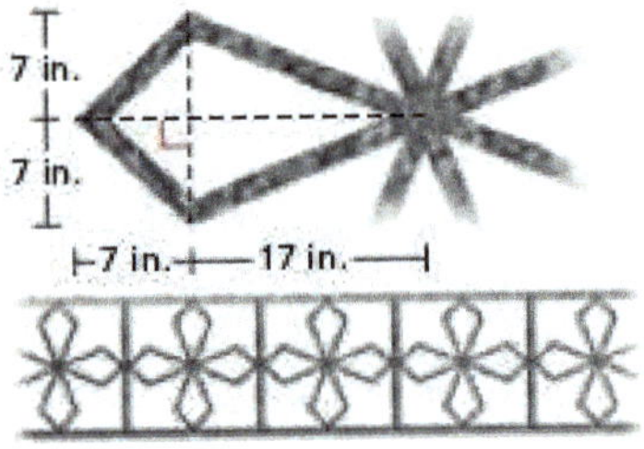

In kite $ABCD$, $m\angle DAX = 32°$, **and** $m\angle XDC = 64°$. **Find each measure.**

15. $m\angle XDA$

16. $m\angle ABC$

17. $m\angle BCD$

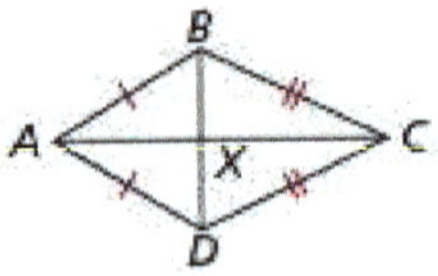

Polygons and Quadrilaterals

Review 6.2 / Other Special Quadrilaterals

18. Find $m\angle Q$.

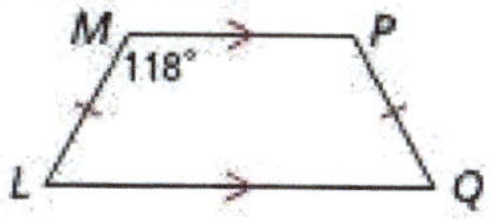

19. $SZ = 62.6$, and $KZ = 34$. Find RJ.

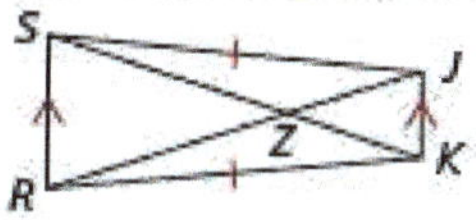

20. **Algebra** Find the value of a so that $XYZW$ is isosceles. Give your answer as a simplified radical.

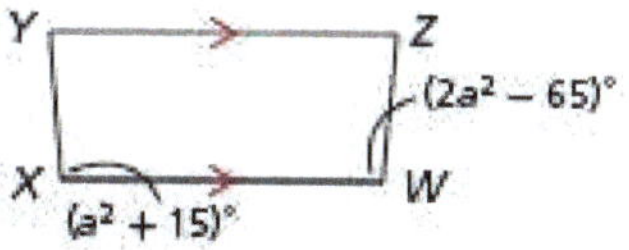

21. **Algebra** $GJ = 4x - 1$, and $FH = 9x - 15$. Find the value of x so that $FGHJ$ is isosceles.

22. Find PQ.

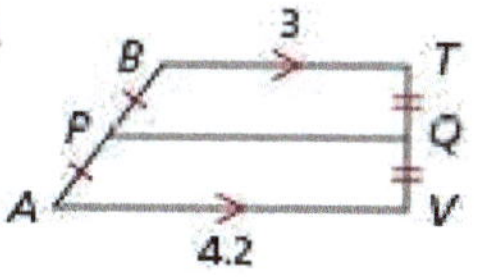

23. Find KR.

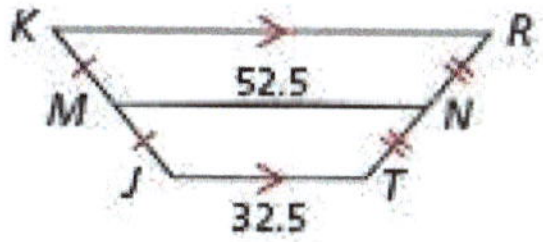

Example 1

Write a ratio expressing the slope of each line.

1. ℓ **2.** m **3.** n

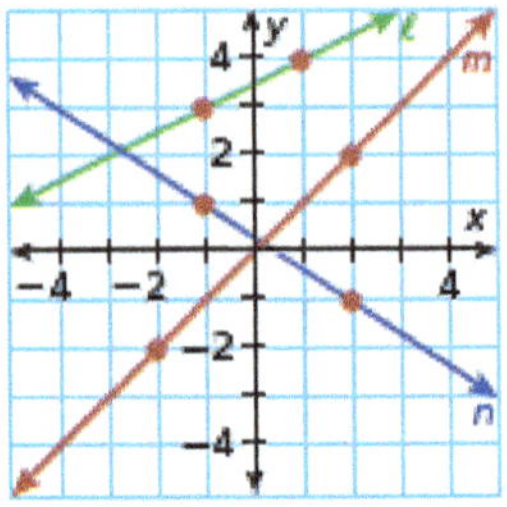

Example 2

4. The ratio of the side lengths of a quadrilateral is 2:4:5:7, and its perimeter is 36 m. What is the length of the shortest side?

5. The ratio of the angle measures in a triangle is 5:12:19. What is the measure of the largest angle?

Example 3

Solve each proportion.

6. $\frac{x}{2} = \frac{40}{16}$ **7.** $\frac{7}{y} = \frac{21}{27}$ **8.** $\frac{6}{58} = \frac{t}{29}$

9. $\frac{y}{3} = \frac{27}{y}$

10. $\frac{16}{x-1} = \frac{x-1}{4}$

11. $\frac{x^2}{18} = \frac{x}{6}$

Example 4

12. Given that $2a = 8b$, find the ratio of a to b in simplest form.

13. Given that $6x = 27y$, find the ratio $y:x$ in simplest form.

Example 5

14. **Architecture** The Arkansas State Capitol Building is a smaller version of the U.S. Capitol Building. The U.S. Capitol is 752 ft long and 288 ft tall. The Arkansas State Capitol is 564 ft long. What is the height of the Arkansas State Capitol?

Example 1

Identify the pairs of congruent angles and corresponding sides.

1.

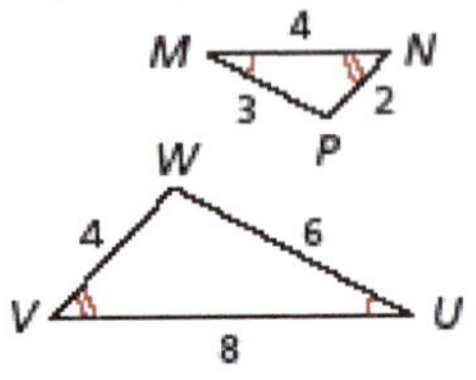

2.

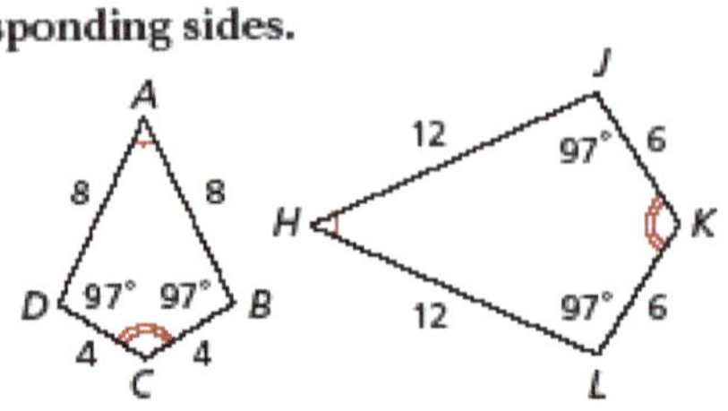

Example 2

Multi-Step **Determine whether the polygons are similar. If so, write the similarity ratio and a similarity statement.**

3. rectangles *ABCD* and *EFGH*

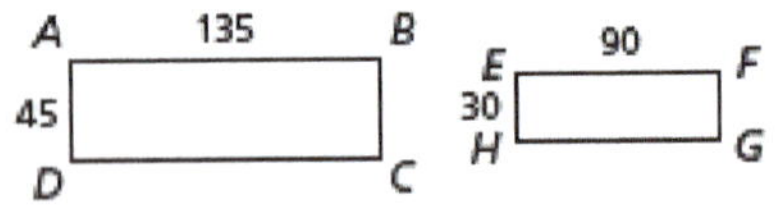

4. $\triangle RMP$ and $\triangle UWX$

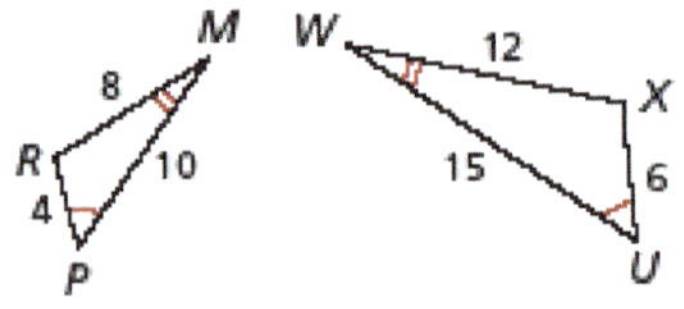

Example 3

5. **Art** The town of Goodland, Kansas, claims that it has one of the world's largest easels. It holds an enlargement of a van Gogh painting that is 24 ft wide. The original painting is 58 cm wide and 73 cm tall. If the reproduction is similar to the original, what is the height of the reproduction to the nearest foot?

Similarity

Similarity Relationships

Triangle Similarity: AA, SSS, and SAS

Example 1

Explain why the triangles are similar and write a similarity statement.

1.

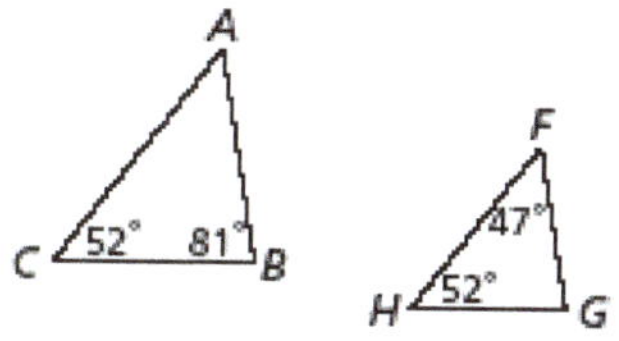

2.

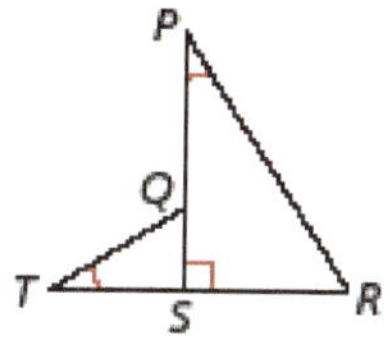

Example 2

Verify that the triangles are similar.

3. $\triangle DEF$ and $\triangle JKL$

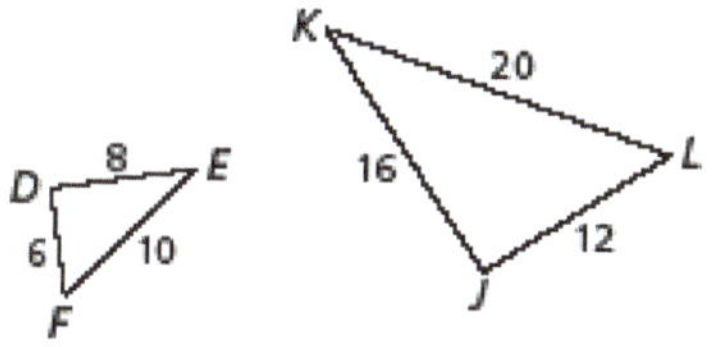

4. $\triangle MNP$ and $\triangle MRQ$

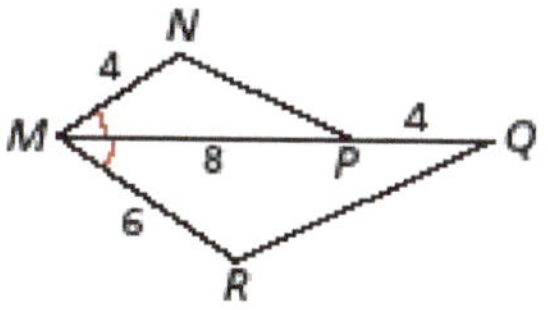

Example 3

Multi-Step **Explain why the triangles are similar and then find each length.**

5. *AB*

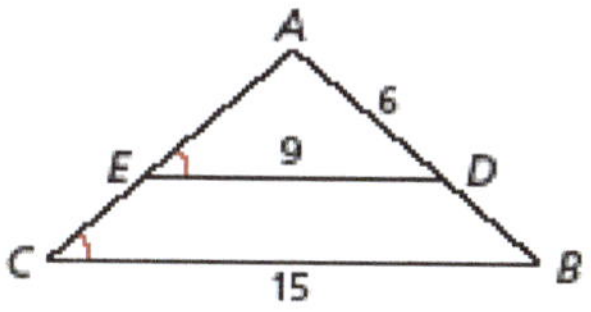

6. *WY*

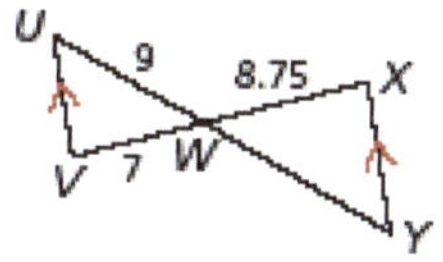

Example 4

7. Given: $\overleftrightarrow{MN} \parallel \overline{KL}$

Prove: $\triangle JMN \sim \triangle JKL$

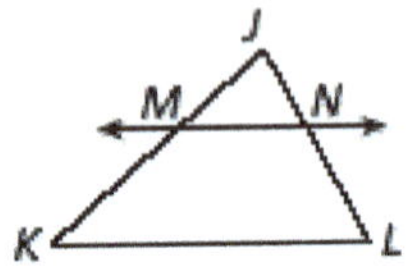

8. Given: $SQ = 2QP$, $TR = 2RP$

Prove: $\triangle PQR \sim \triangle PST$

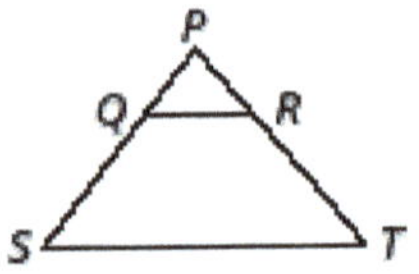

9. The coordinates of A, B, and C are $A(0, 0)$, $B(2, 6)$, and $C(8, -2)$. What theorem or postulate justifies the statement $\triangle ABC \sim \triangle DEF$, if the coordinates of D and E are twice the coordinates of B and C?

Example 5

10. Surveying In order to measure the distance AB across the meteorite crater, a surveyor at S locates points A, B, C, and D as shown. What is AB to the nearest meter? nearest kilometer?

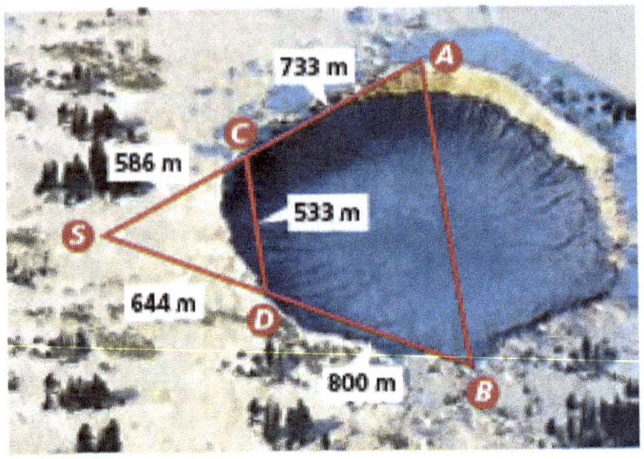

Similarity

Review 7.1 / Similarity Relationships

7.1.1 Ratio and Proportion

Write a ratio expressing the slope of each line.

1. ℓ
2. m
3. n

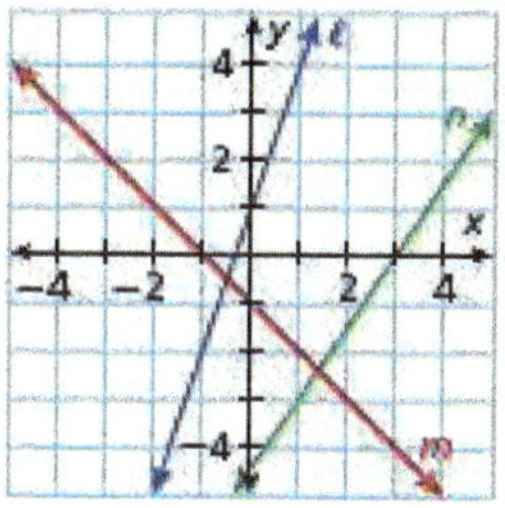

4. The ratio of the side lengths of an isosceles triangle is 4:4:7, and its perimeter is 52.5 cm. What is the length of the base of the triangle?

5. The ratio of the angle measures in a parallelogram is 2:3:2:3. What is the measure of each angle?

Solve each proportion.

6. $\frac{6}{8} = \frac{9}{y}$
7. $\frac{x}{14} = \frac{50}{35}$
8. $\frac{z}{12} = \frac{3}{8}$
9. $\frac{2m+2}{3} = \frac{12}{2m+2}$
10. $\frac{5y}{16} = \frac{125}{y}$
11. $\frac{x+2}{12} = \frac{5}{x-2}$

12. Given that $5y = 25x$, find the ratio of x to y in simplest form.

13. Given that $35b = 21c$, find the ratio $b : c$ in simplest form.

14. **Travel** Madurodam is a park in the Netherlands that contains a complete Dutch city built entirely of miniature models. One of the models of a windmill is 1.2 m tall and 0.8 m wide. The width of the actual windmill is 20 m. What is its height?

7.1.2 Ratios in Similar Polygons

Identify the pairs of congruent angles and corresponding sides.

15.

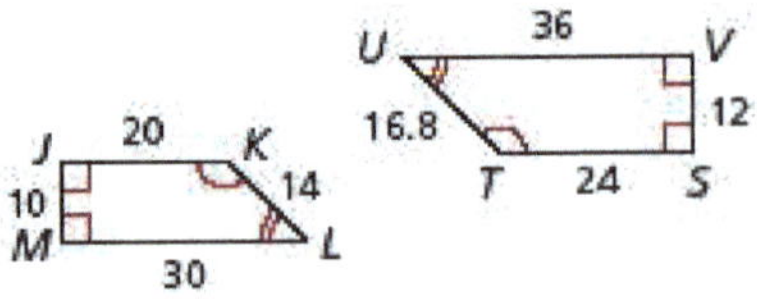

16.

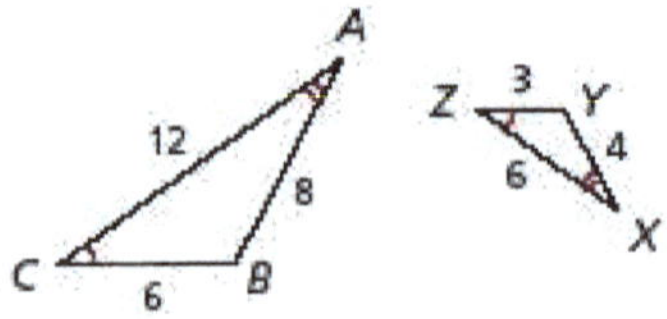

Multi-Step Determine whether the polygons are similar. If so, write the similarity ratio and a similarity statement.

17. $\triangle RSQ$ and $\triangle UXZ$

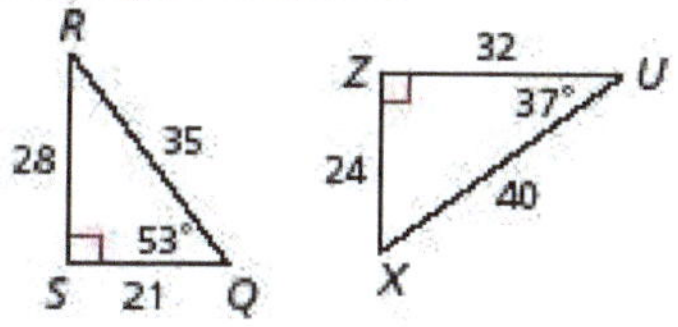

18. rectangles *ABCD* and *JKLM*

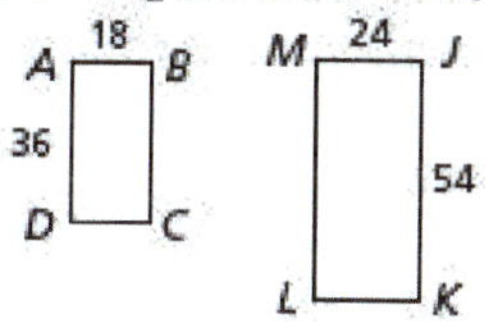

19. **Hobbies** The ratio of the model car's dimensions to the actual car's dimensions is $\frac{1}{56}$. The model has a length of 3 in. What is the length of the actual car?

7.1.3 Triangle Similarity: AA, SSS, and SAS

Explain why the triangles are similar and write a similarity statement.

20.

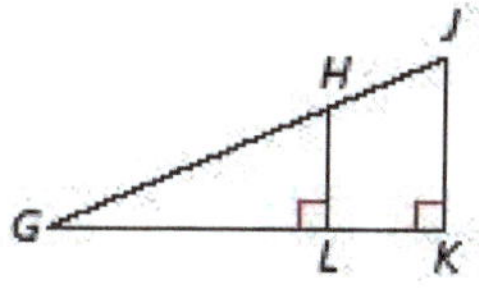

21.

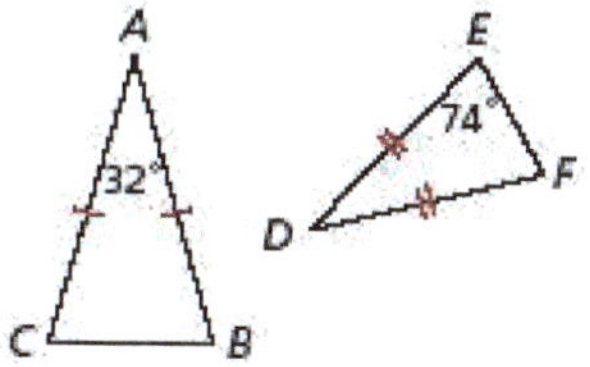

Similarity

Review 7.1 / Similarity Relationships

Verify that the given triangles are similar.

22. $\triangle KLM$ and $\triangle KNL$

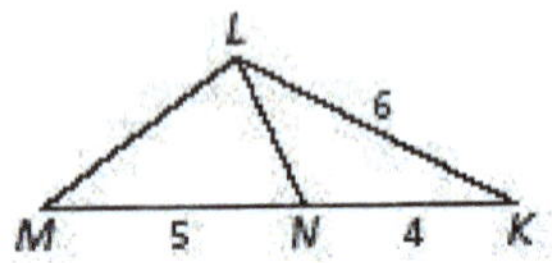

23. $\triangle UVW$ and $\triangle XYZ$

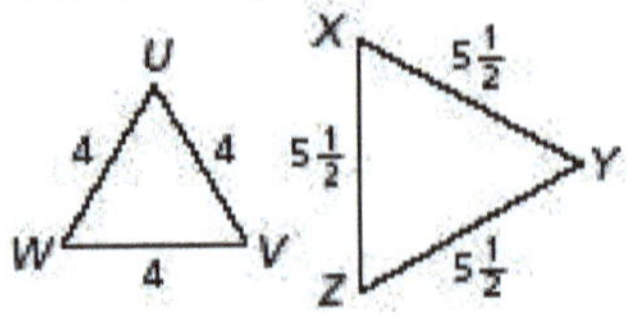

Multi-Step **Explain why the triangles are similar and then find each length.**

24. *AB*

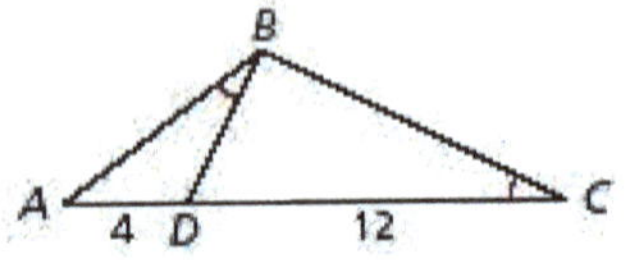

25. *PS*

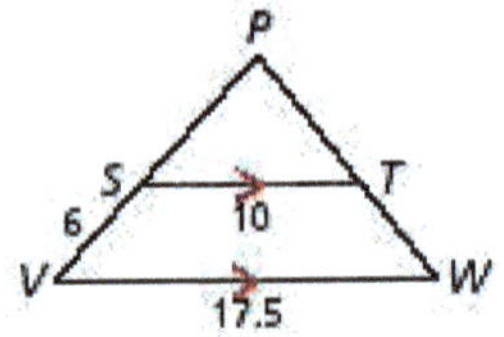

26. Given: $CD = 3AC$, $CE = 3BC$

Prove: $\triangle ABC \sim \triangle DEC$

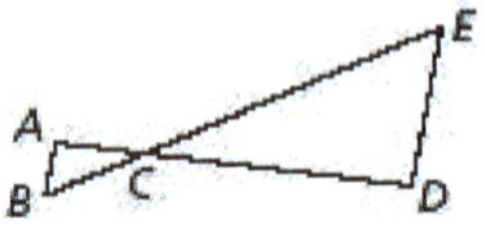

27. Given: $\frac{PR}{MR} = \frac{QR}{NR}$

Prove: $\angle 1 \cong \angle 2$

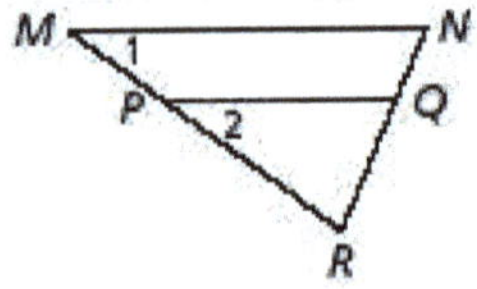

28. **Photography** The picture shows a person taking a pinhole photograph of himself. Light entering the opening reflects his image on the wall, forming similar triangles. What is the height of the image to the nearest tenth of an inch?

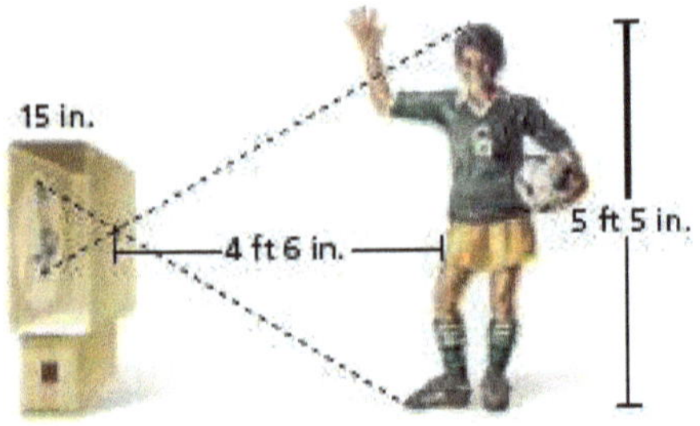

Similarity in Right Triangles

Example 1

Write a similarity statement comparing the three triangles in each diagram.

1. 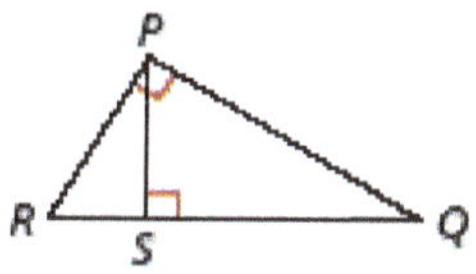

2. 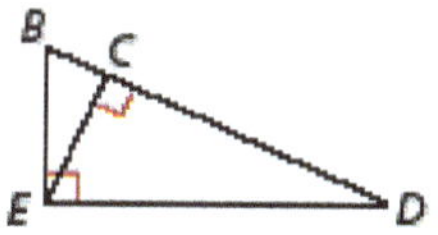

3. 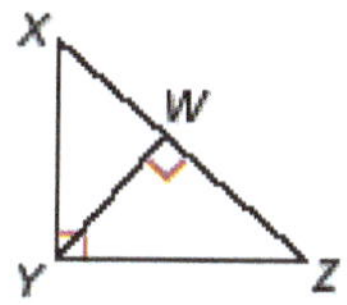

Example 2

Find the geometric mean of each pair of numbers. If necessary, give the answer in simplest radical form.

4. 2 and 50

5. 4 and 16

6. $\frac{1}{2}$ and 8

7. 9 and 12

8. 16 and 25

9. 7 and 11

Example 3

Find x, y, and z.

10. 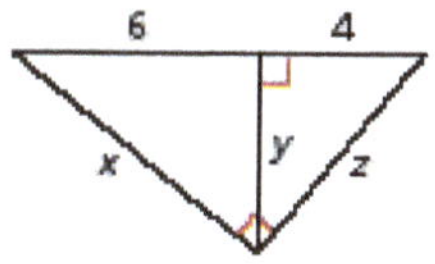

11. 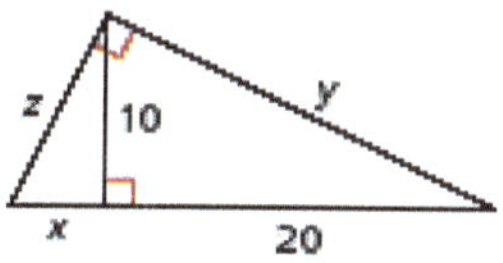

12. 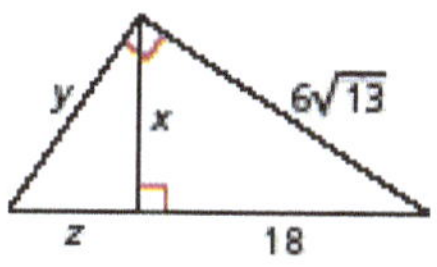

Example 4

13. **Measurement** To estimate the length of the USS *Constitution* in Boston harbor, a student locates points T and U as shown. What is RS to the nearest tenth?

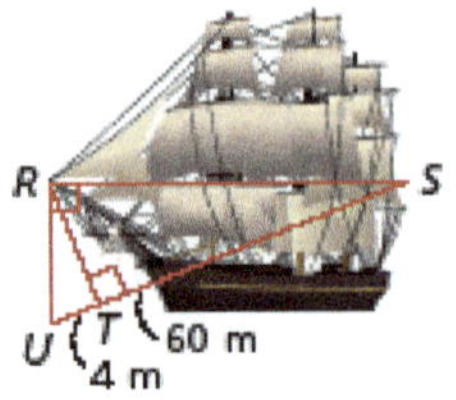

Trigonometric Ratios

Example 1

Write each trigonometric ratio as a fraction and as a decimal rounded to the nearest hundredth.

1. $\sin C$
2. $\tan A$
3. $\cos A$
4. $\cos C$
5. $\tan C$
6. $\sin A$

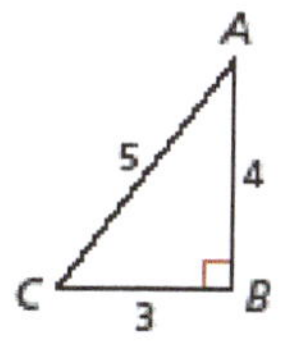

Example 2

Use a special right triangle to write each trigonometric ratio as a fraction.

7. $\cos 60°$
8. $\tan 30°$
9. $\sin 45°$

Example 3

Use your calculator to find each trigonometric ratio. Round to the nearest hundredth.

10. $\tan 67°$
11. $\sin 23°$
12. $\sin 49°$

13. $\cos 88°$

14. $\cos 12°$

15. $\tan 9°$

Example 4

Find each length. Round to the nearest hundredth.

16. *BC*

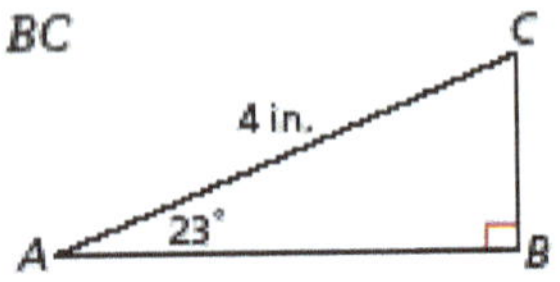

17. *QR*

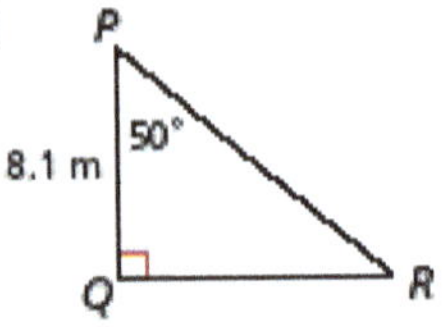

18. *KL*

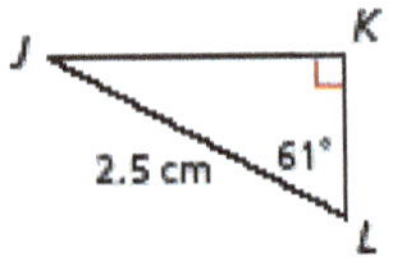

Example 5

19. **Architecture** A pediment has a pitch of 15°, as shown. If the width of the pediment, *WZ*, is 56 ft, what is *XY* to the nearest inch?

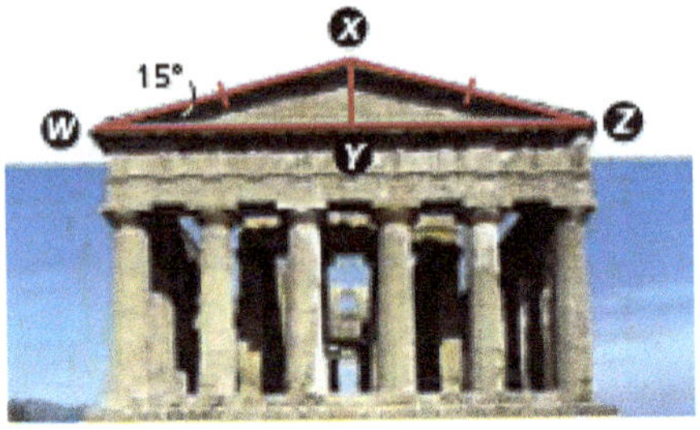

Right Triangles and Trigonometry

Trigonometric Ratios

Solving Right Triangles

Example 1

Use the given trigonometric ratio to determine which angle of the triangle is $\angle A$.

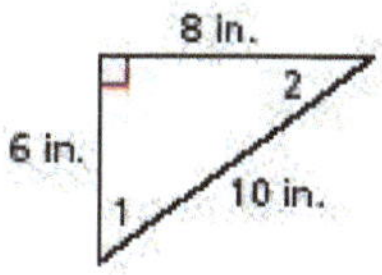

1. $\sin A = \frac{4}{5}$ 2. $\tan A = 1\frac{1}{3}$ 3. $\cos A = 0.6$

4. $\cos A = 0.8$ 5. $\tan A = 0.75$ 6. $\sin A = 0.6$

Example 2

Use your calculator to find each angle measure to the nearest degree.

7. $\tan^{-1}(2.1)$ 8. $\cos^{-1}\left(\frac{1}{3}\right)$ 9. $\cos^{-1}\left(\frac{5}{6}\right)$

10. $\sin^{-1}(0.5)$ 11. $\sin^{-1}(0.61)$ 12. $\tan^{-1}(0.09)$

Example 3

Multi-Step **Find the unknown measures. Round lengths to the nearest hundredth and angle measures to the nearest degree.**

13.
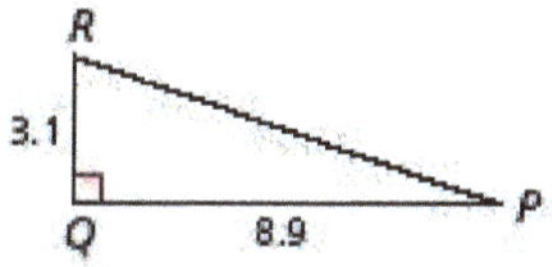

14.
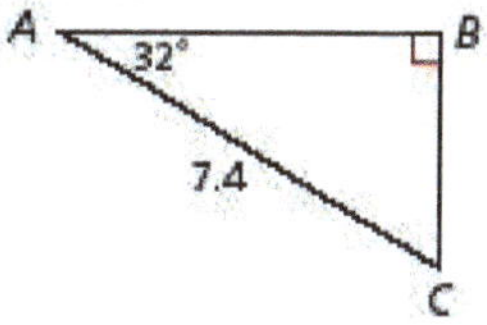

15.
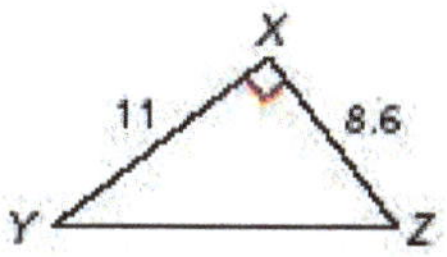

Example 4

Multi-Step For each triangle, find the side lengths to the nearest hundredth and the angle measures to the nearest degree.

16. $D(4, 1)$, $E(4, -2)$, $F(-2, -2)$

17. $R(3, 3)$, $S(-2, 3)$, $T(-2, -3)$

18. $A(-1, 1)$, $B(1, 1)$, $C(1, 5)$

Example 5

19. **Cycling** A hill in the Tour de France bike race has a grade of 8%. To the nearest degree, what is the angle that this hill makes with a horizontal line?

G Thinkwell's Geometry with Edward Burger — Worksheet

Right Triangles and Trigonometry

Review 8.1 / Trigonometric Ratios

8.1.1 Similarity in Right Triangles

Write a similarity statement comparing the three triangles in each diagram.

1. 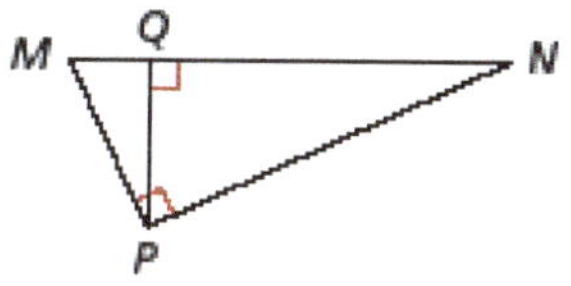

2. 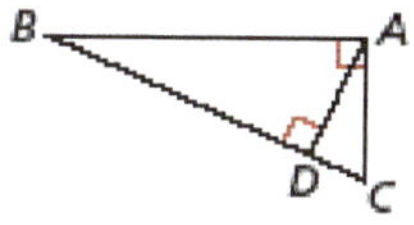

3. 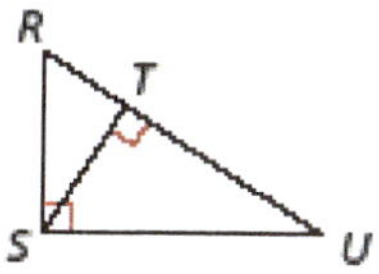

Find the geometric mean of each pair of numbers. If necessary, give the answer in simplest radical form.

4. 5 and 45

5. 3 and 15

6. 5 and 8

7. $\frac{1}{4}$ and 80

8. 1.5 and 12

9. $\frac{2}{3}$ and $\frac{27}{40}$

Find x, y, and z.

10. 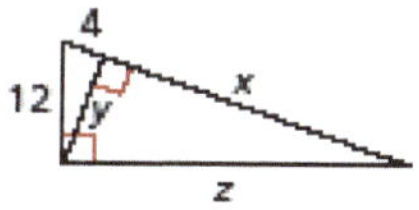

11. 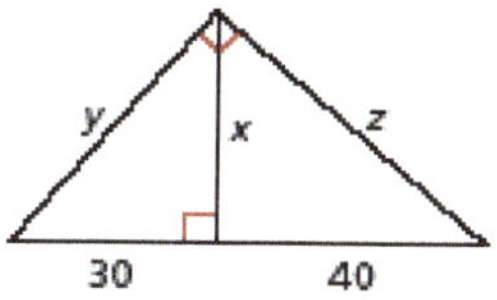

12. 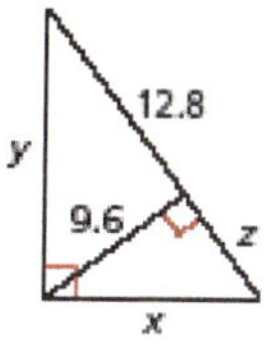

13. **Measurement** To estimate the height of the Taipei 101 tower, Andrew stands so that his lines of sight to the top and bottom of the tower form a 90° angle. What is the height of the tower to the nearest foot?

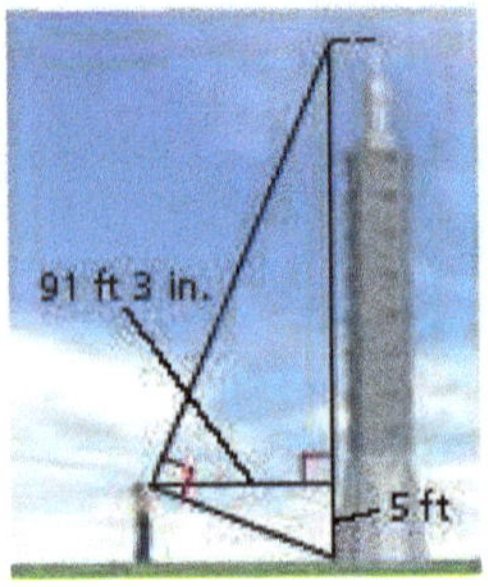

8.1.2 Trigonometric Ratios

Write each trigonometric ratio as a fraction and as a decimal rounded to the nearest hundredth.

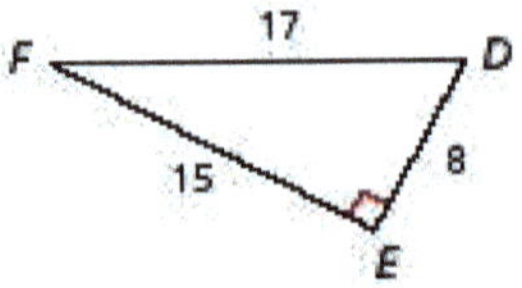

14. $\cos D$

15. $\tan D$

16. $\tan F$

17. $\cos F$

18. $\sin F$

19. $\sin D$

Use a special right triangle to write each trigonometric ratio as a fraction.

20. $\tan 60°$

21. $\sin 30°$

22. $\cos 45°$

Right Triangles and Trigonometry

Review 8.1 / Trigonometric Ratios

Use your calculator to find each trigonometric ratio. Round to the nearest hundredth.

23. $\tan 51°$

24. $\sin 80°$

25. $\cos 77°$

26. $\tan 14°$

27. $\sin 55°$

28. $\cos 48°$

Find each length. Round to the nearest hundredth.

29. *PQ*

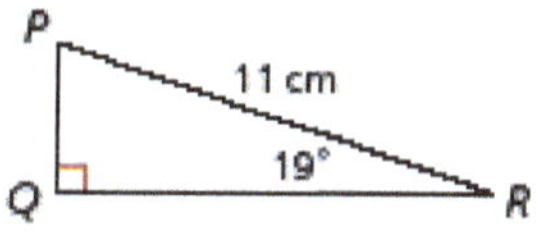

30. *AC*

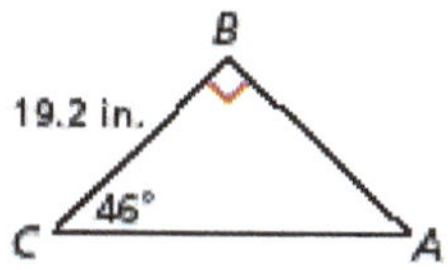

31. *GH*

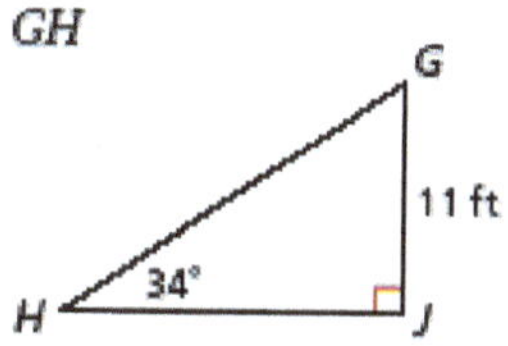

32. *XZ*

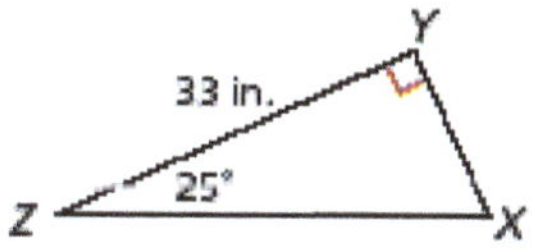

33. *KL*

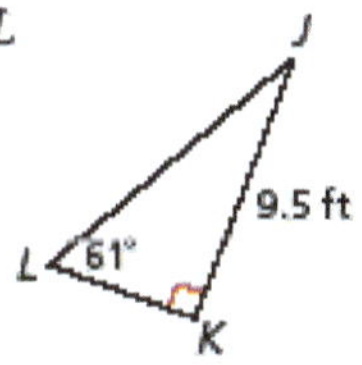

34. *EF*

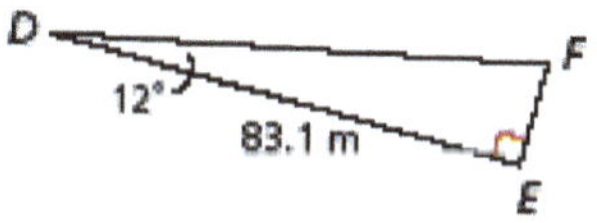

35. **Sports** A jump ramp for waterskiing makes an angle of 15° with the surface of the water. The ramp rises 1.58 m above the surface. What is the length of the ramp to the nearest hundredth of a meter?

8.1.3 Solving Right Triangles

Use the given trigonometric ratio to determine which angle of the triangle is $\angle A$.

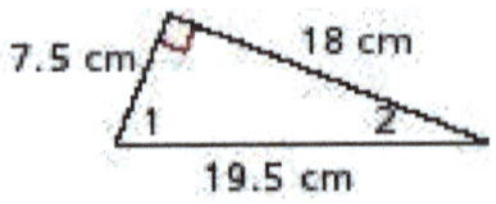

36. $\tan A = \frac{5}{12}$

37. $\tan A = 2.4$

38. $\sin A = \frac{12}{13}$

39. $\sin A = \frac{5}{13}$

40. $\cos A = \frac{12}{13}$

41. $\cos A = \frac{5}{13}$

Use your calculator to find each angle measure to the nearest degree.

42. $\sin^{-1}(0.31)$

43. $\tan^{-1}(1)$

44. $\cos^{-1}(0.8)$

45. $\cos^{-1}(0.72)$

46. $\tan^{-1}(1.55)$

47. $\sin^{-1}\left(\frac{9}{17}\right)$

Multi-Step **Find the unknown measures. Round lengths to the nearest hundredth and angle measures to the nearest degree.**

48.

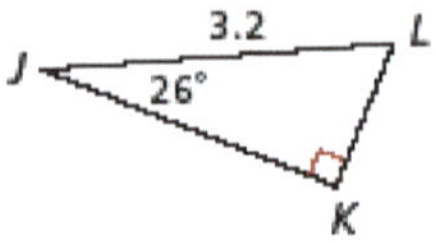

49.

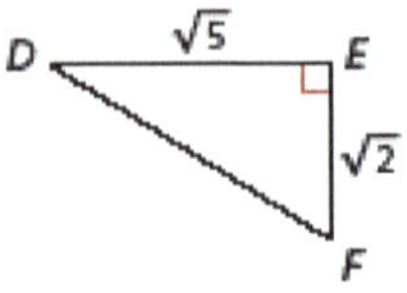

50.

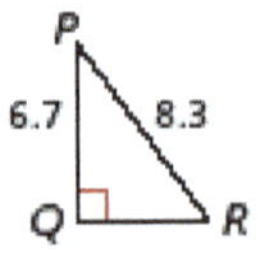

Multi-Step For each triangle, find the side lengths to the nearest hundredth and the angle measures to the nearest degree.

51. $A(2, 0)$, $B(2, -5)$, $C(1, -5)$

52. $M(3, 2)$, $N(3, -2)$, $P(-1, -2)$

53. Building For maximum accessibility, a wheelchair ramp should have a slope between $\frac{1}{16}$ and $\frac{1}{20}$. What is the range of angle measures that a ramp should make with a horizontal line? Round to the nearest degree.

Example 1

Find each measurement.

1. the area of the parallelogram

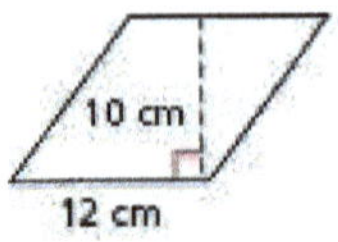

2. the height of the rectangle, in which $A = 10x^2$ ft^2

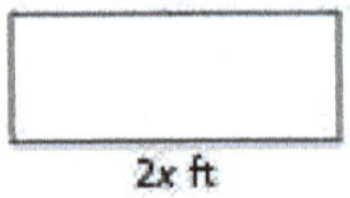

3. the perimeter of a square in which $A = 169$ cm^2

Example 2

Find each measurement.

4. the area of the trapezoid

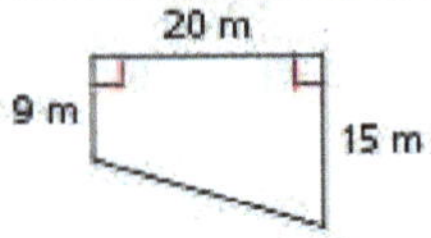

5. the base of the triangle, in which $A = 58.5$ in^2

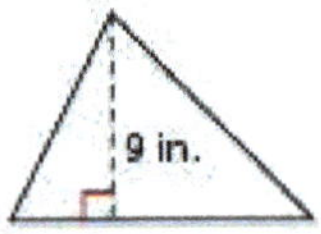

6. b_1 of a trapezoid in which $A = (48x + 68)$ in^2, $h = 8$ in., and $b_2 = (9x + 12)$ in.

Example 3

Find each measurement.

7. the area of the rhombus

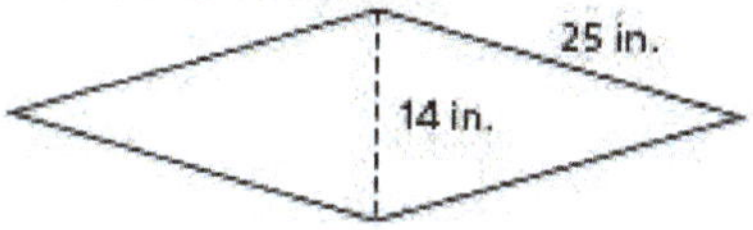

8. d_2 of the kite, in which $A = 187.5\ m^2$

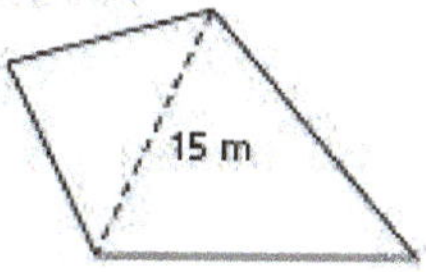

9. d_2 of a kite in which $A = 12x^2y^3\ cm^2$, $d_1 = 3xy$ cm

Example 4

10. **Art** The stained-glass window shown is a rectangle with a base of 4 ft and a height of 3 ft. Use the grid to find the area of each piece.

Developing Formulas for Circles and Regular Polygons

Example 1

Find each measurement.

1. the circumference of $\odot C$

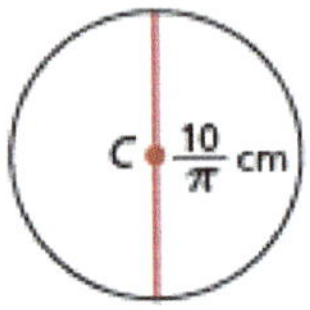

2. the area of $\odot A$ in terms of π

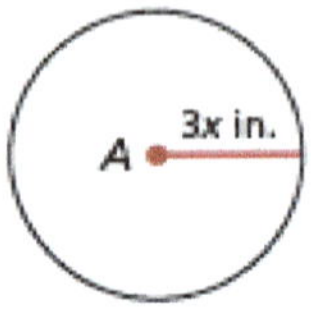

3. the circumference of $\odot P$ in which $A = 36\pi$ in^2

Example 2

4. **Food** A pizza parlor offers pizzas with diameters of 8 in., 10 in., and 12 in. Find the area of each size pizza. Round to the nearest tenth.

Example 3

Find the area of each regular polygon. Round to the nearest tenth.

5. 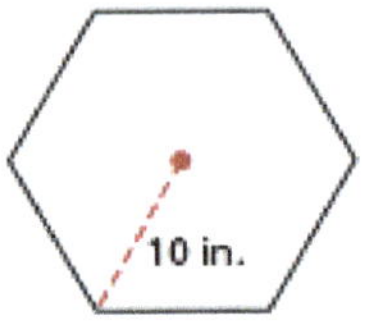

6. 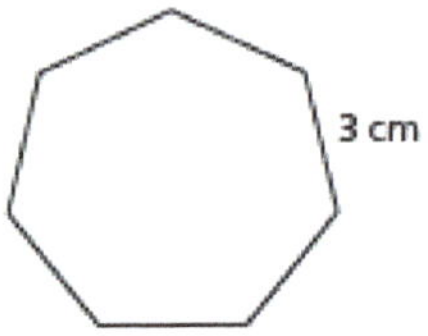

7. an equilateral triangle with an apothem of 2 ft

8. a regular dodecagon with a side length of 5 m

9.1.1 Developing Formulas for Triangles and Quadrilaterals

Find each measurement.

1. the height of the parallelogram, in which $A = 7.5\ \text{m}^2$

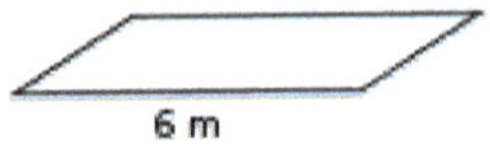

2. the perimeter of the rectangle

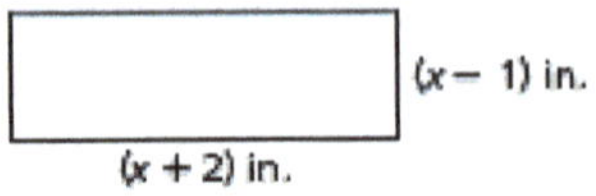

3. the area of a parallelogram in which $b = (3x + 5)$ ft and $h = (7x - 1)$ ft

4. the area of the triangle

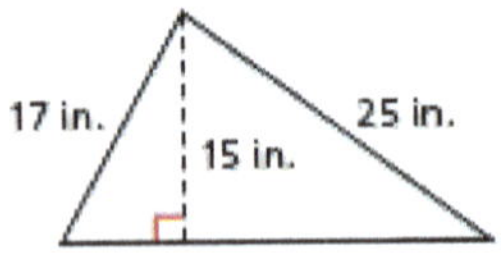

5. the height of the trapezoid, in which $A = 280\ \text{cm}^2$

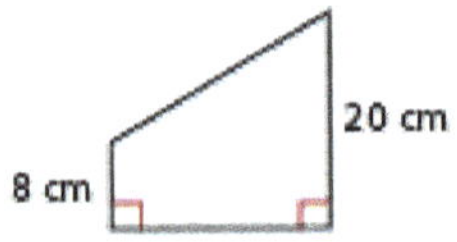

6. the area of a triangle in which $b = (x + 1)$ ft and $h = 8x$ ft

7. the area of the kite

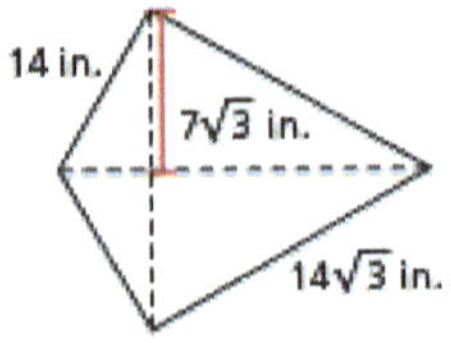

8. d_2 of the rhombus, in which $A = (3x^2 + 6x)\ \text{m}^2$

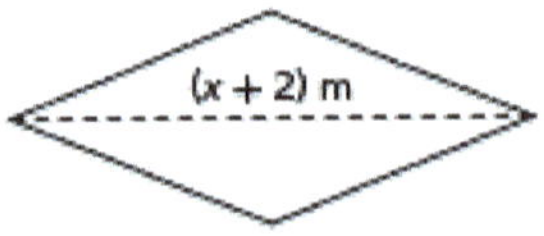

9. the area of a kite in which $d_1 = (6x + 5)$ ft and $d_2 = (4x + 8)$ ft

Crafts **In origami, a *square base* is the starting point for the creation of many figures, such as a crane. In the pattern for the square base, *ABCD* is a square, and *E, F, G,* and *H* are the midpoints of the sides. If *AB* = 6 in., find the area of each shape.**

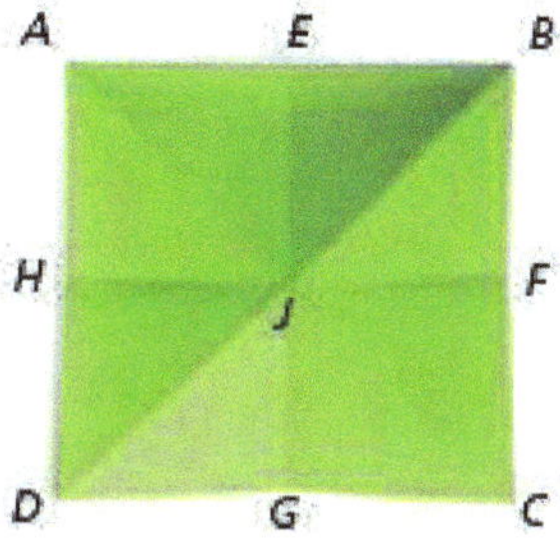

10. rectangle *ABFH*

11. $\triangle AEJ$

12. trapezoid *ABFJ*

9.1.2 Developing Formulas for Circles and Regular Polygons

Find each measurement. Give your answers in terms of π.

13. the area of $\odot M$

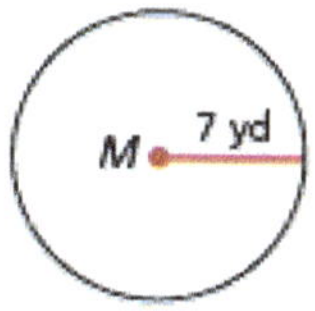

14. the circumference of $\odot Z$

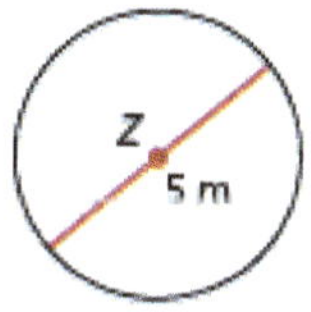

15. the diameter of $\odot G$ in which $C = 10$ ft.

16. **Sports** A horse trainer uses circular pens that are 35 ft, 50 ft, and 66 ft in diameter. Find the area of each pen. Round to the nearest tenth.

Find the area of each regular polygon. Round to the nearest tenth, if necessary.

17.

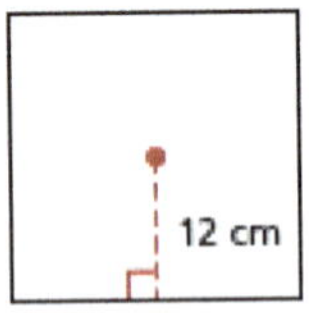

18.

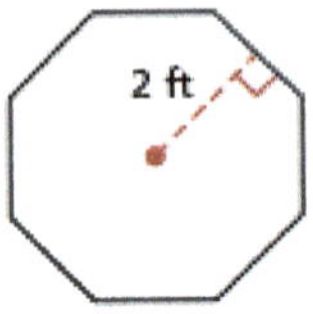

19. a regular nonagon with a perimeter of 144 in.

20. a regular pentagon with an apothem of 2 ft.

Spatial Reasoning

Surface Area and Volume

Surface Area of Prisms and Cylinders

Example 1

Find the lateral area and surface area of each right prism.

1.

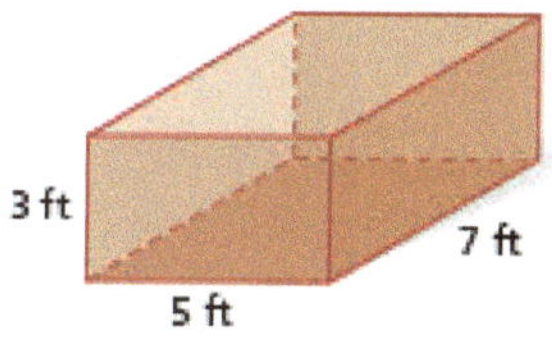

2.

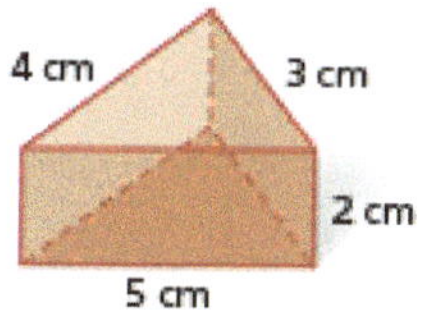

3. a cube with edge length 9 inches

Example 2

Find the lateral area and surface area of each right cylinder. Give your answers in terms of π.

4.

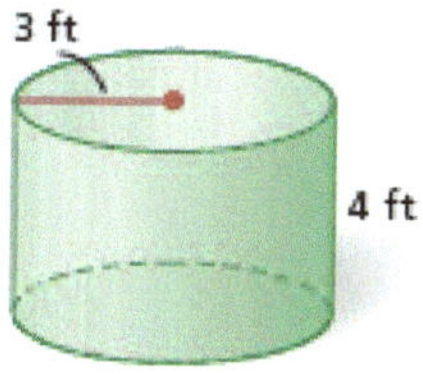

5.

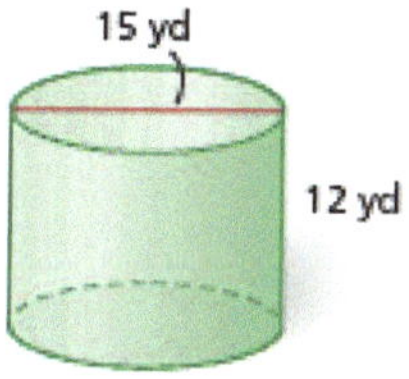

6. a cylinder with base area 64π m^2 and a height 3 meters less than the radius

Example 3

Multi-Step **Find the surface area of each composite figure. Round to the nearest tenth.**

7. 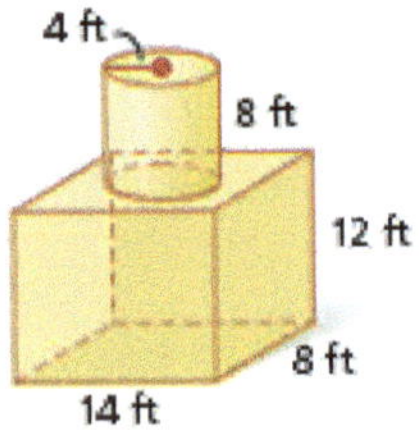

8. 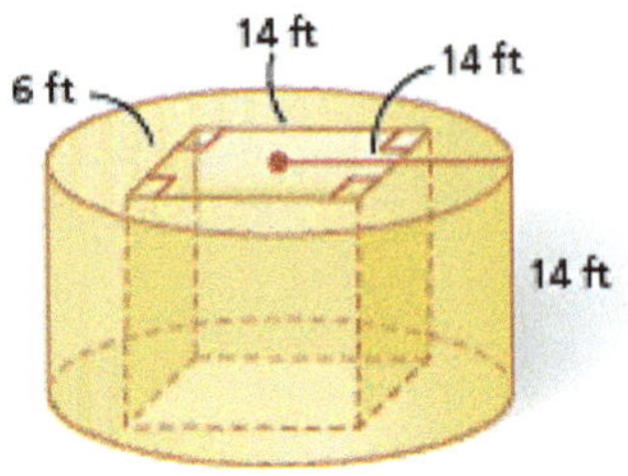

Example 4

Describe the effect of each change on the surface area of the given figure.

9. The dimensions are cut in half.

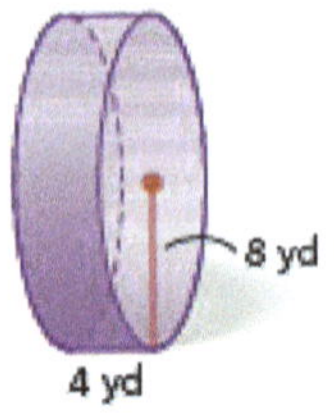

10. The dimensions are multiplied by $\frac{2}{3}$.

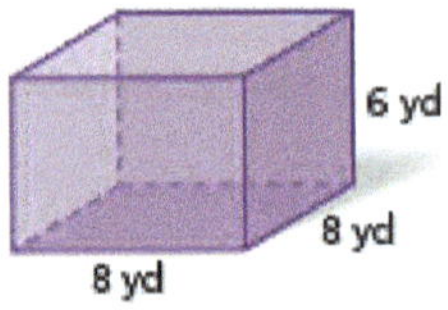

Example 5

11. **Consumer Application** The greater the lateral area of a florescent light bulb, the more light the bulb produces. One cylindrical light bulb is 16 inches long with a 1-inch radius. Another cylindrical light bulb is 23 inches long with a $\frac{3}{4}$-inch radius. Which bulb will produce more light?

Surface Area of Pyramids and Cones

Example 1

Find the lateral area and surface area of each regular pyramid.

1.

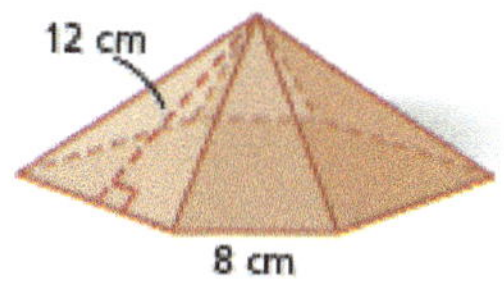

2.

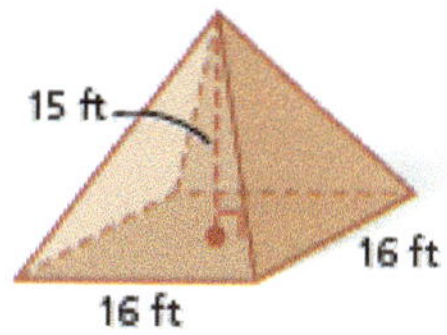

3. a regular triangular pyramid with base edge length 15 in. and slant height 20 in.

Example 2

Find the lateral area and surface area of each right cone. Give your answers in terms of π.

4.

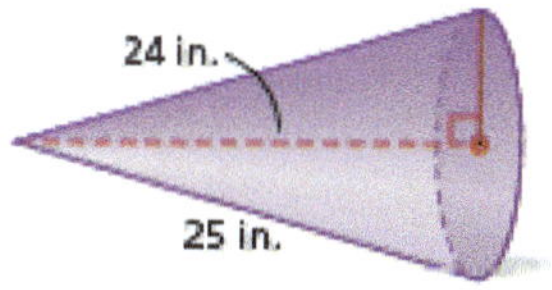

5.

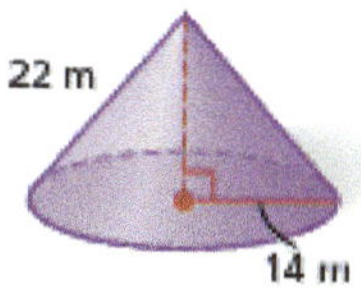

6. a cone with base area 36π ft^2 and slant height 8 ft

Spatial Reasoning

Surface Area and Volume

Surface Area of Pyramids and Cones

Example 3

Describe the effect of each change on the surface area of the given figure.

7. The dimensions are cut in half.

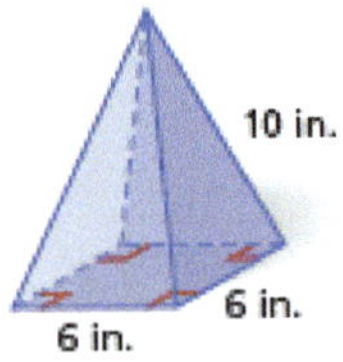

8. The dimensions are tripled.

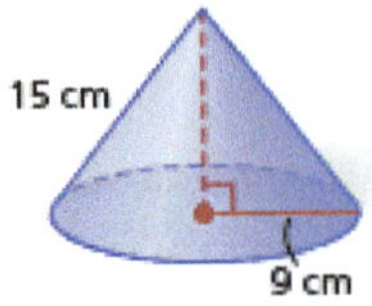

Example 4

Find the surface area of each composite figure.

9.

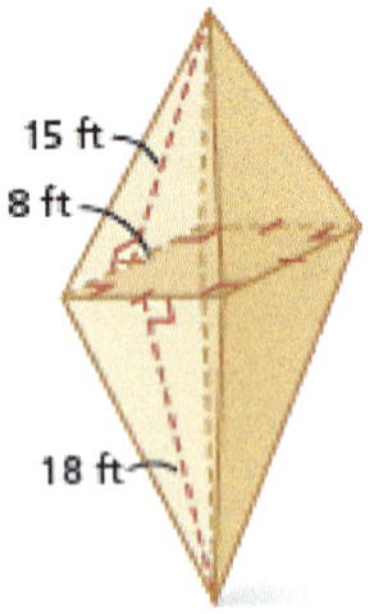

10.

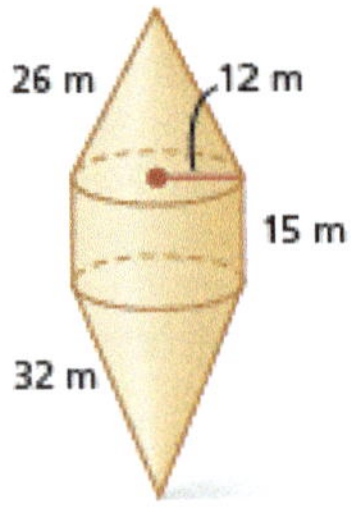

Example 5

11. **Crafts** Anna is making a birthday hat from a pattern that is $\frac{3}{4}$ of a circle of colored paper. If Anna's head is 7 inches in diameter, will the hat fit her? Explain.

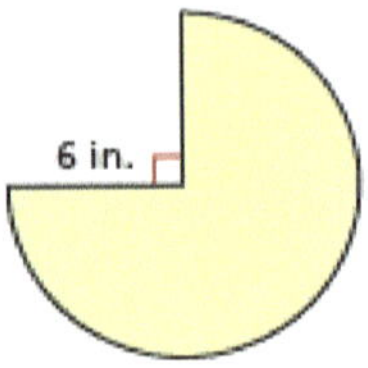

Example 1

Find the volume of each prism.

1.

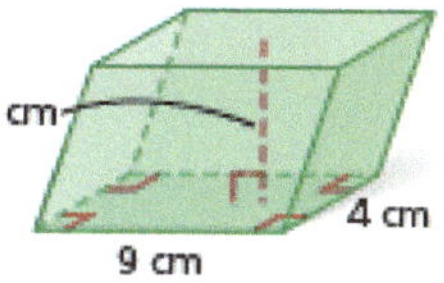

2.

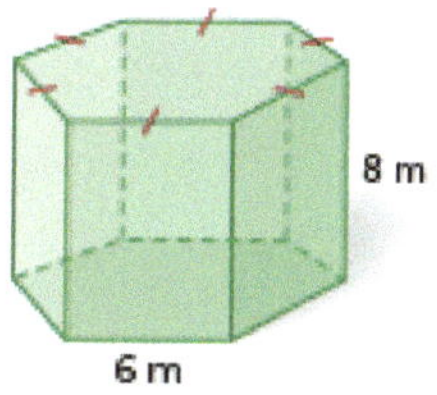

3. a cube with edge length 8 ft

Example 2

4. **Food** The world's largest ice cream cake, built in New York City on May 25, 2004, was approximately a 19 ft by 9 ft by 2 ft rectangular prism. Estimate the volume of the ice cream cake in gallons. If the density of the ice cream was 4.73 pounds per gallon, estimate the weight of the cake. (*Hint:* 1 gallon ≈ 0.134 cubic feet)

Example 3

Find the volume of each cylinder. Give your answers both in terms of π and rounded to the nearest tenth.

5.

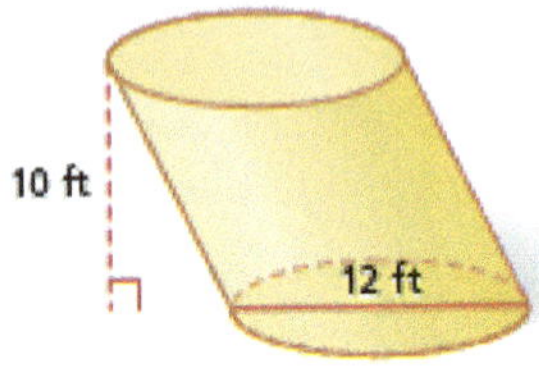

6.

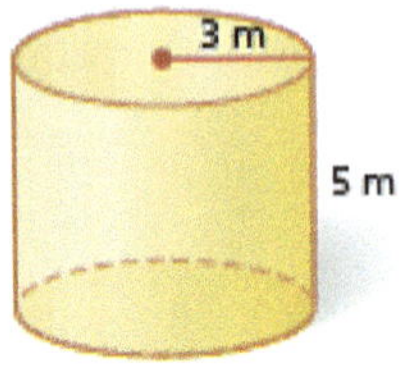

7. a cylinder with base area 25π cm^2 and height 3 cm more than the radius

Example 4

Describe the effect of each change on the volume of the given figure.

8. The dimensions are multiplied by $\frac{1}{4}$.

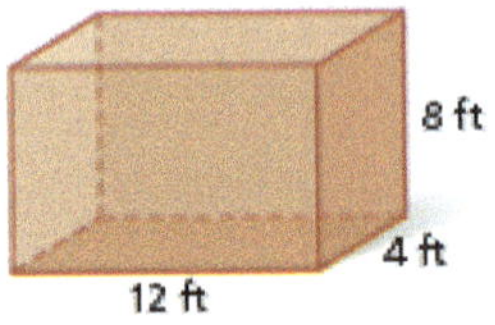

9. The dimensions are tripled.

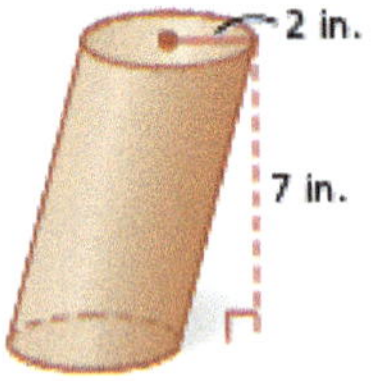

Example 5

Find the volume of each composite figure. Round to the nearest tenth.

10.

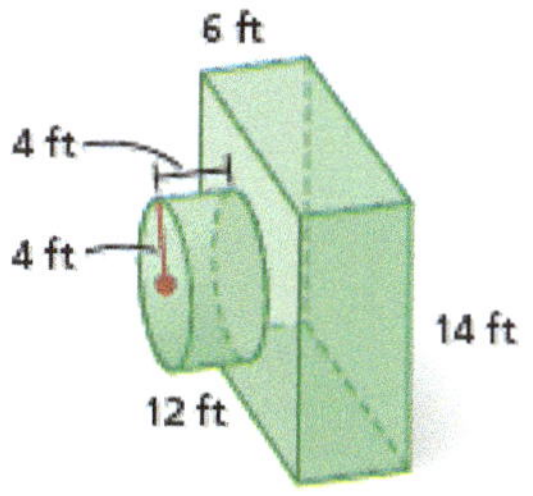

11.

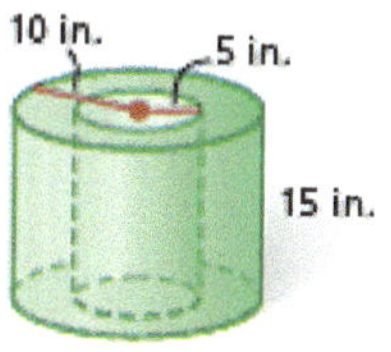

10.1.1 Surface Area of Prisms and Cylinders

Find the lateral area and surface area of each right prism. Round to the nearest tenth, if necessary.

1.

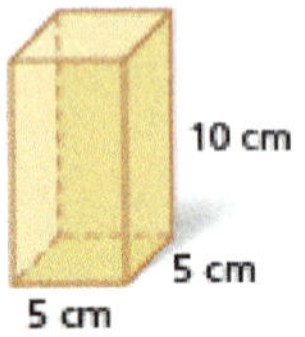

2.

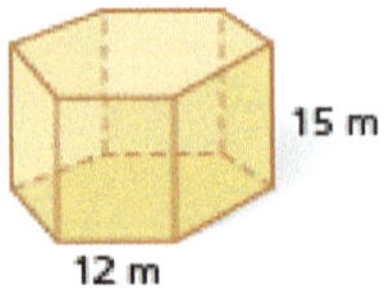

3. a right equilateral triangular prism with base edge length 8 ft and height 14 ft

Find the lateral area and surface area of each right cylinder. Give your answers in terms of π.

4.

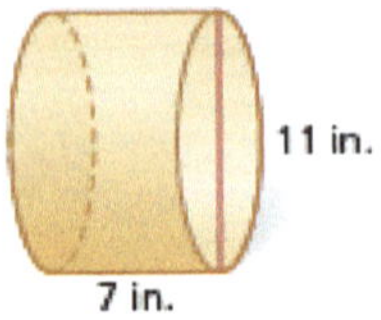

5.

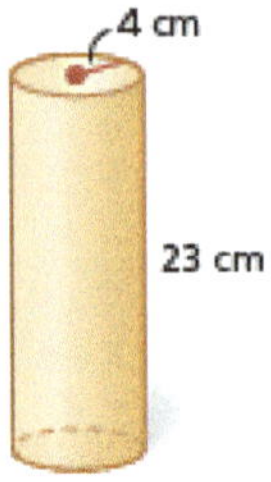

6. a cylinder with base circumference $16\pi\ \text{yd}^2$ and a height equal to 3 times the radius

Multi-Step Find the surface area of each composite figure. Round to the nearest tenth.

7.

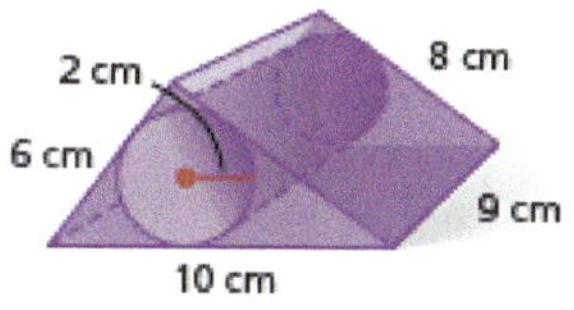

8.

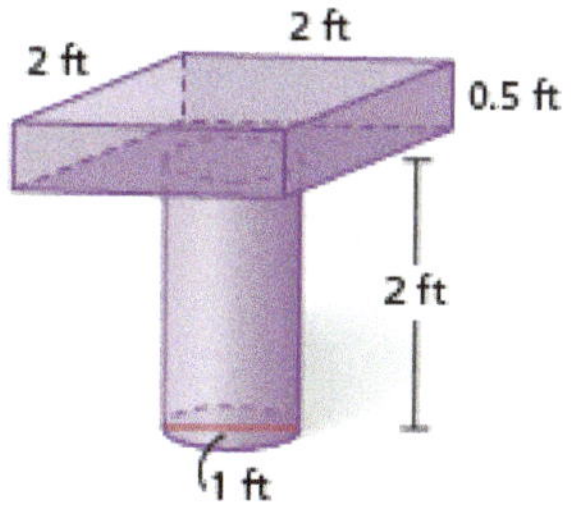

Describe the effect of each change on the surface area of the given figure.

9. The dimensions are tripled.

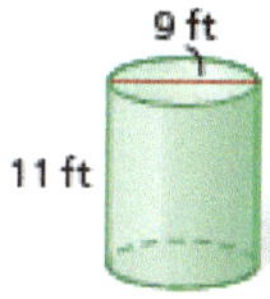

10. The dimensions are doubled.

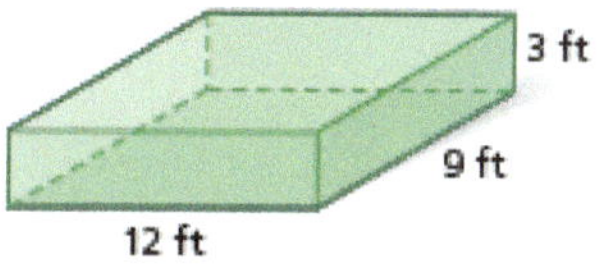

11. **Biology** Plant cells are shaped approximately like a right rectangular prism. Each cell absorbs oxygen and nutrients through its surface. Which cell can be expected to absorb at a greater rate? (*Hint:* 1 μm = 1 micrometer = 0.000001 meter)

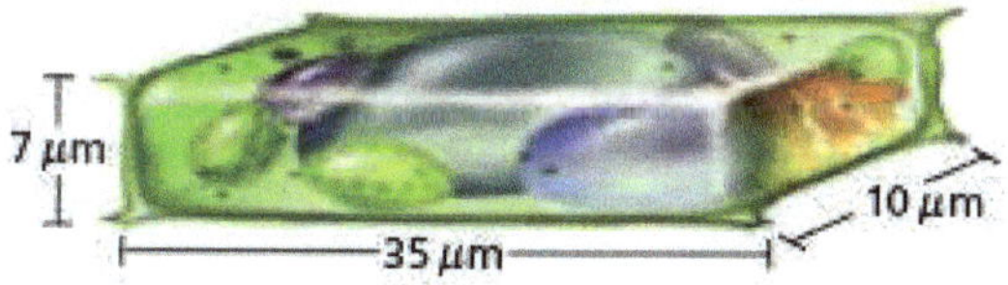

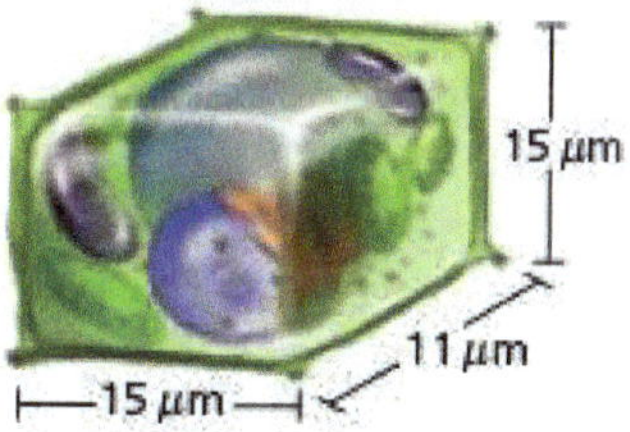

10.1.2 Surface Area of Pyramids and Cones

Find the lateral area and surface area of each regular pyramid.

12.

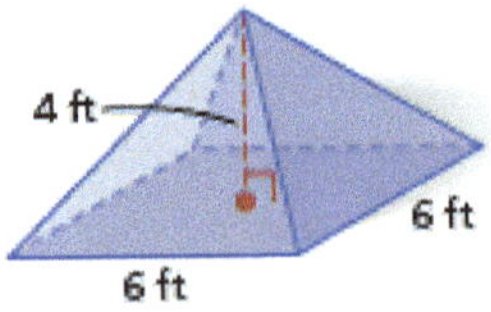

13.

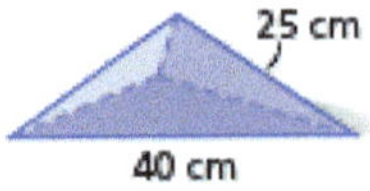

14. a regular hexagonal pyramid with base edge length 7 ft and slant height 15 ft

Find the lateral area and surface area of each right cone. Give your answers in terms of π.

15.

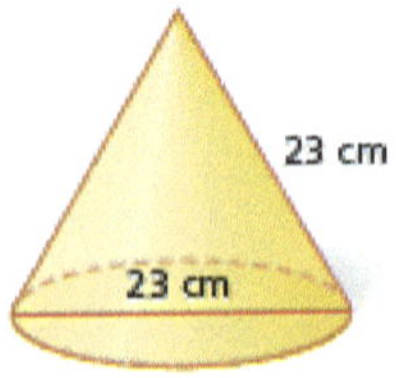

16.

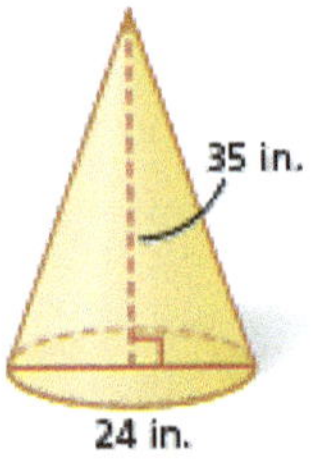

17. a cone with radius 8 m and height that is 1 m less than twice the radius

Describe the effect of each change on the surface area of the given figure.

18. The dimensions are divided by 3.

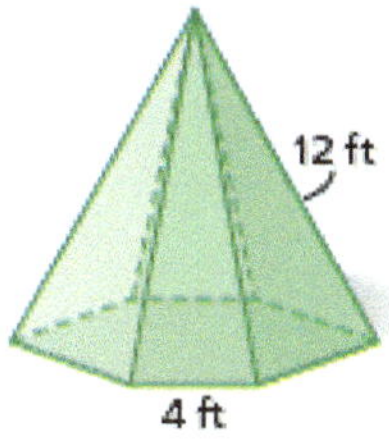

19. The dimensions are doubled.

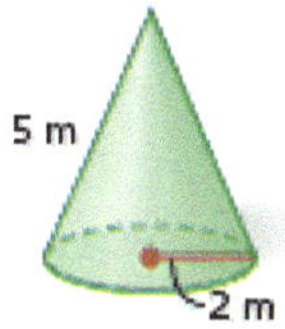

Find the surface area of each composite figure.

20.

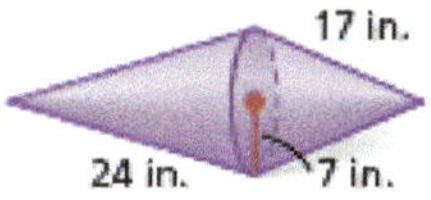

21.

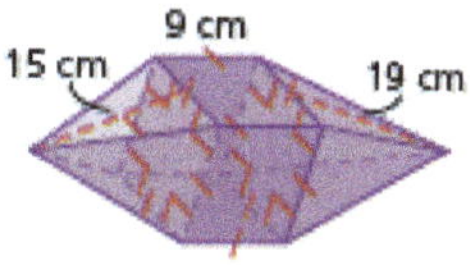

22. It is a tradition in England to celebrate May 1st by hanging cone-shaped baskets of flowers on neighbors' door handles. Addy is making a basket from a piece of paper that is a semicircle with diameter 12 in. What is the diameter of the basket?

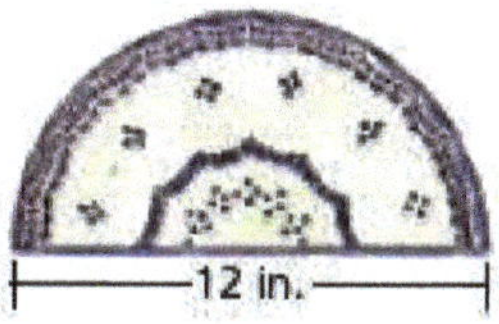

10.1.3 Volume of Prisms and Cylinders

Find the volume of each prism.

23.

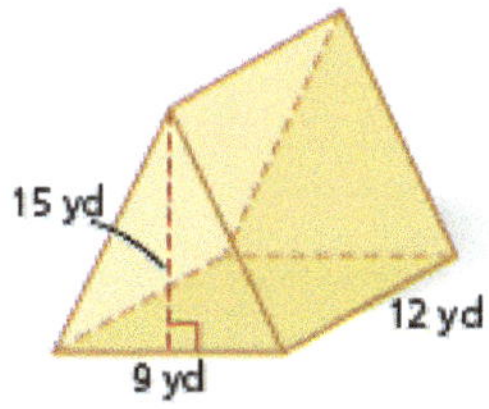

24.

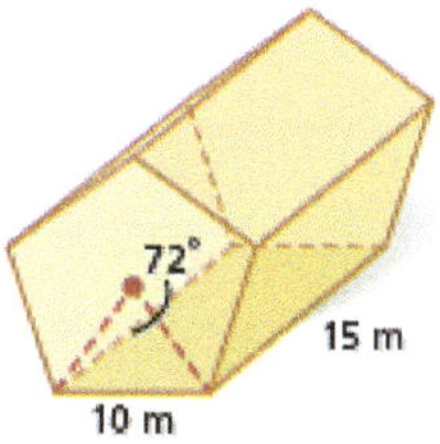

25. a square prism with a base area of 49 ft^2 and a height 2 ft less than the base edge length

26. **Landscaping** Colin is buying dirt to fill a garden bed that is a 9 ft by 16 ft rectangle. If he wants to fill it to a depth of 4 in., how many cubic yards of dirt does he need? If dirt costs \$25 per yd^3, how much will the project cost? (*Hint:* 1 yd^3 = 27 ft^3)

Find the volume of each cylinder. Give your answers both in terms of π and rounded to the nearest tenth.

27.

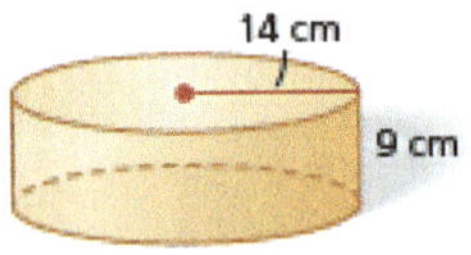

28.

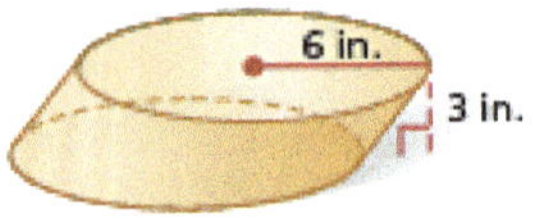

29. a cylinder with base area 24π cm^2 and height 16 cm

Describe the effect of each change on the volume of the given figure.

30. The dimensions are multiplied by 5.

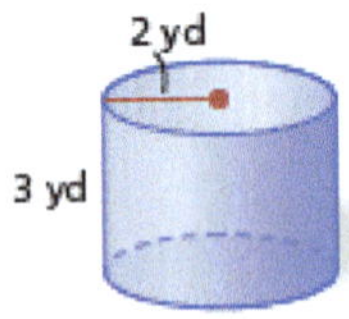

31. The dimensions are multiplied by $\frac{3}{5}$.

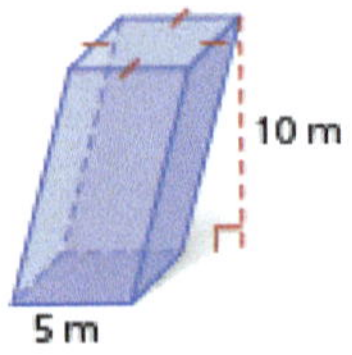

Find the volume of each composite figure.

32.

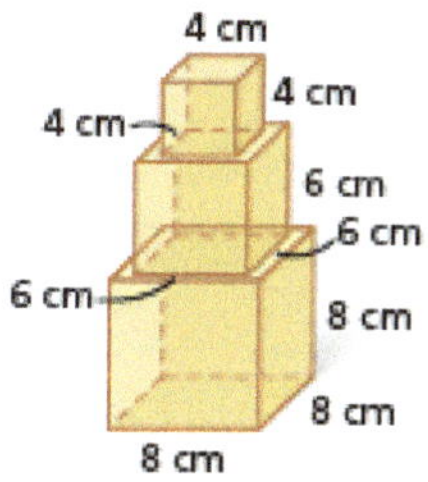

33.

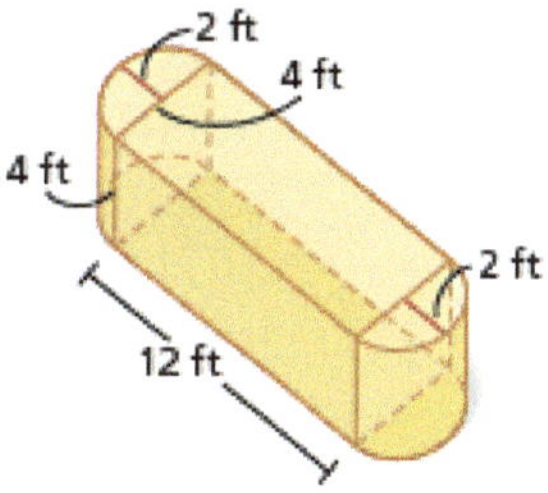

Example 1

Identify each line or segment that intersects each circle.

1. 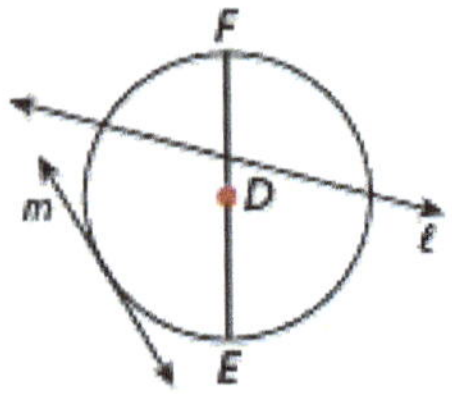

2. 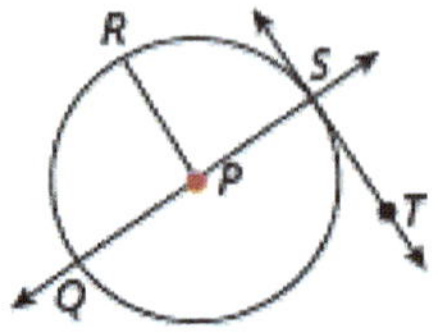

Example 2

Multi-Step **Find the length of each radius. Identify the point of tangency and write the equation of the tangent line at this point.**

3. 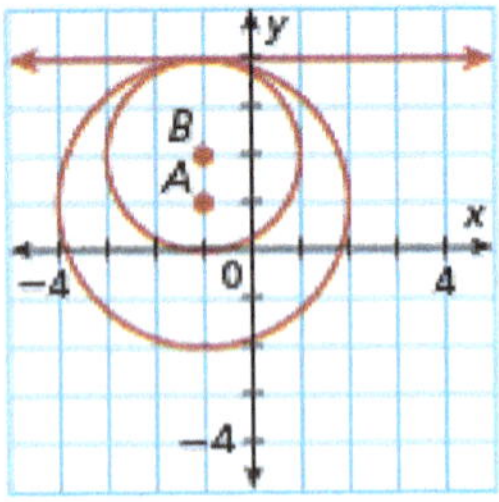

4. 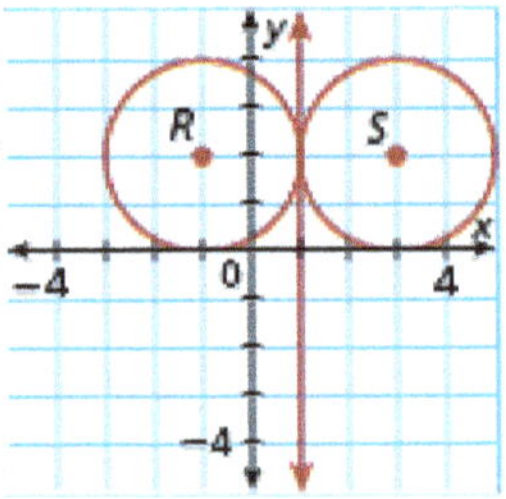

Example 3

5. **Space Exploration** The International Space Station orbits Earth at an altitude of 240 mi. What is the distance from the space station to Earth's horizon to the nearest mile?

Example 4

The segments in each figure are tangent to the circle. Find each length.

6. *JK*

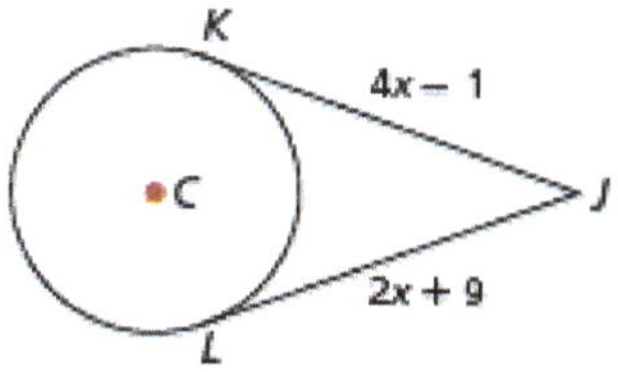

7. *ST*

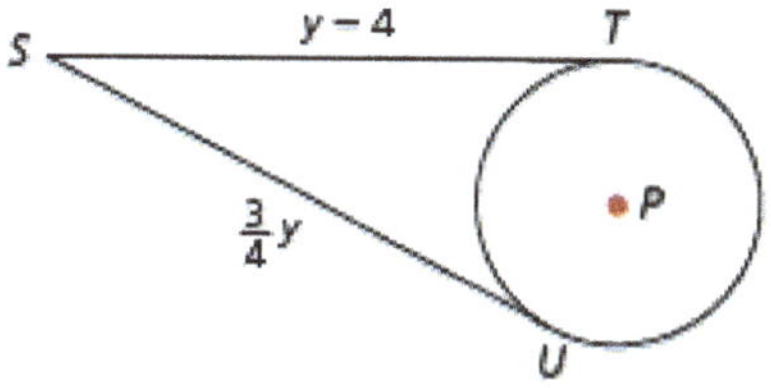

Example 1

Consumer Application Use the following information for Exercises 1–6.
The circle graph shows how a typical household spends money on energy. Find each of the following.

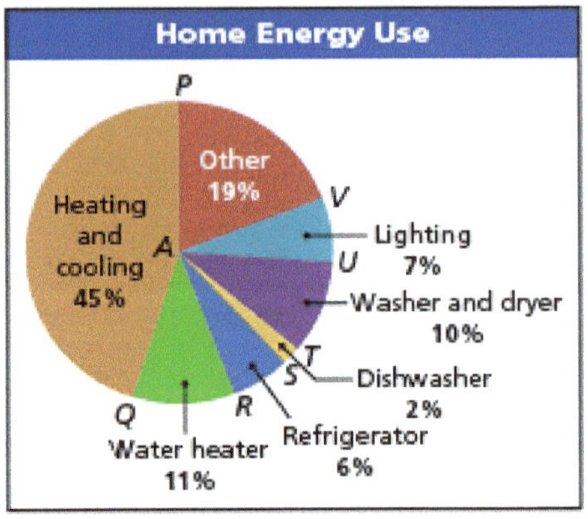

1. m$\angle PAQ$

2. m$\angle VAU$

3. m$\angle SAQ$

4. m$\widehat{UT}$

5. m$\widehat{RQ}$

6. m$\widehat{UPT}$

Example 2

Find each measure.

7. m$\widehat{DF}$

8. m$\widehat{DEB}$

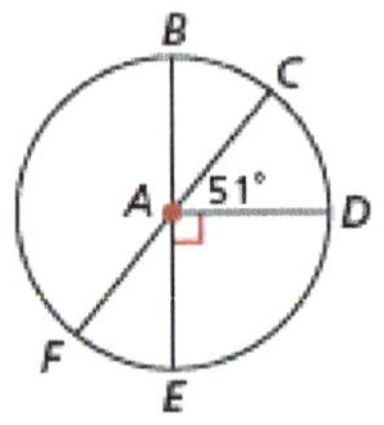

9. m$\widehat{JL}$

10. m$\widehat{HLK}$

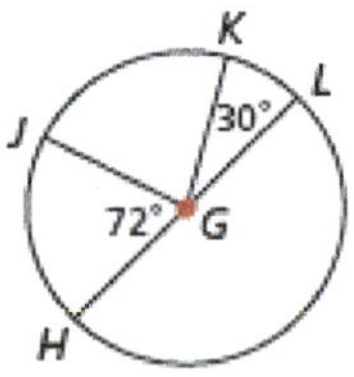

Example 3

11. $\angle QPR \cong \angle RPS$. Find QR.

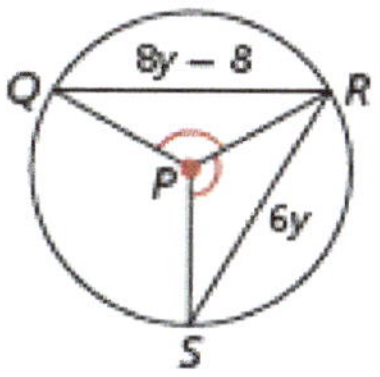

12. $\odot A \cong \odot B$, and $\overset{\frown}{CD} \cong \overset{\frown}{EF}$. Find $m\angle EBF$.

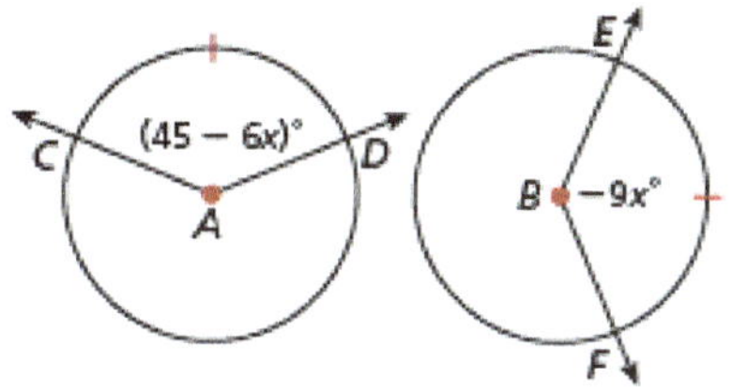

Example 4

Multi-Step Find each length to the nearest tenth.

13. RS

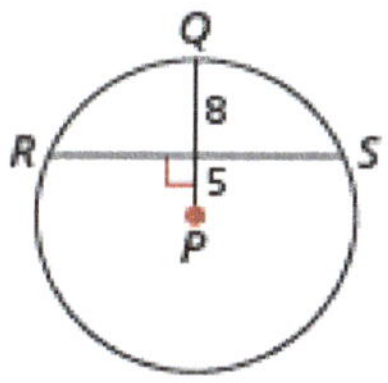

14. EF

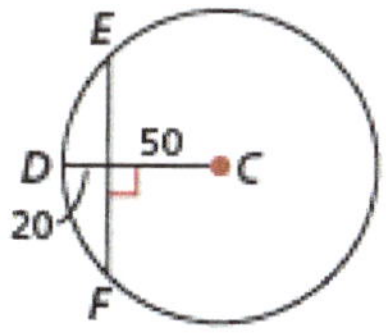

11.1.1 Lines that Intersect Circles

Identify each line or segment that intersects each circle.

1.

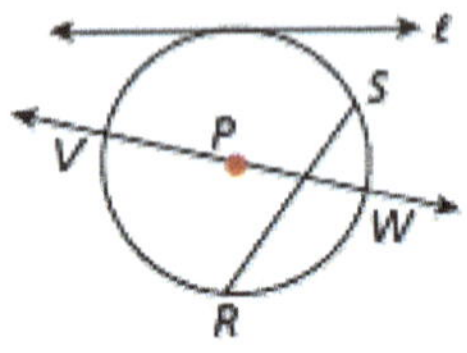

2.

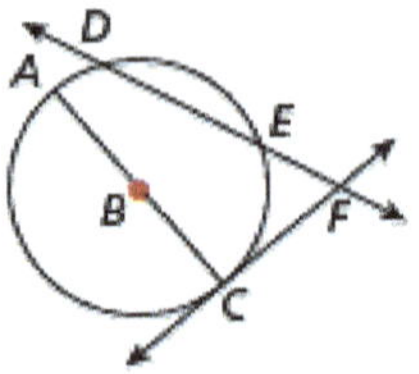

Multi-Step Find the length of each radius. Identify the point of tangency and write the equation of the tangent line at this point.

3.

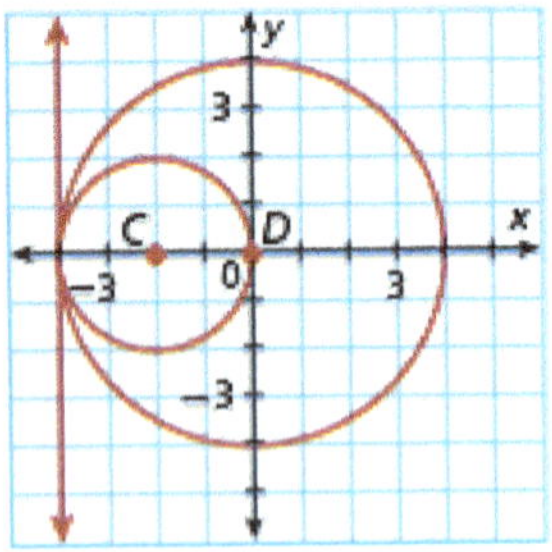

4.

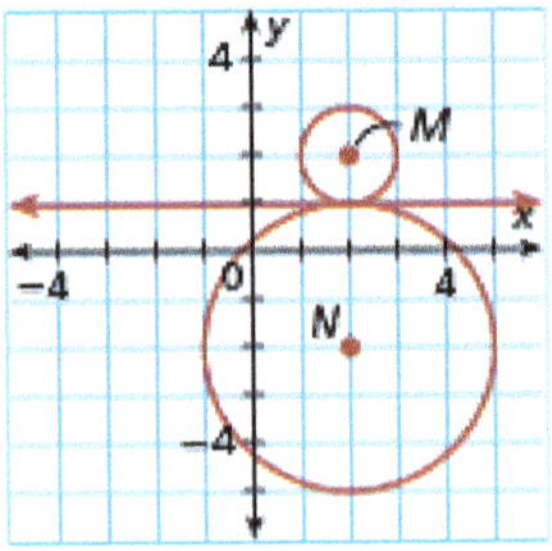

5. **Astronomy** Olympus Mons's peak rises 25 km above the surface of the planet Mars. The diameter of Mars is approximately 6794 km. What is the distance from the peak of Olympus Mons to the horizon to the nearest kilometer?

The segments in each figure are tangent to the circle. Find each length.

6. AB

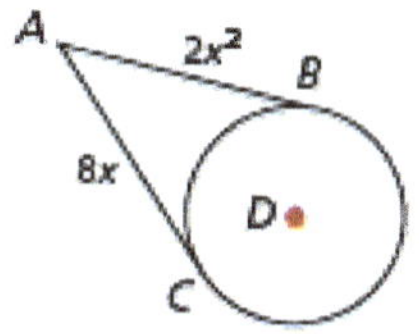

7. RT

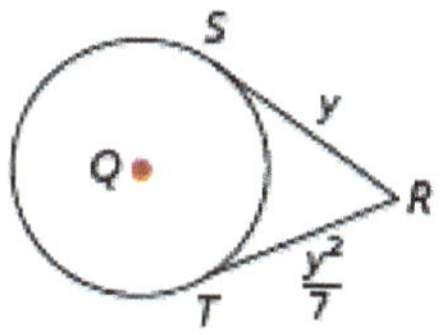

11.1.2 Arcs and Chords

Sports Use the following information for Exercises 8–13.
The key shows the number of medals won by U.S. athletes at the 2004 Olympics in Athens. Find each of the following to the nearest tenth.

8. $m\angle ADB$

Medals	
Gold	35
Silver	39
Bronze	29

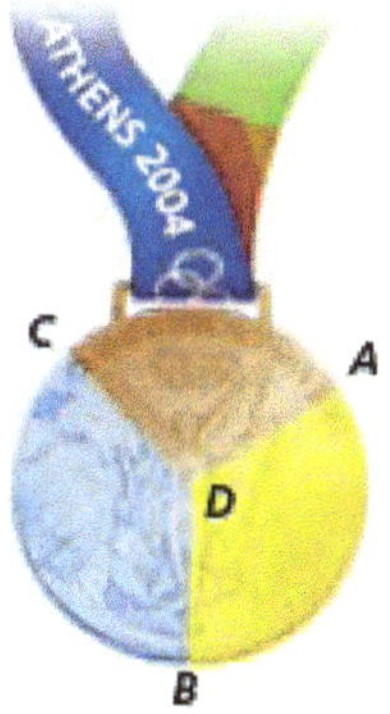

9. $m\angle ADC$

10. $m\widehat{AB}$

11. $m\widehat{BC}$

12. $m\widehat{ACB}$

13. $m\widehat{CAB}$

Find each measure.

14. $m\widehat{MP}$

15. $m\widehat{QNL}$

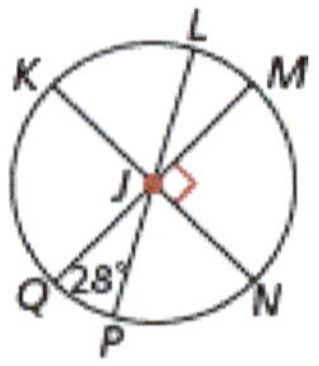

16. $m\widehat{WT}$

17. $m\widehat{WTV}$

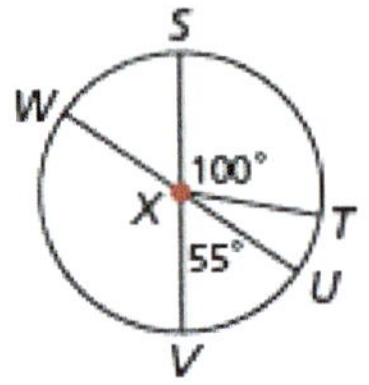

18. $\odot A \cong \odot B$, and $\widehat{CD} \cong \widehat{EF}$. Find $m\angle CAD$.

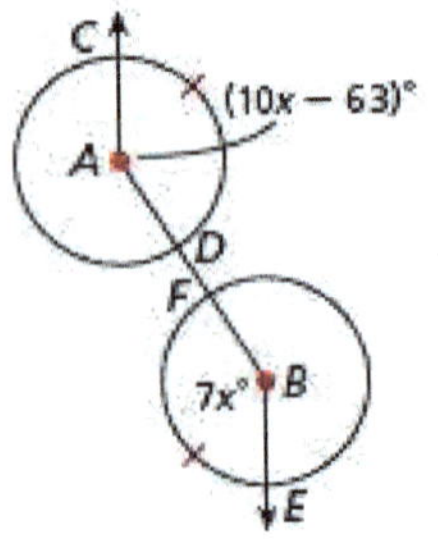

19. $\overline{JK} \cong \overline{LM}$. Find $m\widehat{JK}$.

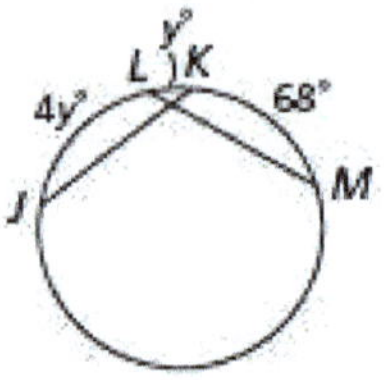

Multi-Step **Find each length to the nearest tenth.**

20. CD

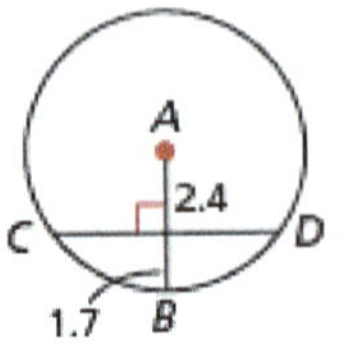

21. RS

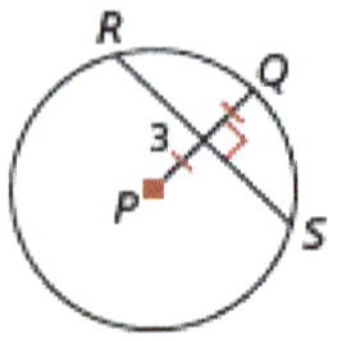

Inscribed Angles

Example 1

Find each measure.

1. m∠DEF

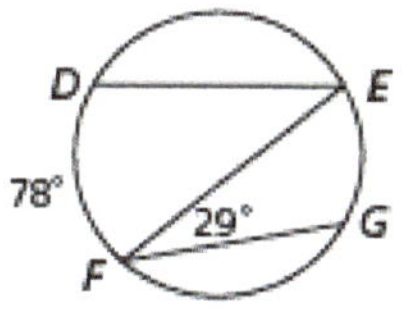

2. m$\widehat{EG}$

3. m$\widehat{JKL}$

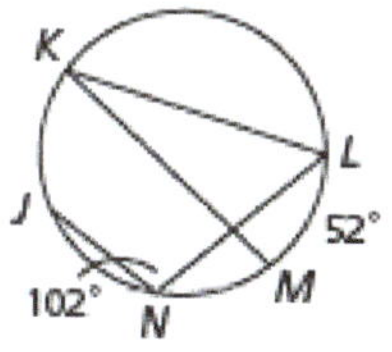

4. m∠LKM

Example 2

5. **Crafts** A circular loom can be used for knitting. What is the m∠QTR in the knitting loom?

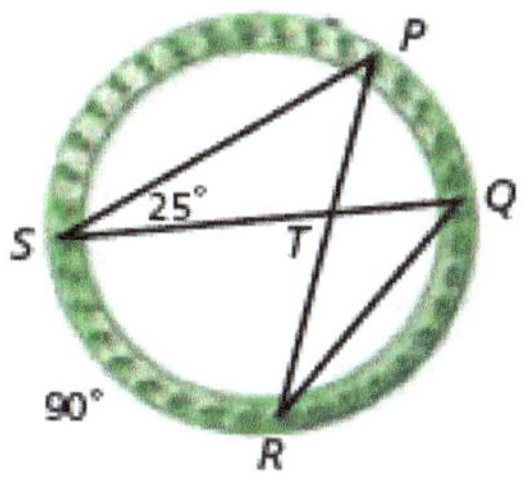

Example 3

Find each value.

6. x

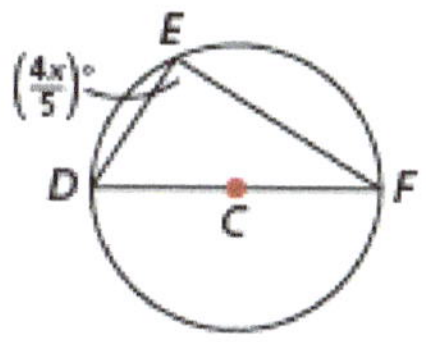

7. y

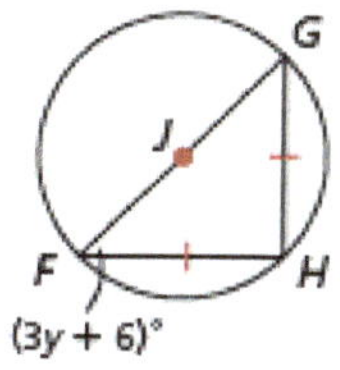

8. m∠XYZ

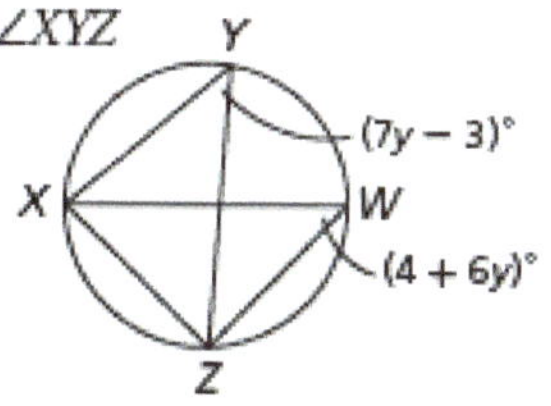

Example 4

Multi-Step **Find the angle measures of each quadrilateral.**

9. *PQRS*

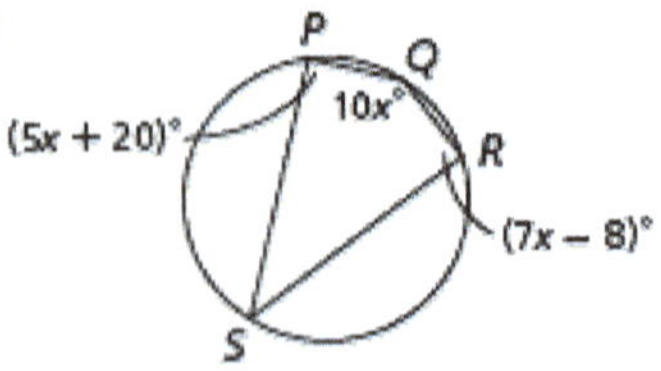

10. *ABCD*

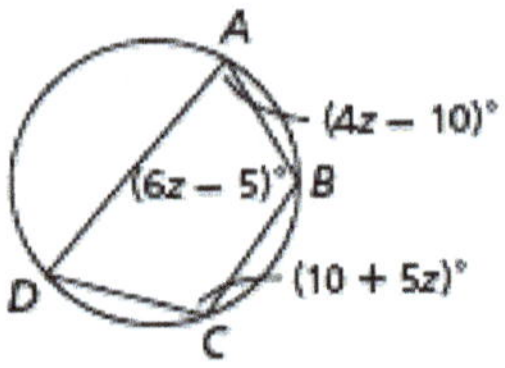

Example 1

Find each measure.

1. m∠*DAB*

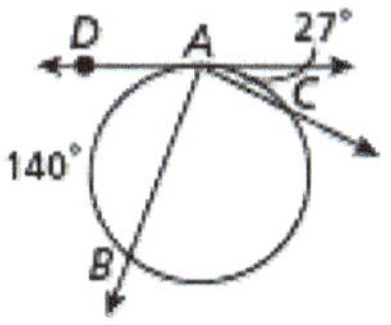

2. m$\widehat{AC}$

3. m$\widehat{PN}$

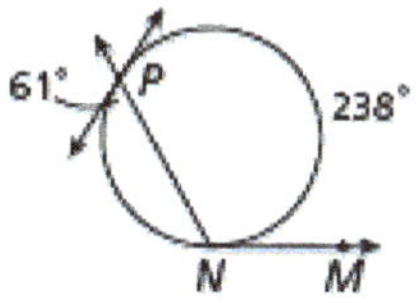

4. m∠*MNP*

Example 2

Find each measure.

5. m∠*STU*

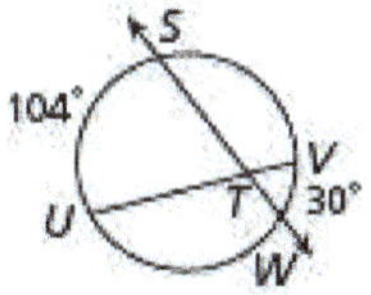

6. m∠*HFG*

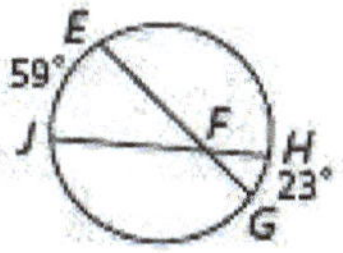

7. m∠*NPK*

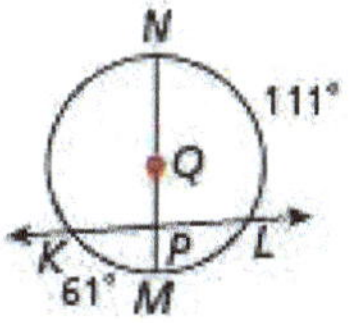

Example 3

Find the value of *x*.

8.

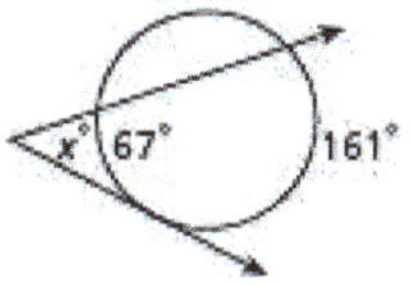

9.

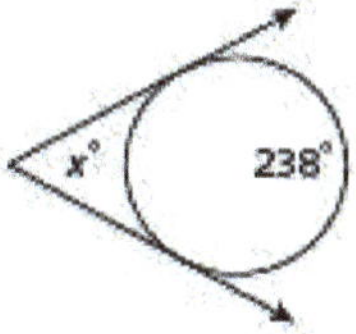

10.

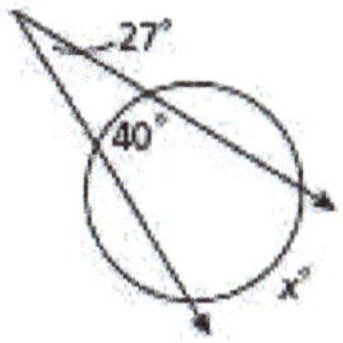

Example 4

11. **Science** A satellite orbits Mars. When it reaches *S* it is about 12,000 km above the planet. How many arc degrees of the planet are visible to a camera in the satellite?

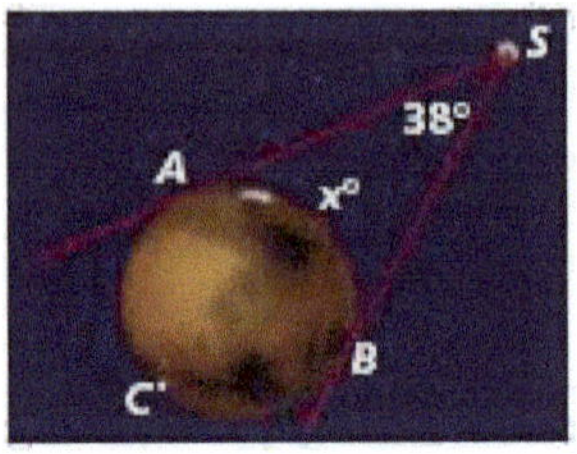

Example 5

Multi-Step **Find each measure.**

12. $m\widehat{DF}$

13. $m\widehat{CD}$

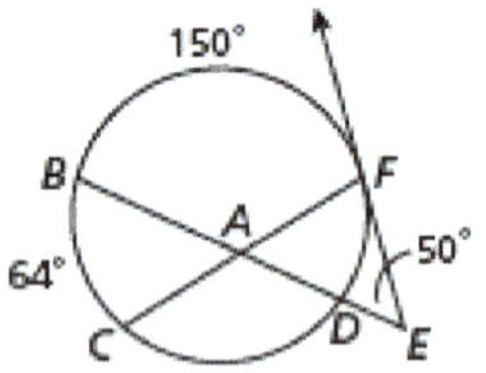

14. $m\widehat{PN}$

15. $m\widehat{KN}$

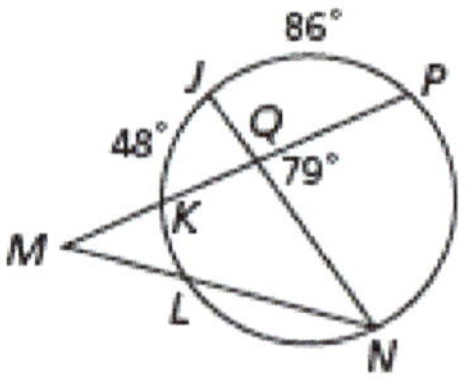

Example 1

Find the value of the variable and the length of each chord.

1.

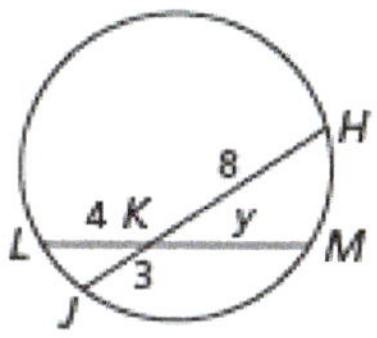

2.

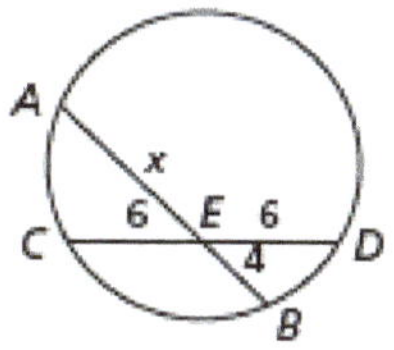

3.

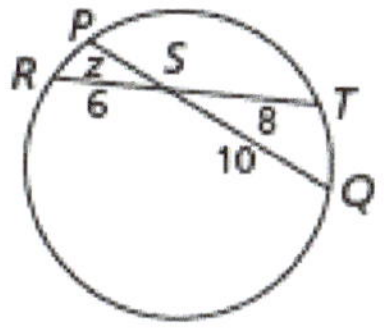

Example 2

4. **Engineering** A section of an aqueduct is based on an arc of a circle as shown. $\overline{EF}$ is the perpendicular bisector of $\overline{GH}$. $GH = 50$ ft, and $EF = 20$ ft. What is the diameter of the circle?

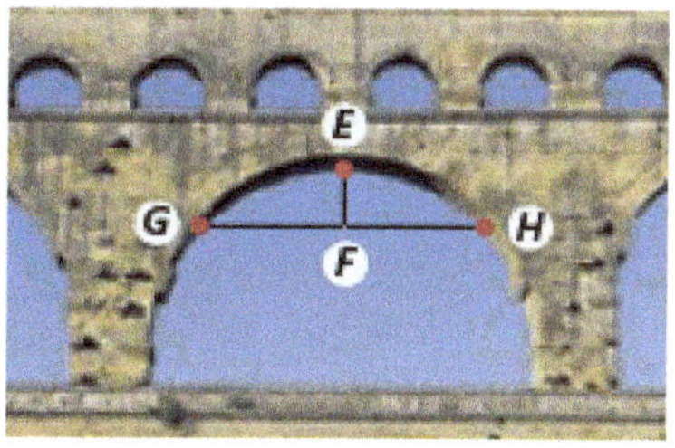

Example 3

Find the value of the variable and the length of each secant segment.

5.

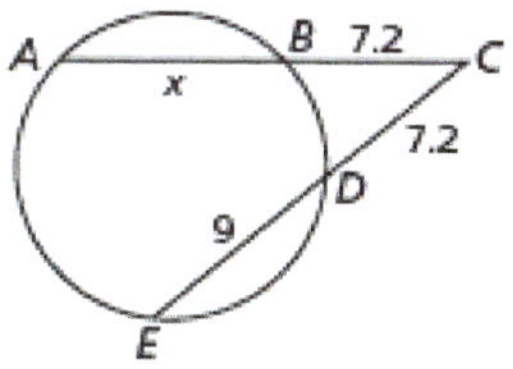

6.

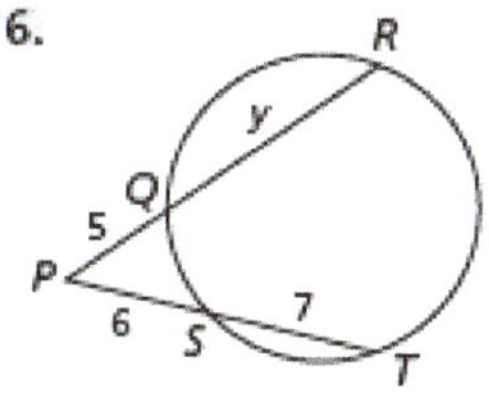

7.

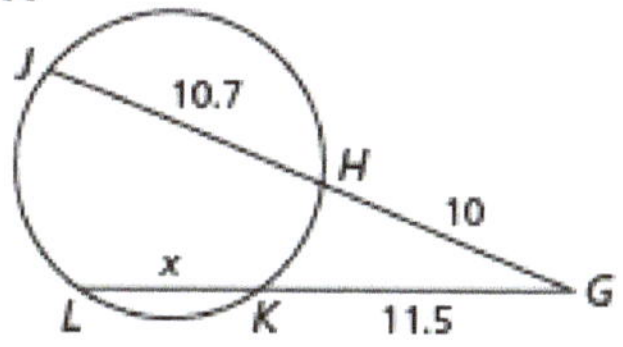

Example 4

Find the value of the variable.

8.

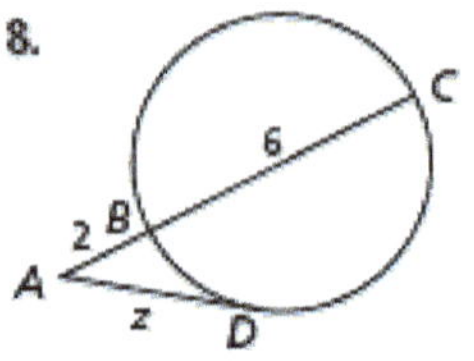

9.

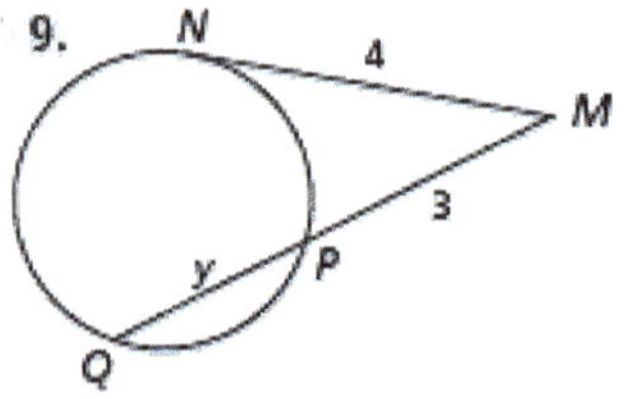

10.

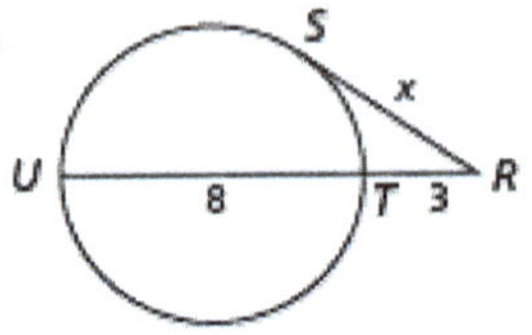

Circles

Review 11.2 / Angles and Segments in Circles

11.2.1 Inscribed Angles

Find each measure.

1. $m\widehat{ML}$

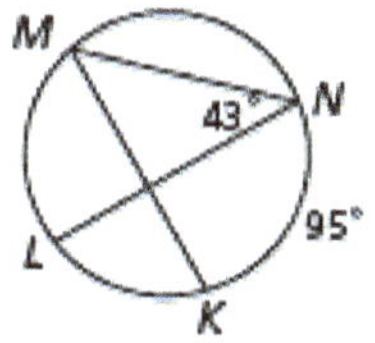

2. $m\angle KMN$

3. $m\widehat{EGH}$

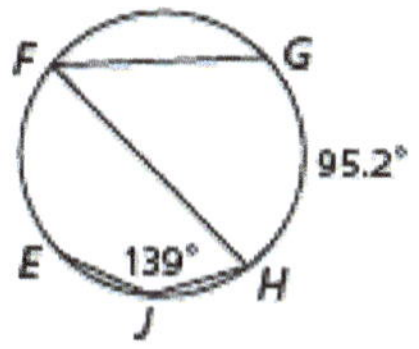

4. $m\angle GFH$

5. **Crafts** An artist created a stained glass window. If $m\angle BEC = 40°$ and $m\widehat{AB} = 44°$, what is $m\angle ADC$?

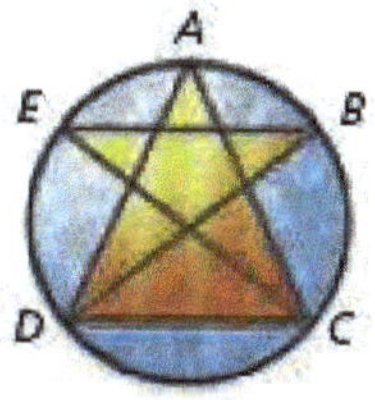

Algebra **Find each value.**

6. y

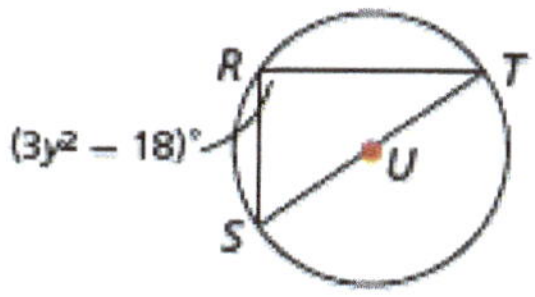

7. z

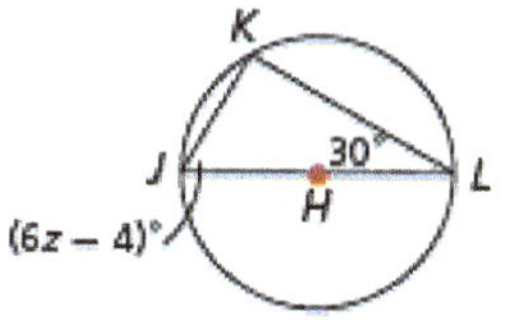

8. $m\widehat{AB}$

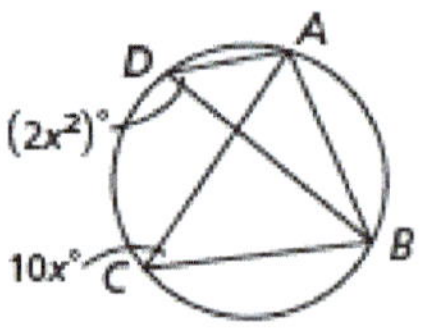

9. $m\angle MPN$

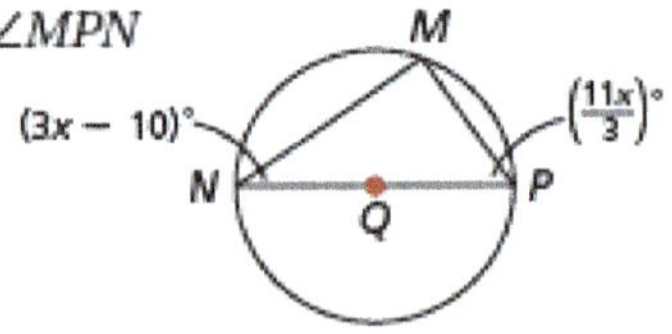

Multi-Step **Find the angle measures of each quadrilateral.**

10. *BCDE*

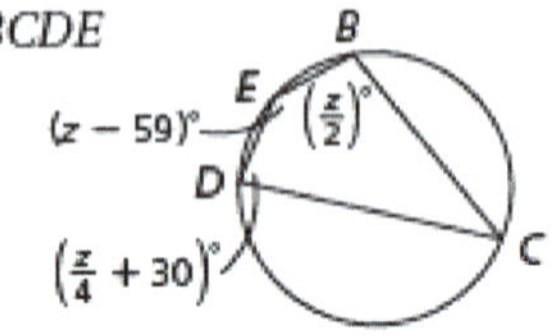

11. *TUVW*

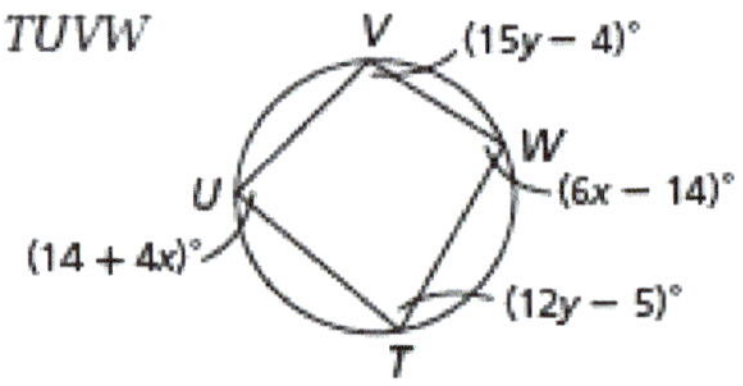

11.2.2 Angle Relationships in Circles

Find each measure.

12. $m\angle BCD$

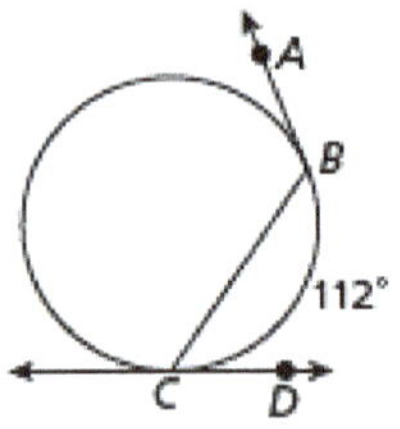

13. $m\angle ABC$

14. $m\angle XZW$

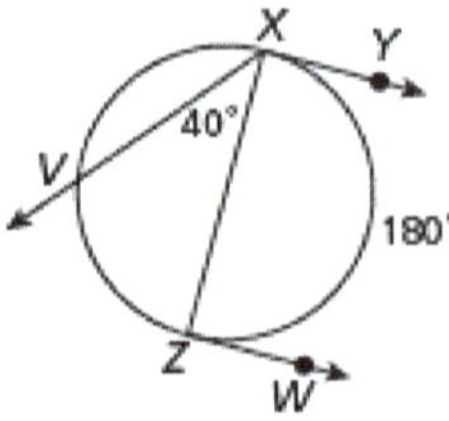

15. $m\widehat{XZV}$

16. m$\angle QPR$

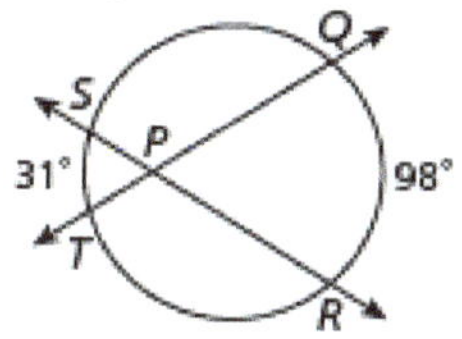

17. m$\angle ABC$

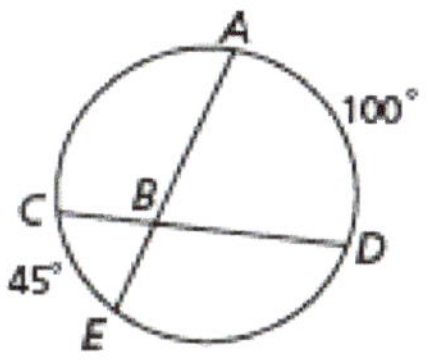

18. m$\angle MKJ$

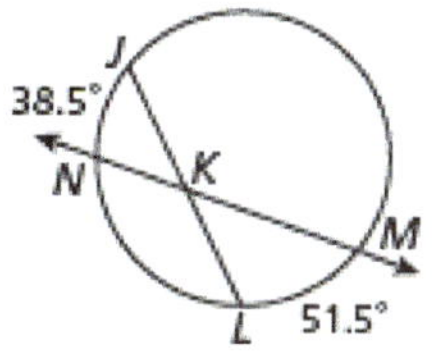

Find the value of x.

19.

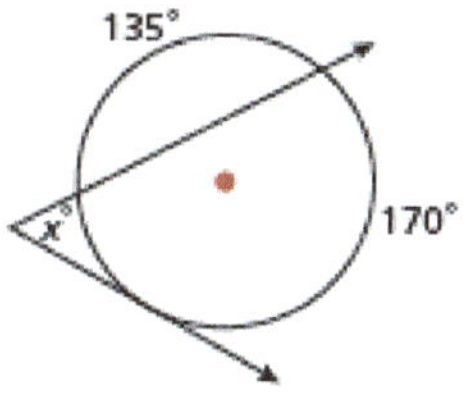

20.

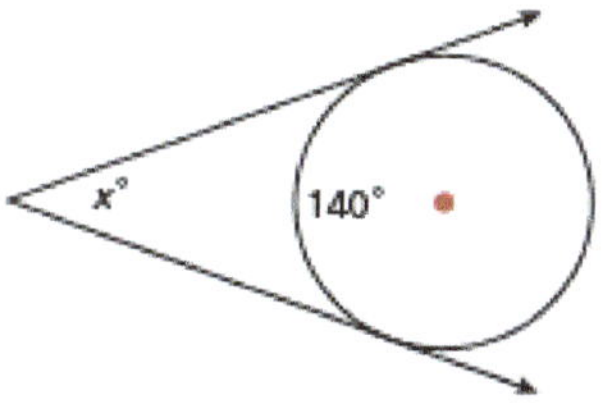

21.

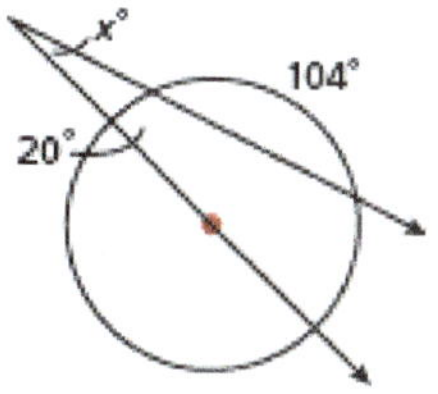

22. **Archaeology** Stonehenge is a circular arrangement of massive stones near Salisbury, England. A viewer at V observes the monument from a point where two of the stones A and B are aligned with stones at the endpoints of a diameter of the circular shape. Given that $m\widehat{AB} = 48°$, what is m$\angle AVB$?

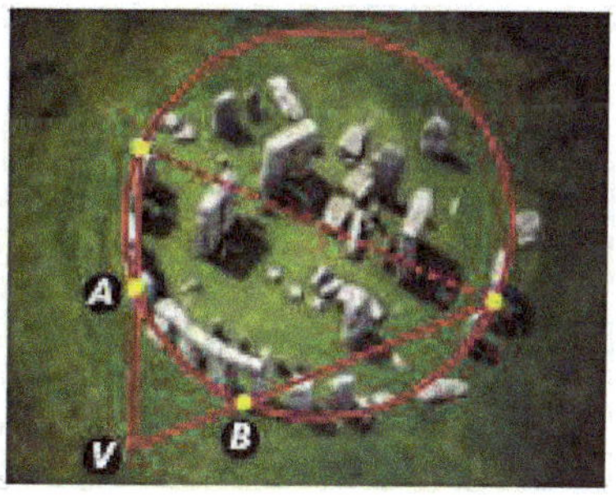

Multi-Step Find each measure.

23. $m\widehat{EG}$

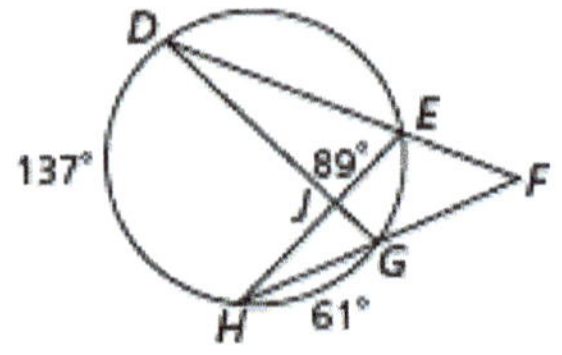

24. $m\widehat{DE}$

25. $m\widehat{PR}$

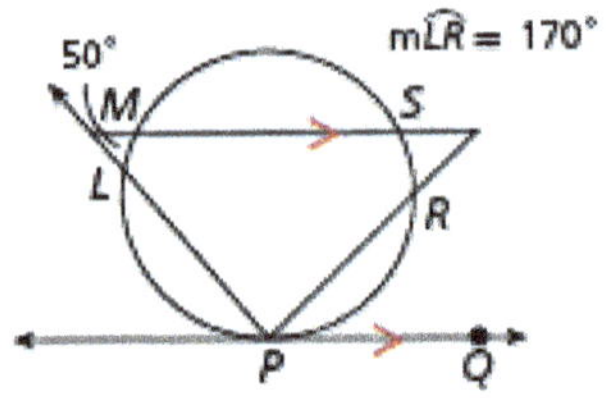

26. $m\widehat{LP}$

11.2.3 Segment Relationships in Circles

Find the value of the variable and the length of each chord.

27.

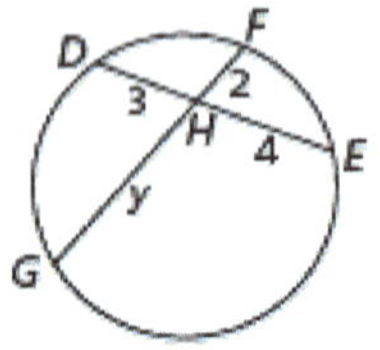

28.

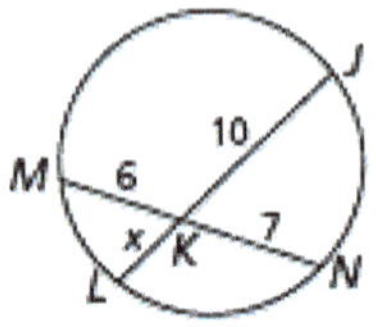

29.

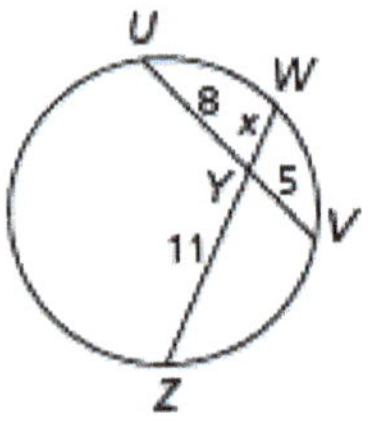

30. **Geology** Molokini is a small, crescent-shaped island $2\frac{1}{2}$ miles from the Maui coast. It is all that remains of an extinct volcano. To approximate the diameter of the mouth of the volcano, a geologist can use a diagram like the one shown. What is the approximate diameter of the volcano's mouth to the nearest foot?

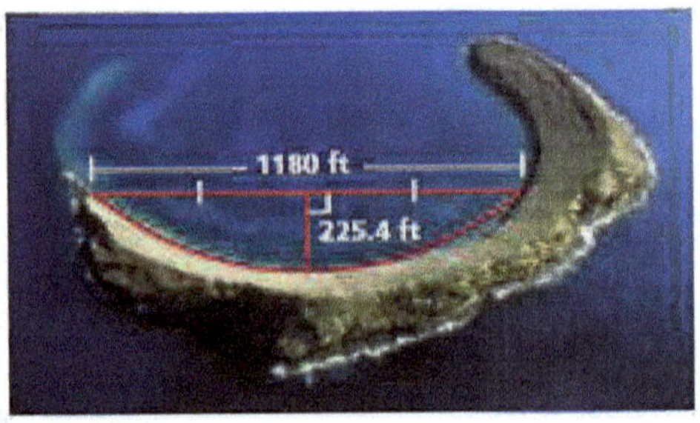

Circles

Review 11.2 / Angles and Segments in Circles

Find the value of the variable and the length of each secant segment.

31.

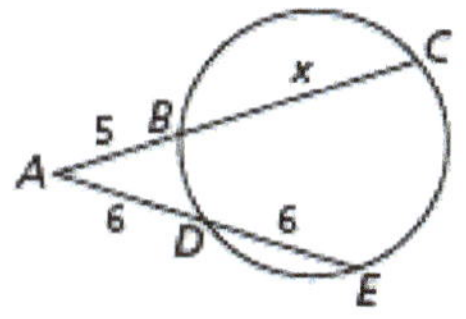

32.

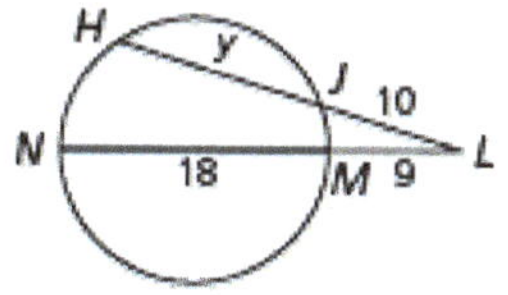

33.

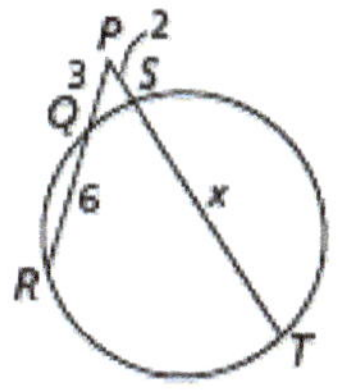

Find the value of the variable.

34.

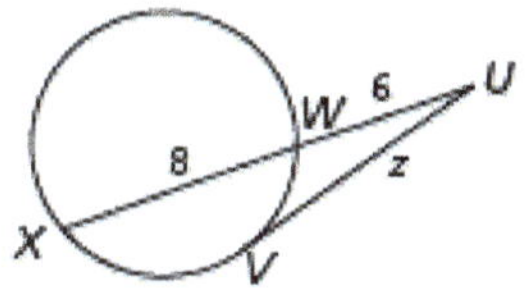

35.

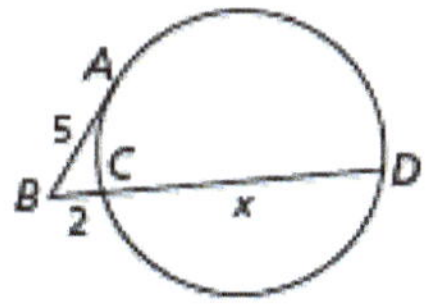

36.

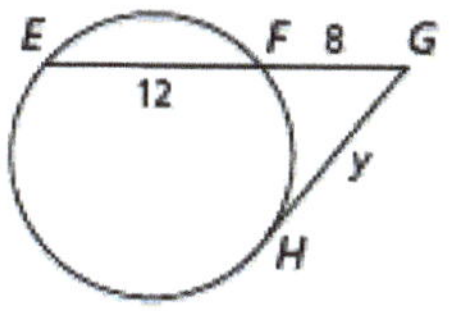

Example 1

Tell whether each transformation appears to be a reflection.

1. 

2. 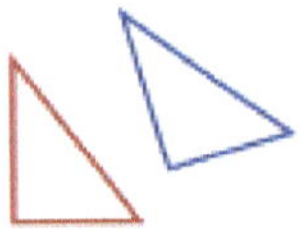

3. 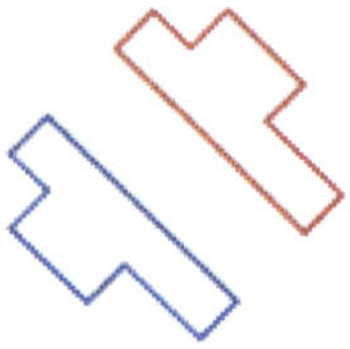

4. 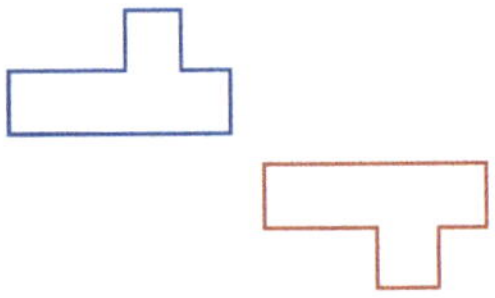

Example 2

Multi-Step Copy each figure and the line of reflection. Draw the reflection of the figure across the line.

5. 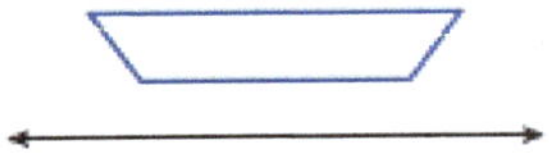

6. 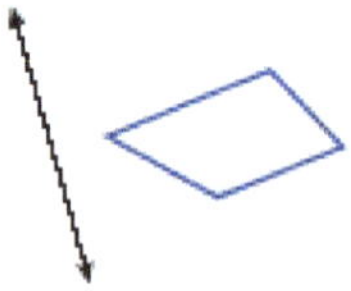

Example 3

7. **City Planning** The towns of San Pablo and Tanner are located on the same side of Highway 105. Two access roads are planned that connect the towns to a point P on the highway. Draw a diagram that shows where point P should be located in order to make the total length of the access roads as short as possible.

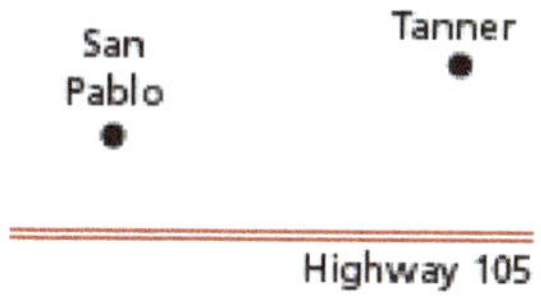

Example 4

Reflect the figure with the given vertices across the given line.

8. $A(-2, 1)$, $B(2, 3)$, $C(5, 2)$; x-axis

9. $R(0, -1)$, $S(2, 2)$, $T(3, 0)$; y-axis

10. $M(2, 1)$, $N(3, 1)$, $P(2, -1)$, $Q(1, -1)$; $y = x$

11. $A(-2, 2)$, $B(-1, 3)$, $C(1, 2)$, $D(-2, -2)$; $y = x$

Example 1

Tell whether each transformation appears to be a translation.

1.

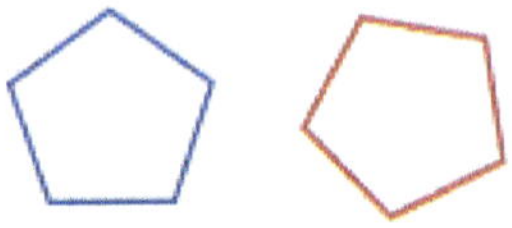

2.

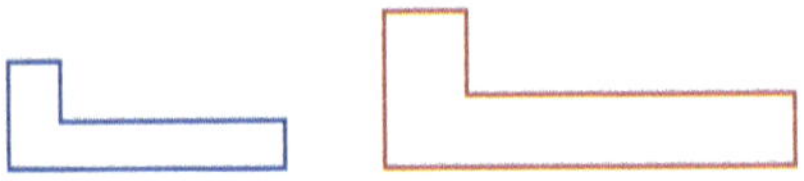

3.

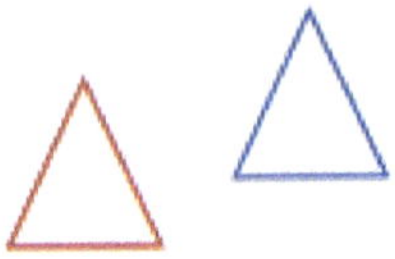

4.

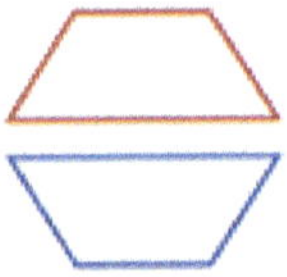

Example 2

Multi-Step Copy each figure and the translation vector. Draw the translation of the figure along the given vector.

5.

6.

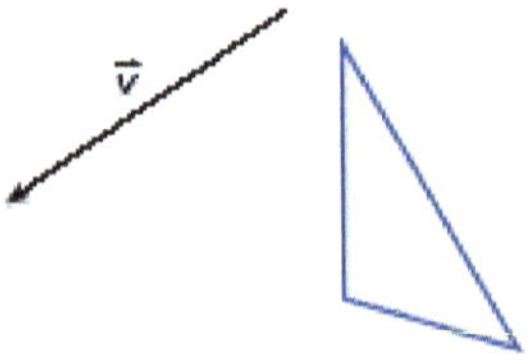

Example 3

Translate the figure with the given vertices along the given vector.

7. $A(-4, -4)$, $B(-2, -3)$, $C(-1, 3)$; $\langle 5, 0\rangle$

8. $R(-3, 1)$, $S(-2, 3)$, $T(2, 3)$, $U(3, 1)$; $\langle 0, -4\rangle$

9. $J(-2, 2)$, $K(-1, 2)$, $L(-1, -2)$, $M(-3, -1)$; $\langle 3, 2\rangle$

Example 4

10. **Art** The Zulu people of southern Africa are known for their beadwork. To create a typical Zulu pattern, translate the polygon with vertices $(1, 5)$, $(2, 3)$, $(1, 1)$, and $(0, 3)$ along the vector $\langle 0, -4\rangle$. Translate the image along the same vector. Repeat to generate a pattern. What are the vertices of the fourth polygon in the pattern?

Example 1

Tell whether each transformation appears to be a rotation.

1.

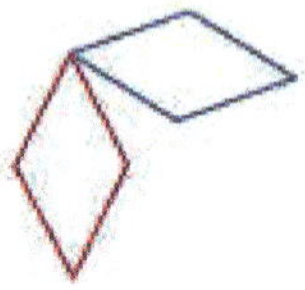

2.

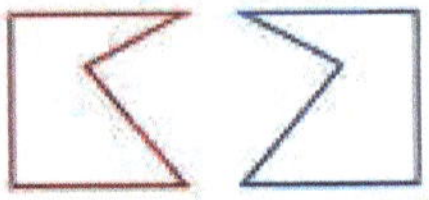

3.

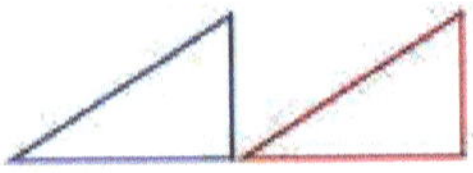

4.

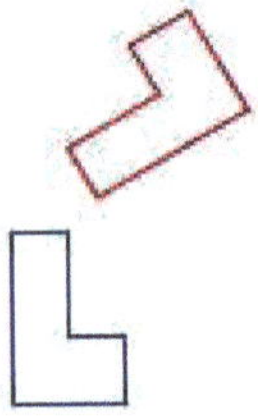

Example 2

Copy each figure and the angle of rotation. Draw the rotation of the figure about point *P* by m∠*A*.

5.

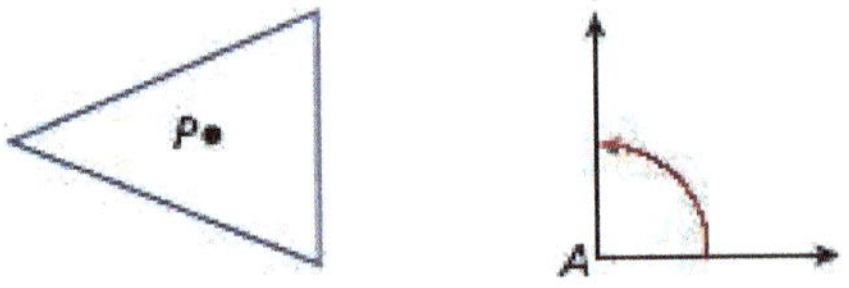

6.

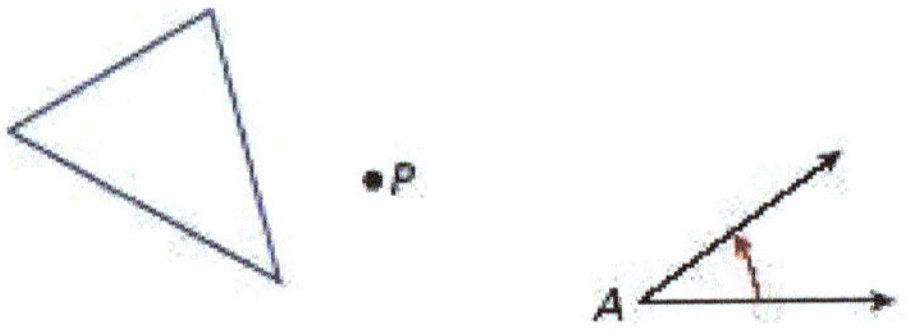

Example 3

Rotate the figure with the given vertices about the origin using the given angle of rotation.

7. $A(1, 0)$, $B(3, 2)$, $C(5, 0)$; 90°

8. $J(2, 1)$, $K(4, 3)$, $L(2, 4)$, $M(-1, 2)$; 90°

9. $D(2, 3)$, $E(-1, 2)$, $F(2, 1)$; 180°

10. $P(-1, -1)$, $Q(-4, -2)$, $R(0, -2)$; 180°

Example 4

11. **Animation** An artist uses a coordinate plane to plan the motion of an animated car. To simulate the car driving around a curve, the artist places the car at the point $(10, 0)$ and then rotates it about the origin by 30°. Give the car's final position, rounding the coordinates to the nearest tenth.

12.1.1 Reflections

Tell whether each transformation appears to be a reflection.

1. 

2. 

3. 

4. 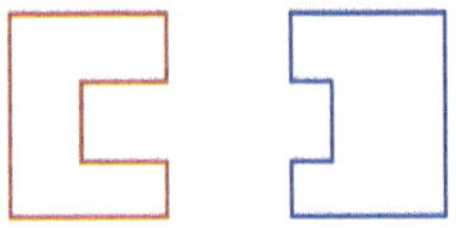

Multi-Step Copy each figure and the line of reflection. Draw the reflection of the figure across the line.

5. 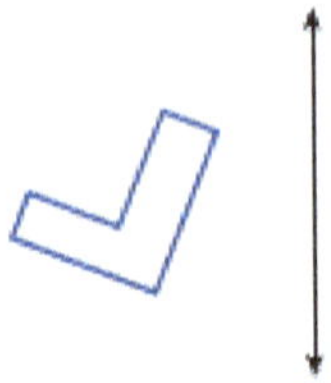

6. 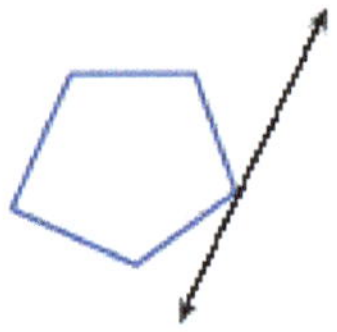

7. **Recreation** Cara is playing pool. She wants to hit the ball at point *A* without hitting the ball at point *B*. She has to bounce the cue ball, located at point *C*, off the side rail and into her ball. Draw a diagram that shows the exact point along the rail that Cara should aim for.

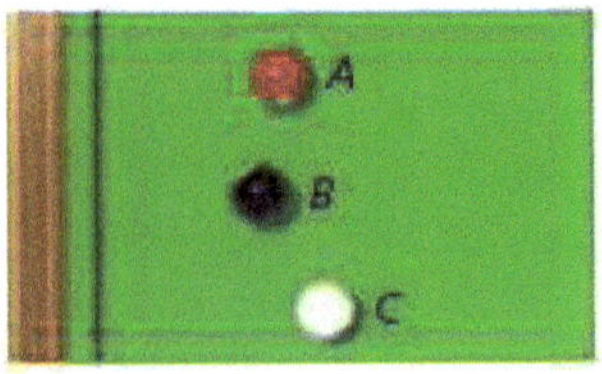

Reflect the figure with the given vertices across the given line.

8. $A(-3, 2)$, $B(0, 2)$, $C(-2, 0)$; y-axis

9. $M(-4, -1)$, $N(-1, -1)$, $P(-2, -2)$; $y = x$

10. $J(1, 2)$, $K(-2, -1)$, $L(3, -1)$; x-axis

11. $S(-1, 1)$, $T(1, 4)$, $U(3, 2)$, $V(1, -3)$; $y = x$

12.1.2 Translations

Tell whether each transformation appears to be a translation.

12. 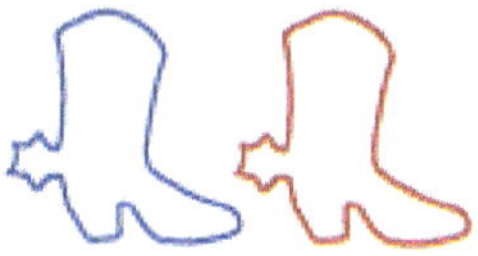

13. 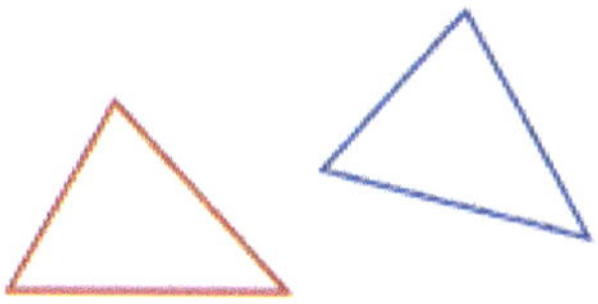

14. 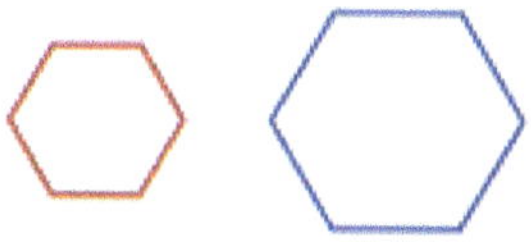

15. 

Multi-Step Copy each figure and the translation vector. Draw the translation of the figure along the given vector.

16.

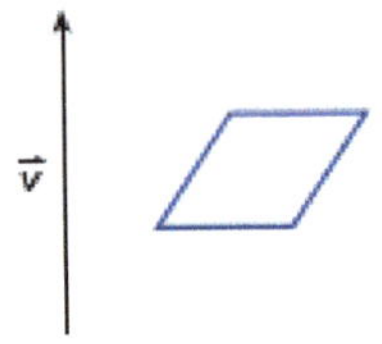

17.

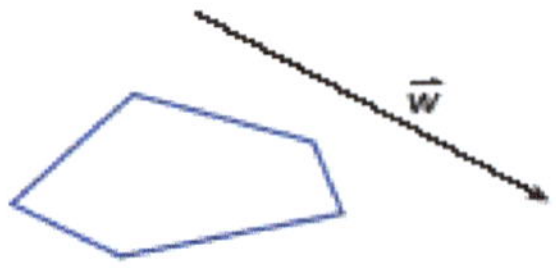

Translate the figure with the given vertices along the given vector.

18. $P(-1, 2)$, $Q(1, -1)$, $R(3, 1)$, $S(2, 3)$; $\langle -3, 0\rangle$

19. $A(1, 3)$, $B(-1, 2)$, $C(2, 1)$, $D(4, 2)$; $\langle -3, -3\rangle$

20. $D(0, 15)$, $E(-10, 5)$, $F(10, -5)$; $\langle 5, -20\rangle$

21. **Animation** An animator draws the ladybug shown and then translates it along the vector $\langle 1, 1 \rangle$, followed by a translation of the new image along the vector $\langle 2, 2 \rangle$, followed by a translation of the second image along the vector $\langle 3, 3 \rangle$.

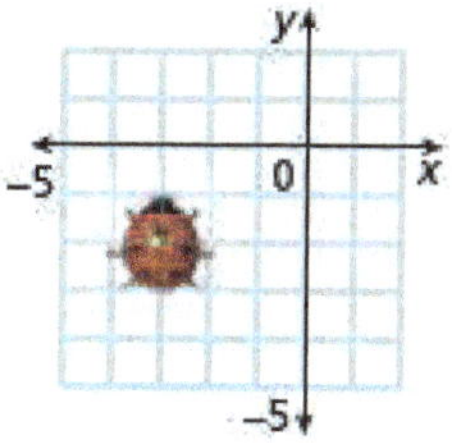

a. Sketch the ladybug's final position.

b. What single vector moves the ladybug from its starting position to its final position?

12.1.3 Rotations

Tell whether each transformation appears to be a rotation.

22. 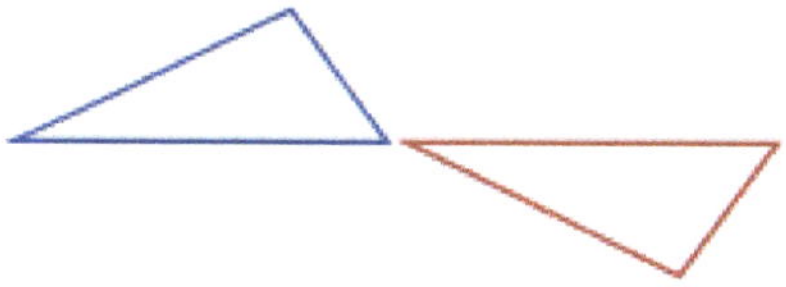

23. 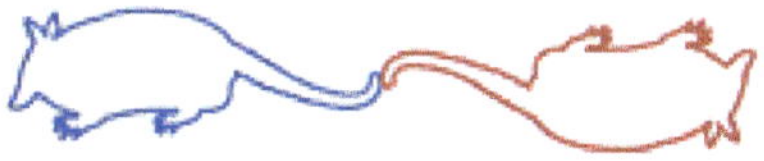

24. 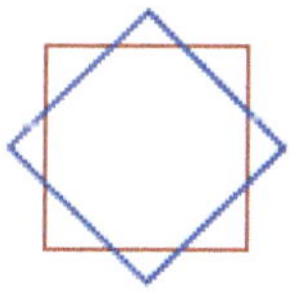

25. 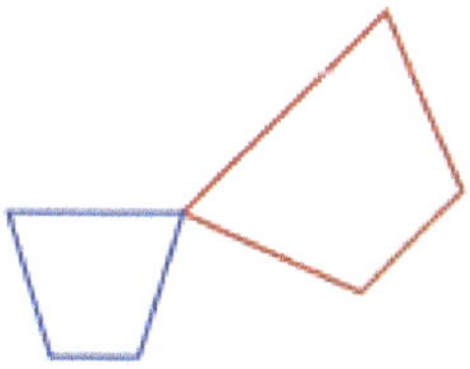

Copy each figure and the angle of rotation. Draw the rotation of the figure about point P by m∠A.

26. 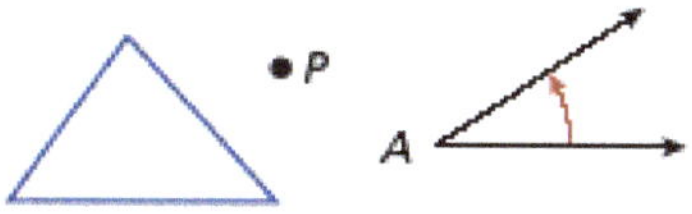

27. 

Rotate the figure with the given vertices about the origin using the given angle of rotation.

28. $E(-1, 2)$, $F(3, 1)$, $G(2, 3)$; 90°

29. $A(-1, 0)$, $B(-1, -3)$, $C(1, -3)$, $D(1, 0)$; 90°

30. $P(0, -2)$, $Q(2, 0)$, $R(3, -3)$; 180°

31. $L(2, 0)$, $M(-1, -2)$, $N(2, -2)$; 180°

32. **Architecture** The CN Tower in Toronto, Canada, features a revolving restaurant that takes 72 minutes to complete a full rotation. A table that is 50 feet from the center of the restaurant starts at position $(50, 0)$. What are the coordinates of the table after 6 minutes? Round coordinates to the nearest tenth.

Simplifying Expressions

Example 1

Simplify each expression.

1. $-12 + 67 + 12 + 23$
2. $16 + 2\frac{1}{2} + 4 + 1\frac{1}{2}$
3. $27 + 98 + 73$
4. $\frac{1}{3} \cdot 8 \cdot 21$
5. $2 \cdot 38 \cdot 50$
6. $50 \cdot 118 \cdot 20$

Example 2

Write each product using the Distributive Property. Then simplify.

7. $14(1002)$
8. $16(19)$
9. $9(38)$
10. $8(57)$
11. $12(112)$
12. $7(109)$

Example 3

Simplify each expression by combining like terms.

13. $6x + 10x$
14. $35x - 15x$
15. $-3a + 9a$

16. $-8r - r$

17. $17x^2 + x$

18. $3.2x + 4.7x$

Example 4

Simplify each expression. Justify each step with an operation or property.

19. $5(x+3) - 7x$

20. $9(a-3) - 4$

21. $5x^2 - 2(x - 3x^2)$

22. $6x - x - 3x^2 + 2x$

23. $12x + 8x + t - 7x$

24. $4a - 2(a-1)$

Writing and Solving Two-Step Equations

Example 1

Translate each sentence into an equation.

1. 20 increased by the product of −1 and a number is 7.
2. Zero is the sum of 4 and the quotient of a number and 3.

Example 2 and 3

Solve each equation. Check your answer.

3. $4a + 3 = 11$
4. $15y + 31 = 61$
5. $8 = 3r - 1$
6. $42 = -2d + 6$
7. $9 - c = -13$
8. $\frac{x}{6} + 4 = 15$
9. $15 = \frac{a}{3} - 2$
10. $4 - \frac{m}{2} = 10$

Example 4

11. **Transportation** Paul bought a student discount card for the bus. The card cost \$7 and allows him to buy daily bus passes for \$1.50. After one month, Paul spent \$29.50. How many daily bus passes did Paul buy?

Example 5

Solve each equation. Check your answer.

12. $12.3 = \frac{m}{0.4} - 0.2$

13. $\frac{2.5 + d}{5} = 1.1$

Solving Two-Step Inequalities

Example 1

Solve. Then graph each solution set on a number line.

1. $5x + 3 < 18$
2. $-19 \geq \frac{z}{7} + 23$
3. $3y - 4 \geq 14$
4. $\frac{m}{4} - 2 > -3$
5. $42 \leq -11p - 13$
6. $\frac{n}{-3} - 4 > 4$

Example 2

7. Three students collected more than \$93 washing cars. They used \$15 to reimburse their parents for cleaning supplies. Then they divided the remaining money equally. How much did each student earn?

Writing and Solving Multi-Step Equations

Example 1

Solve.

1. $14n + 2 - 7n = 37$
2. $10x - 11 - 4x = 43$
3. $1 = -3 + 4p - 2p$

Example 2

Solve.

4. $12 - (x + 3) = 10$
5. $15 = 2(q + 4) + 3$
6. $5(m - 2) + 36 = -4$

Example 3

7. Keisha read twice as many books this year as Ben read. Subtracting 4 from the number of books Keisha read and dividing by 2 gives the number of books Sheldon read. Sheldon read 10 books. How many books did Ben read?

Solving Equations with Variables on Both Sides

Example 1

Solve each equation. Check your answer.

1. $2c - 5 = c + 4$
2. $8r + 4 = 10 + 2r$
3. $2x - 1 = x + 11$
4. $28 - 0.3y = 0.7y - 12$

Example 2

Solve each equation. Check your answer.

5. $-2(x + 3) = 4x - 3$
6. $3c - 4c + 1 = 5c + 2 + 3$
7. $5 + 3(q - 4) = 2(q + 1)$
8. $5 - (t + 3) = -1 + 2(t - 3)$

9. $7x - 4 = -2x + 1 + 9x - 5$

10. $8x + 6 - 9x = 2 - x - 15$

11. $6y = 8 - 9 + 6y$

12. $6 - 2x - 1 = 4x + 8 - 6x - 3$

Example 3

13. **Consumer Economics** A house-painting company charges \$376 plus \$12 per hour. Another painting company charges \$280 plus \$15 per hour.
    a. How long is a job for which both companies will charge the same amount?
    b. What will that cost be?

Systems of Equations

Example 1

Solve each system of equations.

1. $y = x + 1$
   $y = 2x - 1$

2. $y = -2x + 3$
   $y = 5x - 4$

3. $y = 3x - 5$
   $y = 6x + 7$

4. $y = 6x - 12$
   $y = -9x + 3$

5. $y = 5x + 7$
   $y = -3x + 7$

6. $y = 3x + 5$
   $y = 3x - 10$

Example 2

Solve each system of equations.

7. $2x + 2y = 16$
   $2x + 6y = 28$

8. $x + y = 20$
   $x = y - 4$

9. $x + 2y = 21$
   $-x + 3y = 29$

10. $5x - 2y = 4$
    $11x + 4y = -8$

11. $x = -3y$
    $7x - 2y = -69$

12. $-4x - 5y = -7$
    $11y = 2x + 37$

13.1.1 Simplifying Expressions

Simplify each expression.

1. $53 + 28 + 17 + 12$
2. $5 \cdot 14 \cdot 20$
3. $6 \cdot 3 \cdot 5$
4. $4.5 + 7.1 + 8.5 + 3.9$

Write each product using the Distributive Property. Then simplify.

5. $9(62)$
6. $8(29)$
7. $11(25)$
8. $6(53)$

Simplify each expression by combining like terms.

9. $3x + 9x$
10. $14x^2 - 5x^2$
11. $-7x + 8x$
12. $3x^2 - 4$

Simplify each expression. Justify each step with an operation or property.

13. $4(y + 6) + 9$
14. $-7(x + 2) + 4x$
15. $3x + 2 - 2x - 1$
16. $5x - 3x + 3x^2 + 9x$
17. $8x + 2x - 3y - 9x$
18. $7y - 3 + 6y - 7$

13.1.2 Writing and Solving Two-Step Equations

Solve each equation. Check your answer.

19. $x - 8 = 17$

20. $10x = 30$

21. $5h - 20 = 8$

22. $6 = -14 + 2c$

23. $2x + 6 = 10$

24. $17 = 3p - 7$

25. $2d + \frac{1}{5} = \frac{3}{5}$

26. $1 = 2x + \frac{1}{2}$

27. $\frac{z}{2} + 1 = \frac{3}{2}$

28. $\frac{2}{3} = \frac{4j}{6}$

29. $\frac{3}{4} = \frac{3}{8}x - \frac{3}{2}$

30. $\frac{1}{5} - \frac{x}{5} = -\frac{2}{5}$

31. **Consumer Economics** Jennifer is saving money to buy a bike. The bike costs $245. She has $125 saved, and each week she adds $15 to her savings. How long will it take her to save enough money to buy the bike?

32. **Consumer Economics** Amanda and Casey's total restaurant bill, including tip, was \$34. Amanda's portion of the bill was twice as much as Casey's, and they each left a tip of \$2. How much did each person pay?

33. Marissa is buying a shirt for each of her brothers and each of her 2 sisters. Each shirt costs \$7.50, and she spends a total of \$30. How many brothers does Marissa have?

13.1.3 Solving Two-Step Inequalities

Solve. Then graph each solution set on a number line.

34. $5s - 7 > -42$

35. $\frac{b}{2} + 3 < 9$

36. $19 \leq -2q + 5$

37. $-8c - 11 \leq 13$

38. $\frac{y}{-4} + 6 > 10$

39. $\frac{x}{9} - 5 \leq -8$

40. $\frac{r}{-2} - 9 > -14$

41. $44 \geq 13j + 18$

42. $\frac{d}{13} - 12 > 27$

43. Rico has \$5.00. Bagels cost \$0.65 each, and a small container of cream cheese costs \$1.00. What is the greatest number of bagels Rico can buy if he also buys one small container of cream cheese?

44. The 35 members of a drill team are trying to raise at least \$1,200 to cover travel costs to a training camp. They have already raised \$500. How much should each member still raise, on average, to meet the goal?

13.1.4 Writing and Solving Multi-Step Equations

Solve.

1. $b + 18 + 3b = 74$

2. $10x - 3 - 2x = 4$

3. $18w - 10 - 6w = 50$

4. $19 = 5n + 7 - 3n$

5. $-27 = -3p + 15 - 3p$

6. $-x - 8 + 14x = -34$

7. $2(x + 4) + 6 = 22$

8. $1 - 3(n + 5) = -8$

9. $4.3 - 1.4(p + 7) = -9.7$

10. $1.8 + 6n - 3.2 = 7.6$

11. $0 = 9\left(k - \frac{2}{3}\right) + 33$

12. $6(t - 2) - 76 = -142$

13. Abby ran 3 times as many laps as Karen. Adding 4 to the number of laps Abby ran and then dividing by 7 gives the number of laps Jill ran. Jill ran 1 lap. How many laps did Karen run?

13.1.5 Solving Equations with Variables on Both Sides

Solve each equation. Check your answer.

14. $7a - 17 = 4a + 1$

15. $2b - 5 = 8b + 1$

16. $4x - 2 = 3x + 4$

17. $2x - 5 = 4x - 1$

18. $8x - 2 = 3x + 12.25$

19. $5x + 2 = 3x$

20. $3c - 5 = 2c + 5$

21. $-17 - 2x = 6 - x$

22. $3(t - 1) = 9 + t$

23. $5 - x - 2 = 3 + 4x + 5$

24. $2(x + 4) = 3(x - 2)$

25. $3m - 10 = 2(4m - 5)$

26. $5 - (n - 4) = 3(n + 2)$

27. $6(x + 7) - 20 = 6x$

28. $8(x + 1) = 4x - 8$

29. $x - 4 - 3x = -2x - 3 - 1$

30. $-2(x + 2) = -2x + 1$

31. $2(x + 4) - 5 = 2x + 3$

32. **Sports** Justin and Tyson are beginning an exercise program to train for football season. Justin weighs 150 lb and hopes to gain 2 lb per week. Tyson weighs 195 lb and hopes to lose 1 lb per week.
   a. If the plan works, in how many weeks will the boys weigh the same amount?
   b. What will that weight be?

13.1.6 Systems of Equations

Solve each system of equations.

33. $y = -2x - 1$
    $y = 2x + 3$

34. $y = 3x + 6$
    $y = x + 2$

35. $y = 5x - 3$
    $y = -3x + 13$

36. $y = x + 6$
    $y = -2x - 12$

37. $y = 3x - 1$
    $y = -2x + 9$

38. $y = -2x - 6$
    $y = 3x + 29$

39. $3x + 3y = 15$
    $3x - 6y = -12$

40. $2x + y = 11$
    $-x + 2y = 2$

41. $y = 5x - 2$
    $4x + 3y = 13$

42. $5x - 9y = 11$
    $3x + 7y = 19$

43. $12x + 18y = 30$
    $4x - 13y = 67$

44. $-14x - 11y = 97$
    $-12y + 11x = 27$

Example 1

Identify the quadrant that contains each point.

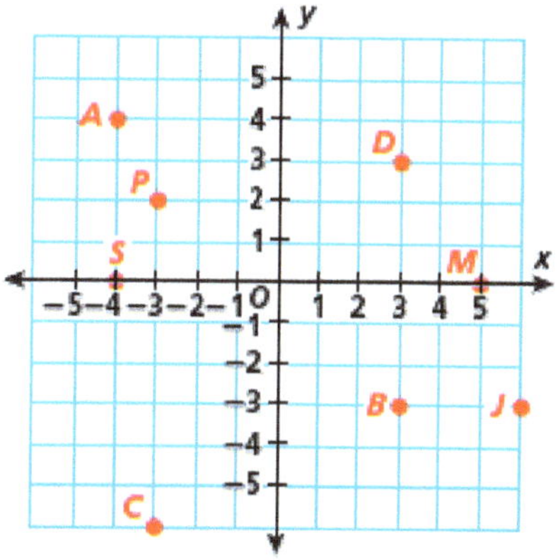

1. *A*
2. *B*
3. *C*
4. *D*

Example 2

Plot each point on a coordinate plane.

5. $E(-1, 2)$
6. $N(2, -4)$
7. $H(-3, -4)$
8. $T(5, 0)$

Example 3

Give the coordinates of each point.

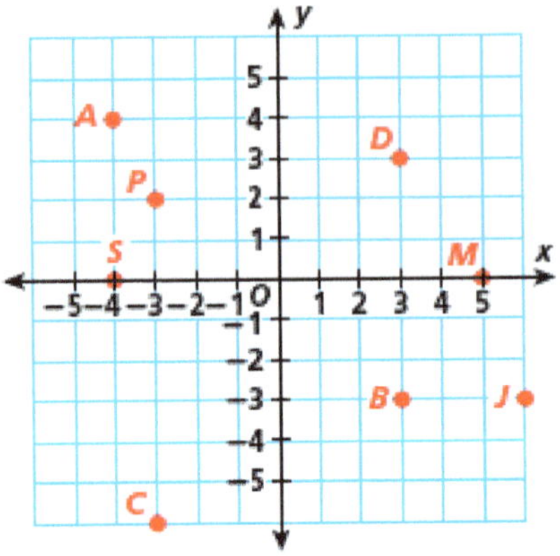

9. J

10. P

11. S

12. M

Equations, Tables, and Graphs

Example 1

1. The amount of water in a pool being filled is represented by the equatic $g = 15m$, where g is the number of gallons of water in the pool and m is the number of minutes since filling began. Make a table and sketch a graph of the equation.

Example 2

2. Use the table to make a graph and to write an equation.

x	0	2	5	9	12
y	3	5	8	12	15

Example 3

3. Use the graph to make a table and to write an equation.

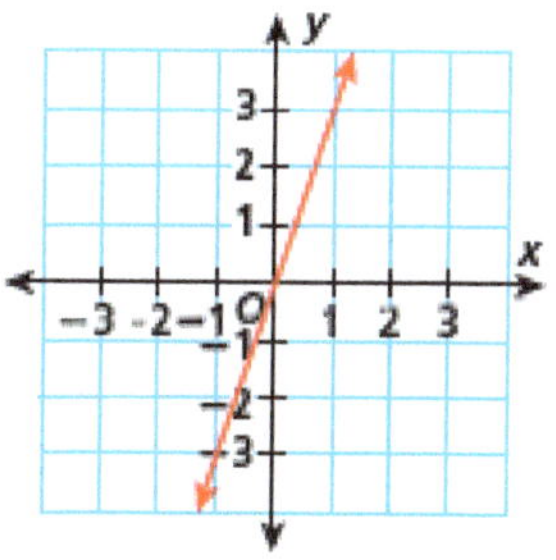

Graphing Using Intercepts

Example 1

Find the x- and y-intercepts.

1.

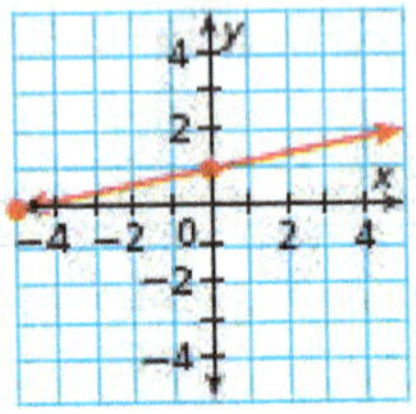

2.

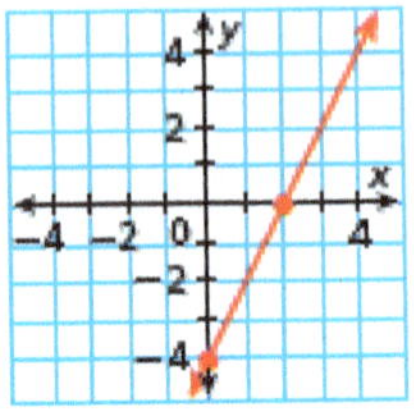

3.

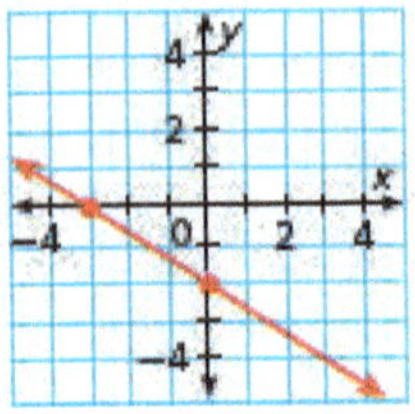

4. $2x - 4y = 4$

5. $-2y = 3x - 6$

6. $4y + 5x = 2y - 3x + 16$

13.2.1 The Coordinate Plane

Identify the quadrant that contains each point.

1. *F*
2. *J*
3. *K*
4. *E*

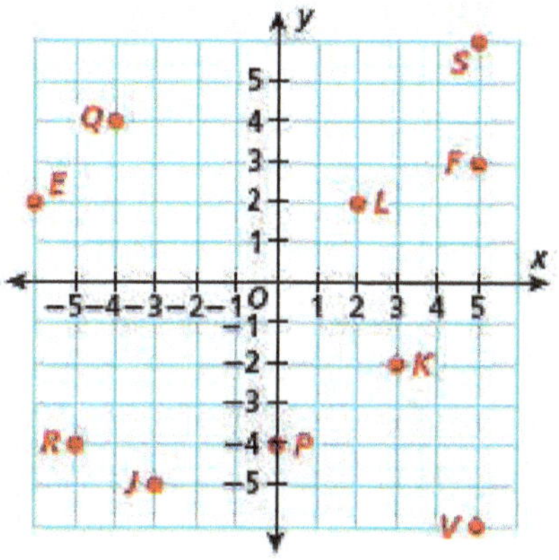

Plot each point on a coordinate plane.

5. $A(-1, 1)$
6. $M(2, -2)$
7. $W(-5, -5)$
8. $G(0, -3)$

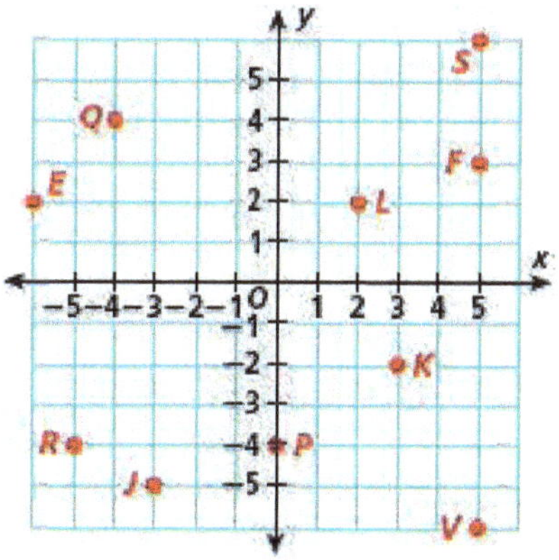

Give the coordinates of each point.

9. *Q*
10. *V*
11. *R*
12. *P*
13. *S*
14. *L*

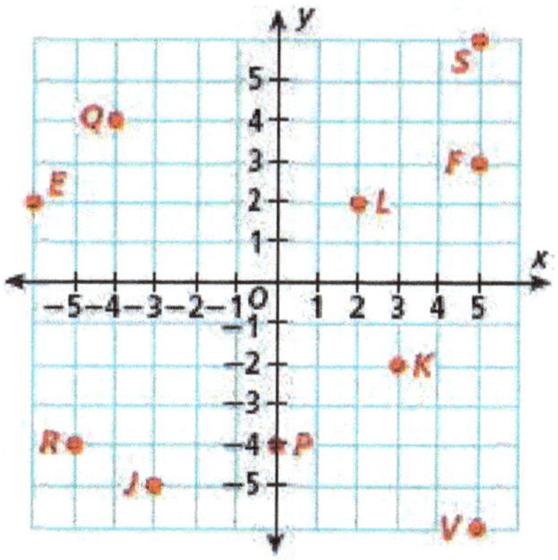

13.2.2 Equations, Tables, and Graphs

15. The amount of sand in the top half of an hourglass is represented by the equation $h = -0.5s$, where h is the height of the sand in centimeters and s is the number of seconds since the top half began draining. Make a table and sketch a graph of the equation.

16. Use the table to make a graph and to write an equation.

x	0	2	4	6	8
y	12	10	8	6	4

17. Use the graph to make a table and to write an equation.

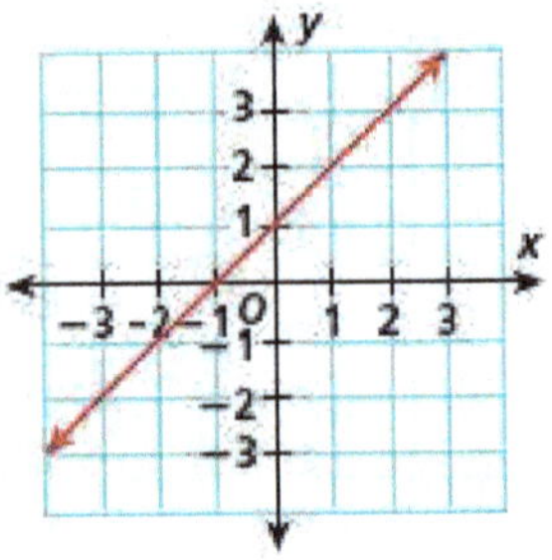

13.2.3 Graphing Using Intercepts

Find the *x*- and *y*-intercepts.

18.

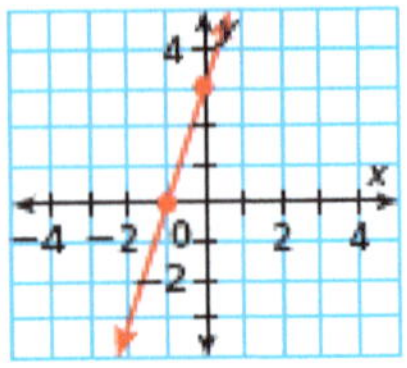

19.

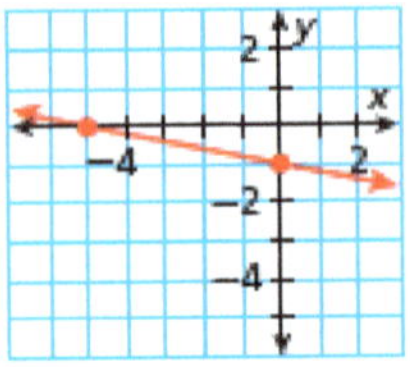

20.

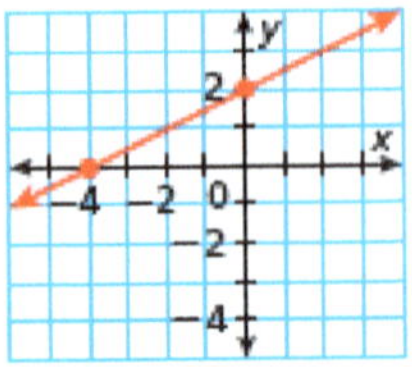

21. $6x + 3y = 12$

22. $4y - 8 = 2x$

23. $-2y + x = 2y - 8$

24. $4x + y = 8$

25. $y - 3x = -15$

26. $2x + y = 10x - 1$

27. **Environmental Science** A fishing lake was stocked with 300 bass. Each year, the population decreases by 25. The population of bass in the lake after x years is represented by the function $f(x) = 300 - 25x$.

a. Graph the function and find its intercepts.

b. What does each intercept represent?

28. **Sports** Julie is running a 5-kilometer race. She ran 1 kilometer every 5 minutes. Julie's distance from the finish line after x minutes is represented by the function $f(x) = 5 - \frac{1}{5}x$.

a. Graph the function and find its intercepts.

b. What does each intercept represent?

Use intercepts to graph the line described by each equation.

29. $4x - 6y = 12$

30. $2x + 3y = 18$

31. $\frac{1}{2}x - 4y = 4$

32. $y - x = -1$

33. $5x + 3y = 15$

34. $x - 3y = -1$

Rates, Ratios, and Proportions

Example 1

1. The ratio of the sale price of a jacket to the original price is 3 : 4. The original price is \$64. What is the sale price?

2. **Chemistry** The ratio of hydrogen atoms to oxygen atoms in water is 2 : 1. If an amount of water contains 341 trillion atoms of oxygen, how many hydrogen atoms are there?

Example 2

Find each unit rate.

3. A computer's fan rotates 2000 times in 40 seconds.

4. Twelve cows produce 224,988 pounds of milk.

5. A yellow jacket can fly 4.5 meters in 9 seconds.

Example 3

6. Lydia wrote $4\frac{1}{2}$ pages of her science report in one hour. What was her writing rate in pages per minute?

7. A model airplane flies 18 feet in 2 seconds. What is the airplane's speed in miles per hour? Round your answer to the nearest hundredth.

8. A vehicle uses 1 tablespoon of gasoline to drive 125 yards. How many miles can the vehicle travel per gallon? Round your answer to the nearest mile. (*Hint:* There are 256 tablespoons in a gallon.)

Example 4

Solve each proportion.

9. $\frac{3}{z} = \frac{1}{8}$

10. $\frac{x}{3} = \frac{1}{5}$

11. $\frac{b}{4} = \frac{3}{2}$

12. $\frac{f+3}{12} = \frac{7}{2}$

13. $\frac{-1}{5} = \frac{3}{2d}$

14. $\frac{3}{14} = \frac{s-2}{21}$

15. $\frac{-4}{9} = \frac{7}{x}$

16. $\frac{3}{s-2} = \frac{1}{7}$

17. $\frac{10}{h} = \frac{52}{13}$

Example 5

18. **Archaeology** Stonehenge II in Hunt, Texas, is a scale model of the ancient construction in Wiltshire, England. The scale of the model to the original is 3:5. The Altar Stone of the original construction is 4.9 meters tall. Write and solve a proportion to find the height of the model of the Altar Stone.

Alfred Sheppard, one of the builders of Stonehenge II.

Applications of Proportions

Example 1

1. Colin bought 3 vintage T-shirts for $45. How much would Colin pay for 6 vintage T-shirts?

2. Amanda purchased 2 movie tickets for $14. How many movie tickets can Amanda buy with $42?

Example 2

Find the missing value in each proportion.

3. $\frac{12}{9} = \frac{n}{3}$

4. $\frac{t}{5} = \frac{28}{20}$

5. $\frac{1}{c} = \frac{6}{12}$

6. $\frac{6}{7} = \frac{30}{b}$

Example 3

7. Ursula is entering a bicycle race for charity. Her mother pledges $0.75 for every 0.5 mile she bikes. If Ursula bikes 17.5 miles, how much will her mother donate?

Finding and Using Percents

Example 1

1. What percent of 91 is 45?

2. What percent of 1270 is 375?

3. What percent of 240 is 180?

4. What percent of 186 is 75?

Example 2

5. Elijah walks 2 miles to school. If Bailey's walk is 80% of the length of Elijah's walk, find the length of Bailey's walk.

6. Jay's term paper is 18 pages long. If Madison's paper is 175% of the length of Jay's paper, find the length of Madison's paper.

7. Of 109.6 million households, 19,508,800 watched the television show *CSI* during the week of October 3, 2005. What percent of American households watched *CSI* this week?

Example 3

Find each number to the nearest tenth.

8. 6.9 is $11\frac{1}{2}$% of what number?

9. 92 is $66\frac{2}{3}$% of what number?

10. 12% of what number is 20?

11. 30% of what number is 96?

Example 4 and 5

12. How much water can a 7.4 oz piece of chalk absorb if it can absorb 32% of its weight?

13. At 2 P.M., a flag pole casts a shadow that is 155% of its actual height. If the shadow is 23.25 ft, what is the actual height of the pole?

13.3.1 Rates, Ratios, and Proportions

1. **Gardening** The ratio of the height of a bonsai ficus tree to the height of a full-size ficus tree is 1:9. The bonsai ficus is 6 inches tall. What is the height of a full-size ficus?

2. **Manufacturing** At one factory, the ratio of defective light bulbs produced to total light bulbs produced is about 3:500. How many light bulbs are expected to be defective when 12,000 are produced?

Find each unit rate.

3. Four gallons of gasoline weigh 25 pounds.

4. Fifteen ounces of gold cost $6058.50.

5. **Biology** The tropical giant bamboo can grow 11.9 feet in 3 days. What is this rate of growth in inches per hour? Round your answer to the nearest hundredth, and show that your answer is reasonable.

Appendix: Essential Algebra Review

Review 13.3 / Ratio and Proportion

6. **Transportation** The maximum speed of the Tupolev Tu-144 airliner is 694 m/s. What is this speed in kilometers per hour?

Solve each proportion.

7. $\frac{v}{6} = \frac{1}{2}$

8. $\frac{2}{5} = \frac{4}{y}$

9. $\frac{2}{h} = \frac{-5}{6}$

10. $\frac{3}{10} = \frac{b+7}{20}$

11. $\frac{5t}{9} = \frac{1}{2}$

12. $\frac{2}{3} = \frac{6}{q-4}$

13. $\frac{x}{8} = \frac{7.5}{20}$

14. $\frac{3}{k} = \frac{45}{18}$

15. $\frac{6}{a} = \frac{15}{17}$

16. $\frac{9}{2} = \frac{5}{x+1}$

17. $\frac{3}{5} = \frac{x}{100}$

18. $\frac{38}{19} = \frac{n-5}{20}$

19. **Science** The image shows a dust mite as seen under a microscope. The actual length of this dust mite is 0.3 millimeter. Use a ruler to measure the length of the dust mite in the image in millimeters. What is the scale of the drawing?

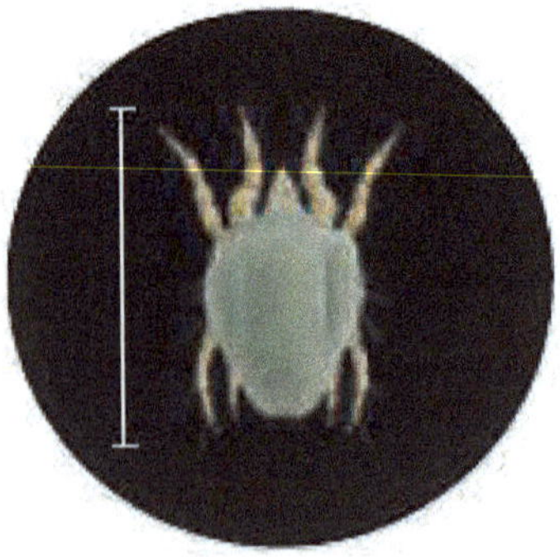

13.3.2 Applications of Proportions

20. Zed buys a sports magazine each month for 3 months and spends $18. How much will Zed spend if he buys the same sports magazine every month for 12 months?

21. Morgan bought 3 pounds of cheese for a party platter for $27. How many pounds of the same cheese can Morgan buy with $54?

Find the missing value in each proportion.

22. $\frac{3}{2} = \frac{24}{d}$

23. $\frac{p}{40} = \frac{3}{8}$

24. $\frac{6}{14} = \frac{x}{7}$

25. $\frac{5}{p} = \frac{7}{77}$

26. According to Ty's study guidelines, how many minutes of science reading should he do if his science class is 90 minutes long?

Ty's Study Guidelines	
Class	Reading Time
Literature	35 minutes for every 50 minutes of class time
Science	20 minutes for every 60 minutes of class time
History	30 minutes for every 55 minutes of class time

13.3.3 Finding and Using Percents

Find the percent of each number.

27. 16% of 70
28. 118% of 3,000
29. 5% of 58
30. 103% of 50

31. In 2004, there were 19,396 bulldogs registered by the American Kennel Club. Approximately 86% of this number were registered in 2003. About how many bulldogs were registered in 2003?

Solve.

32. 56 is 140% of what number?
33. 45 is 20% of what number?
34. 18 is 15% of what number?

35. What percent of 60 is 40?
36. What percent of 48 is 16?
37. What percent of 6 is 18?

38. 210% of what number is 147?

39. A 12-pack of cinnamon-scented pencils sells for $3.00 at a school booster club sale. What is the sales tax rate if the total cost of the pencils is $3.21?

Example 1

Add.

1. $(5n^3 + 3n + 6) + (18n^3 + 9)$

2. $(3.7q^2 - 8q + 3.7) + (4.3q^2 - 2.9q + 1.6)$

3. $(-3x + 12) + (9x^2 + 2x - 18)$

4. $(9x^4 + x^3) + (2x^4 + 6x^3 - 8x^4 + x^3)$

Example 2

5. **Geometry** Write a polynomial that represents the measure of angle *ABD*.

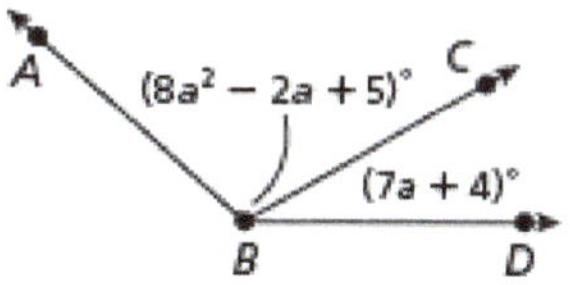

Example 3

Find the opposite of each polynomial.

6. $0.32r^3 + 0.19r^4$

7. $9y^2 - 6x^2$

8. $-4 + w - 2z$

Example 4

Subtract.

9. $(6c^4 + 8c + 6) - (2c^4)$

10. $(16y^2 - 8y + 9) - (6y^2 - 2y + 7y)$

11. $(2r + 5) - (5r - 6)$

12. $(-7k^2 + 3) - (2k^2 + 5k - 1)$

Example 5

13. **Geometry** The length of line segment *AC* is $(15x - y)$ cm and the length of line segment *AB* is $(8x + 3y - 1)$ cm. Write a polynomial that represents the length of line segment *BC*.

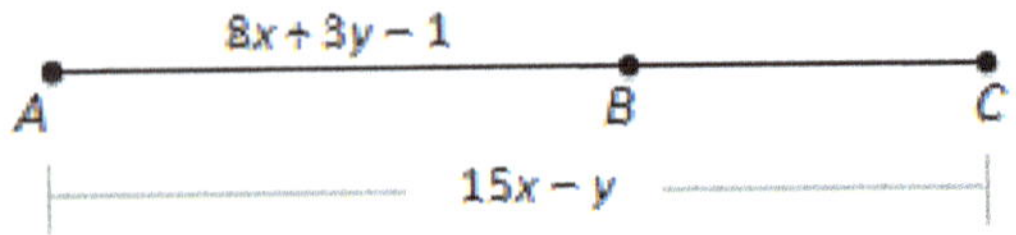

Appendix: Essential Algebra Review

Operations with Polynomials and Factoring

Multiplying and Dividing Monomials

Example 1

Multiply.

1. $(2x^2)(7x^4)$
2. $(-5mn^3)(4m^2n^2)$
3. $(6rs^2)(s^3t^2)\left(\frac{1}{2}r^4t^3\right)$
4. $\left(\frac{1}{3}a^5\right)(12a)$
5. $(-3x^4y^2)(-7x^3y)$
6. $(-2pq^3)(5p^2q^2)(-3q^4)$

Example 2

Divide. Assume that no denominator equals zero.

7. $\frac{15n^5}{5n^2}$
8. $\frac{-4x^7}{2x^6}$
9. $\frac{18a^2b^8}{12ab^5}$
10. $\frac{24s^{10}t^{12}}{18s^5}$
11. $\frac{42a^4c^7}{7a^4c^2}$
12. $\frac{16m^5p}{8m^3p}$

Example 3

Simplify.

13. $(5a)^3$
14. $(9m^3)^2$
15. $(6x^2y^5)^3$
16. $(-7p^5q^3)^2$
17. $(2m^3n^2)^5$
18. $(-3x^4yz)^4$

Example 1

Multiply.

1. $4(x^2 + 2x + 1)$
2. $3ab(2a^2 + 3b^3)$
3. $2a^3b(3a^2b + ab^2)$
4. $-3x(x^2 - 4x + 6)$
5. $5x^2y(2xy^3 - y)$
6. $5m^2n^3 \cdot mn^2(4m - n)$

Example 2

7. **Photography** The length of a rectangular photograph is 3 inches less than twice the width.
   a. Write a polynomial that represents the area of the photograph.
   b. Find the area of the photograph when the width is 4 inches.

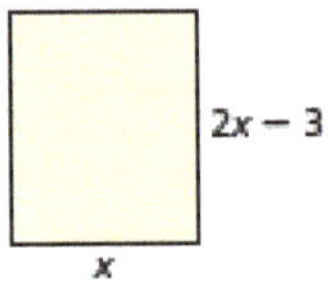

Multiplying Binomials

Example 1

Multiply.

1. $(x + 1)(x - 2)$
2. $(y - 3)(y - 5)$
3. $(4a^3 - 2b)(a - 3b^2)$
4. $(m^2 - 2mn)(3mn + n^2)$

Example 2

5. **Geometry** Write a polynomial that represents the area of the figure.

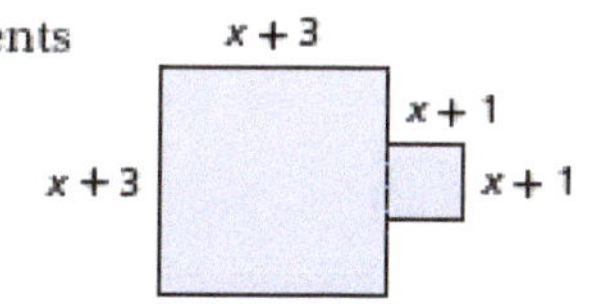

Example 3

Multiply.

6. $(x + 7)^2$
7. $(2 + x)^2$
8. $(5x + 9)^2$

9. $(2a + 7b)^2$

10. $(x - 6)^2$

11. $(2x - 1)^2$

12. $(8 - x)^2$

13. $(6p - q)^2$

14. $(7a - 2b)^2$

15. $(x + 5)(x - 5)$

16. $(x + 6)(x - 6)$

17. $(5x + 1)(5x - 1)$

18. $(2x^2 + 3)(2x^2 - 3)$

19. $(9 - x^3)(9 + x^3)$

20. $(2x - 5y)(2x + 5y)$

Factoring with the GCF

Example 1

Find the GCF of each pair of monomials.

1. $6x^2$ and $5x^2$
2. $15y^3$ and $-20y$
3. $13q^4$ and $2p^2$

Example 2

Factor each polynomial. Check your answer.

4. $15a - 5a^2$
5. $10g^3 - 3g$
6. $-35x + 42$
7. $-4x^2 - 6x$
8. $12h^4 + 8h^2 - 6h$
9. $3x^2 - 9x + 3$
10. $9m^2 + m$
11. $14n^3 + 7n + 7n^2$
12. $36f + 18f^2 + 3$
13. $-15b^2 + 7b$

Example 3

14. **Physics** A model rocket is fired vertically into the air at 320 ft/s. The expression $-16t^2 + 320t$ gives the rocket's height after t seconds. Factor this expression.

Example 4

Factor each expression.

15. $5(m-2) - m(m-2)$

16. $2b(b+3) + 5(b+3)$

17. $4(x-3) - x(y+2)$

Appendix: Essential Algebra Review

Operations with Polynomials and Factoring

Factoring $x^2 + bx + c$

Example 1

Factor each trinomial by guess and check.

1. $x^2 + 13x + 36$ **2.** $x^2 + 11x + 24$ **3.** $x^2 + 14x + 40$

Example 2

Factor each trinomial. Check your answer.

4. $x^2 + 4x + 3$ **5.** $x^2 + 10x + 16$ **6.** $x^2 + 15x + 44$

7. $x^2 - 7x + 6$ **8.** $x^2 - 9x + 14$ **9.** $x^2 - 11x + 24$

Example 3

Factor each trinomial. Check your answer.

10. $x^2 + 6x - 7$ **11.** $x^2 + 6x - 27$ **12.** $x^2 + x - 30$

13. $x^2 - x - 2$ **14.** $x^2 - 3x - 18$ **15.** $x^2 - 4x - 45$

Factoring $ax^2 + bx + c$

Example 1

Factor each trinomial by guess and check.

1. $2x^2 + 9x + 10$
2. $5x^2 + 31x + 6$
3. $5x^2 + 7x - 6$
4. $6x^2 + 37x + 6$
5. $3x^2 - 14x - 24$
6. $6x^2 + x - 2$

Example 2

Factor each trinomial. Check your answer.

7. $5x^2 + 11x + 2$
8. $2x^2 + 11x + 5$
9. $4x^2 - 9x + 5$
10. $2y^2 - 11y + 14$
11. $5x^2 + 9x + 4$
12. $3x^2 + 7x + 2$

Example 3

Factor each trinomial. Check your answer.

13. $4a^2 + 8a - 5$
14. $15x^2 + 4x - 3$
15. $2x^2 + x - 6$

16. $6n^2 - 11n - 10$

17. $10x^2 - 9x - 1$

18. $7x^2 - 3x - 10$

Example 4

Factor each trinomial. Check your answer.

19. $-2x^2 + 5x + 12$

20. $-4n^2 - 16n + 9$

21. $-5x^2 + 7x + 6$

22. $-6x^2 + 13x - 2$

23. $-4x^2 - 8x + 5$

24. $-5x^2 + x + 18$

13.4.1 Adding and Subtracting Polynomials

Add.

1. $(2t^2 - 8t) + (8t^2 + 9t)$

2. $(-7x^2 - 2x + 3) + (4x^2 - 9x)$

3. $(x^5 - x) + (x^4 + x)$

4. $(-2z^3 + z + 2z^3 + z) + (3z^3 - 5z^2)$

Subtract.

5. $(t^3 + 8t^2) - (3t^3)$

6. $(3x^2 - x) - (x^2 + 3x - x)$

7. $(5m + 3) - (6m^3 - 2m^2)$

8. $(3s^2 + 4s) - (-10s^2 + 6s)$

9. **Photography** The measurements of a photo and its frame are shown in the diagram. Write a polynomial that represents the width of the photo.

10. **Geometry** The length of a rectangle is represented by $4a + 3b$, and its width is represented by $7a - 2b$. Write a polynomial for the perimeter of the rectangle.

13.4.2 Multiplying and Dividing Monomials

Multiply.

11. $(3x^2)(8x^5)$

12. $(-2r^3s^4)(6r^2s)$

13. $(15xy^2)\left(\frac{1}{3}x^2z^3\right)(y^3z^4)$

14. $(-2a^3)(-5a)$

15. $(6x^3y^2)(-2x^2y)$

16. $(-3a^2b)(-2b^3)(-a^3b^2)$

17. $(7x^2)(xy^5)(2x^3y^2)$

18. $(-4a^3bc^2)(a^3b^2c)(3ab^4c^5)$

19. $(12mn^2)(2m^2n)(mn)$

Divide. Assume that no denominator equals zero.

20. $\frac{35x^9}{7x^7}$

21. $\frac{20n^{12}}{5n^4}$

22. $\frac{8pq^8}{4pq^5}$

23. $\frac{12a^5b^4}{8a^4b^{11}}$

24. $\frac{15x^2y^7}{10xy^2}$

25. $\frac{-25a^{10}b^3}{5a^3}$

Simplify.

26. $(2y)^8$

27. $(7x^3)^3$

28. $(4b^2)^4$

29. $(5x^4y^5)^3$

30. $(2a^3b^{10})^5$

31. $(-2p^3r)^4$

13.4.3 Multiplying Monomials by Polynomials

Multiply.

32. $9s(s+6)$

33. $9(2x^2-5x)$

34. $3x(9x^2-4x)$

35. $3(2x^2+5x+4)$

36. $5s^2t^3(2s-3t^2)$

37. $x^2y^3 \cdot 5x^2y(6x+y^2)$

38. $-5x(2x^2 - 3x - 1)$

39. $-2a^2b^3(3ab^2 - a^2b)$

40. $-7x^3y \cdot x^2y^2(2x - y)$

41. **Geometry** The length of the rectangle at right is 3 feet longer than its width.

a. Write a polynomial that represents the area of the rectangle.

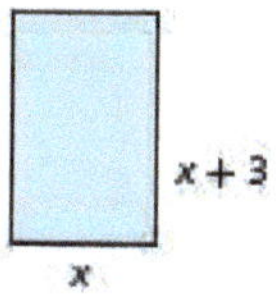

b. Find the area of the rectangle when the width is 5 feet.

13.4.4 Multiplying Binomials

Multiply.

42. $(x + 5)(x - 3)$

43. $(5x - 2)(x + 3)$

44. $(5x + 2)(2x - 1)$

45. $(x - 1)(x - 2)$

46. $(x - 8)(7x + 4)$

47. $(2x + 7)(3x + 7)$

48. $(x + 3)^2$

49. $(2 + 3x)^2$

50. $(r^2 + 5t)^2$

Appendix: Essential Algebra Review

Review 13.4 / Operations with Polynomials and Factoring, Part 1

51. $(s^2 - 7)^2$

52. $(2c - d^3)^2$

53. $(1 - x^2)^2$

54. $(a - 10)(a + 10)$

55. $(5a^2 + 9)(5a^2 - 9)$

56. $(x^3 + y^2)(x^3 - y^2)$

57. **Entertainment** Write a polynomial that represents the area of the circular puzzle. Remember that the formula for area of a circle is $A = \pi r^2$, where r is the radius of the circle. Leave the symbol π in your answer.

13.4.5 Factoring with the GCF

Find the GCF of each pair of monomials.

1. $8a^2$ and 11
2. $9s$ and $63s^3$
3. $-64n^4$ and $24n^2$

Factor each polynomial. Check your answer.

4. $9y^2 + 45y$
5. $36d^3 + 24$
6. $-14x^4 + 5x^2$
7. $-15f - 10f^2$
8. $-4d^4 + d^3 - 3d^2$
9. $14x^3 + 63x^2 - 7x$
10. $21c^2 + 14c$
11. $33d^3 + 22d + 11$
12. $-5g^3 - 15g^2$
13. **Finance** After t years, the amount of money in a savings account that earns simple interest is $P + Prt$, where P is the starting amount and r is the yearly interest rate. Factor this expression.

Factor each expression.

14. $6a(a-2) - 5b(b+4)$
15. $-4x(x+2) + 9(x+2)$
16. $6y(y-7) + (y-7)$

17. $a(x-3)+2b(x-3)$

18. $-3(2+b)+4b(b+2)$

19. $5(3x-2)+x(3x-2)$

13.4.6 Factoring x^2+bx+c

Factor each trinomial by guess and check.

20. $x^2+13x+30$

21. $x^2+11x+28$

22. $x^2+16x+48$

Factor each trinomial. Check your answer.

23. $x^2+12x+11$

24. $x^2+16x+28$

25. $x^2+15x+36$

26. x^2-6x+5

27. $x^2-9x+18$

28. $x^2-12x+32$

29. x^2+x-12

30. $x^2+4x-21$

31. $x^2+9x-36$

32. $x^2-12x-13$

33. $x^2-10x-24$

34. $x^2-2x-35$

35. Factor $n^2 - 12n - 45$. Show that the original polynomial and the factored form have the same value for $n = 0, 1, 2, 3,$ and 4.

13.4.7 Factoring $ax^2 + bx + c$

Factor each trinomial by guess and check.

36. $9x^2 + 9x + 2$

37. $2x^2 + 7x + 5$

38. $10d^2 + 17d + 7$

39. $6x^2 + 14x + 4$

40. $6x^2 - 13x + 6$

41. $5x^2 + 9x - 18$

Factor each trinomial. Check your answer.

42. $6x^2 + 23x + 7$

43. $10n^2 - 17n + 7$

44. $3x^2 + 11x + 6$

45. $7x^2 + 15x + 2$

46. $3n^2 + 4n + 1$

47. $3x^2 - 19x + 20$

48. $6x^2 + 11x + 4$

49. $4x^2 - 31x + 21$

50. $10x^2 + 31x + 15$

Appendix: Essential Algebra Review

Review 13.4 / Operations with Polynomials and Factoring, Part 2

51. $2n^2 - 7n - 4$

52. $3x^2 - 4x - 15$

53. $3n^2 - n - 4$

54. $-4x^2 - 4x + 15$

55. $-3x^2 + 16x - 16$

56. $-3x^2 - x + 2$

Radical Expressions

Example 1

Simplify each expression.

1. $\sqrt{81}$
2. $\sqrt{\frac{98}{2}}$
3. $\sqrt{(a+7)^2}$

Example 2

Simplify. All variables represent nonnegative numbers.

4. $\sqrt{180}$
5. $\sqrt{40}$
6. $\sqrt{648}$
7. $\sqrt{m^5n^3}$
8. $\sqrt{32x^4y^3}$
9. $\sqrt{200a^2b}$

Example 3

Simplify. All variables represent nonnegative numbers.

10. $\sqrt{\frac{17}{25}}$
11. $\sqrt{\frac{7}{16}}$
12. $\sqrt{\frac{6}{49}}$
13. $\sqrt{\frac{b}{c^2}}$
14. $\sqrt{\frac{4x^2}{36x}}$
15. $\sqrt{\frac{7a^4}{9a^3}}$

Radical Expressions

Example 4

Simplify. All variables represent nonnegative numbers.

16. $\sqrt{\frac{108}{49}}$

17. $\sqrt{\frac{204}{25}}$

18. $\sqrt{\frac{512}{81}}$

19. $\sqrt{\frac{1}{36x^2}}$

20. $\sqrt{\frac{50x^2}{169}}$

21. $\sqrt{\frac{72x^7}{4x^4}}$

Example 5

22. **Recreation** Your boat is traveling due north from a dock. Your friend's boat left at the same time from the same dock and is headed due east. After an hour, your friend calls and tells you that he has just stopped because of engine trouble. How far must you travel to meet your friend? Give your answer as a radical expression in simplest form. Then estimate the distance to the nearest mile.

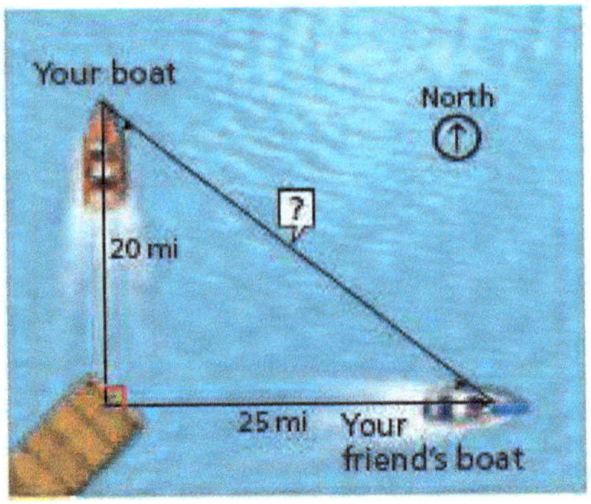

Solving Quadratic Equations by Factoring

Example 1

Use the Zero Product Property to solve each equation. Check your answer.

1. $(x+2)(x-8)=0$
2. $(x-6)(x-5)=0$
3. $(x+7)(x+9)=0$
4. $(x)(x-1)=0$
5. $(x)(x+11)=0$
6. $(3x+2)(4x-1)=0$

Example 2

Solve each quadratic equation by factoring. Check your answer.

7. $x^2+4x-12=0$
8. $x^2-8x-9=0$
9. $x^2-5x+6=0$
10. $x^2-3x=10$
11. $x^2+10x=-16$
12. $x^2+2x=15$
13. $x^2-8x+16=0$
14. $-3x^2=18x+27$
15. $x^2+36=12x$
16. $x^2+14x+49=0$
17. $x^2-16x+64=0$
18. $2x^2+6x=-18$

Example 3

19. **Games** A group of friends tries to keep a beanbag from touching the ground without using their hands. Once the beanbag has been kicked, its height can be modeled by $h = -16t^2 + 14t + 2$, where h is the height in feet above the ground and t is the time in seconds. Find the time it takes the beanbag to reach the ground.

Solving Quadratic Equations by Using Square Roots

Example 1

Solve using square roots. Check your answer.

1. $x^2 = 225$ **2.** $x^2 = 49$ **3.** $x^2 = -100$

4. $x^2 = 400$ **5.** $-25 = x^2$ **6.** $36 = x^2$

Example 2

Solve using square roots. Check your answer.

7. $3x^2 - 75 = 0$ **8.** $0 = 81x^2 - 25$ **9.** $49x^2 + 64 = 0$

10. $16x^2 + 10 = 131$ **11.** $0 = 4x^2 - 16$ **12.** $100x^2 + 26 = 10$

Example 3

Solve. Round to the nearest hundredth.

13. $3x^2 = 81$ **14.** $0 = x^2 - 60$ **15.** $100 - 5x^2 = 0$

Example 4

16. **Geometry** The length of a rectangle is 3 times its width. The area of the rectangle is 170 square meters. Find the width. Round to the nearest tenth of a meter. (*Hint:* Use $A = bh$.)

13.5.1 Radical Expressions

Simplify.

1. $\sqrt{100}$
2. $\sqrt{\frac{800}{2}}$
3. $\sqrt{3^2 + 4^2}$
4. $\sqrt{3 \cdot 27}$
5. $\sqrt{a^4}$
6. $\sqrt{(x+1)^2}$
7. $\sqrt{(5-x)^2}$
8. $\sqrt{(x-3)^2}$

Simplify. All variables represent nonnegative numbers.

9. $\sqrt{125}$
10. $\sqrt{4000}$
11. $\sqrt{216a^2b^2}$
12. $\sqrt{320r^2s^2}$
13. $\sqrt{\frac{15}{64}}$
14. $\sqrt{\frac{45}{4}}$
15. $\sqrt{\frac{64a^4}{4a^6}}$
16. $\sqrt{\frac{14z^3}{9z^3}}$
17. $\sqrt{\frac{128}{81}}$
18. $\sqrt{\frac{x^3}{y^6}}$
19. $\sqrt{\frac{150}{196x^2}}$
20. $\sqrt{\frac{192s^3}{49s}}$

21. **Amusement Parks** A thrill ride at an amusement park carries riders 160 feet straight up and then releases them for a free fall. The time t in seconds that it takes an object in free fall to reach the ground is $t = \sqrt{\frac{d}{16}}$, where d is the distance in feet that it falls. How long does it take the riders to reach the ground? Give your answer as a radical expression in simplest form. Then estimate the answer to the nearest tenth of a second.

13.5.2 Solving Quadratic Equations by Factoring

Use the Zero Product Property to solve each equation. Check your answer.

22. $(x-8)(x+6)=0$

23. $(x+4)(x+7)=0$

24. $(x-2)(x-5)=0$

25. $(x-9)(x)=0$

26. $(x)(x+25)=0$

27. $(2x+1)(3x-1)=0$

Solve each quadratic equation by factoring. Check your answer.

28. $x^2+8x+15=0$

29. $x^2-2x-8=0$

30. $x^2-4x+3=0$

31. $x^2+10x+25=0$

32. $x^2-x=12$

33. $-x^2=4x+4$

34. **Multi-Step** The height of a flare can be approximated by the function $h = -16t^2 + 95t + 6$, where h is the height in feet and t is the time in seconds. Find the time it takes the flare to hit the ground.

13.5.3 Solving Quadratic Equations by Using Square Roots

Solve using square roots. Check your answer.

35. $x^2 = 169$

36. $x^2 = 25$

37. $x^2 = -36$

38. $x^2 = 10{,}000$

39. $-121 = x^2$

40. $625 = x^2$

41. $4 - 81x^2 = 0$

42. $-4x^2 - 49 = 0$

43. $64x^2 - 5 = 20$

44. $9x^2 + 9 = 25$

45. $49x^2 + 1 = 170$

46. $81x^2 + 17 = 81$

Solve. Round to the nearest hundredth.

47. $4x^2 = 88$

48. $x^2 - 29 = 0$

49. $x^2 + 40 = 144$

50. $3x^2 - 84 = 0$

51. $50 - x^2 = 0$

52. $2x^2 - 10 = 64$

53. **Entertainment** For a scene in a movie, a sack of money is dropped from the roof of a 600 ft skyscraper. The height of the sack above the ground is given by $h = -16t^2 + 600$, where t is the time in seconds. How long will it take the sack to reach the ground? Round to the nearest tenth of a second.

Answer Keys

Understanding Points, Lines, and Planes

1. *A, B, C, D, E*

2. Possible answer: $\overleftrightarrow{AC}$, $\overleftrightarrow{BD}$

3. Possible answer: *EBC* and $\mathcal{N}$

4. Possible answer: *B*, *C*, or *D*

5. 

6. 

7. Possible answer: $\overleftrightarrow{AB}$

8. Possible answer: plane *ABD*

9. 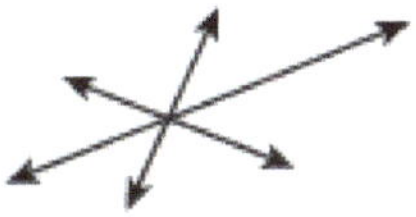

10. 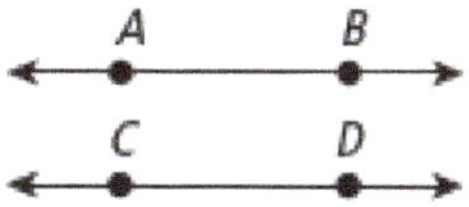

Fundamentals of Geometry

Points, Lines, Planes, and Angles

Measuring and Constructing Segments

1. $$\begin{aligned} AB &= |1-(-2.5)| \\ &= |3.5| \\ &= 3.5 \end{aligned}$$

2. $$\begin{aligned} BC &= |3.5-1| \\ &= |2.5| \\ &= 2.5 \end{aligned}$$

3. Answers vary.

4. $$\begin{aligned} AC &= AB + BC \\ 15.8 &= 9.9 + BC \\ -9.9 &\quad -9.9 \\ 5.9 &= BC \end{aligned}$$

5. $$\begin{aligned} MP &= MN + NP \\ 5y + 9 &= 17 + 3y \\ -3y &\quad\;\; -3y \\ 2y + 9 &= 17 \\ -9 &\quad -9 \\ \frac{2y}{2} &= \frac{8}{2} \\ y &= 4 \\ MP &= 5y + 9 \\ &= 5(4) + 9 = 29 \end{aligned}$$

6. 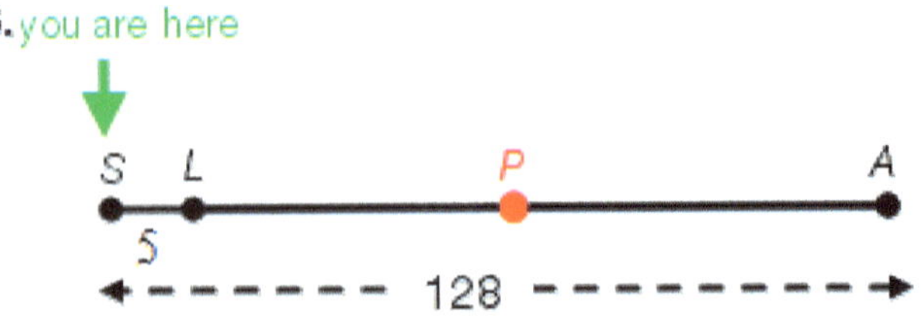

First find the distance from Lubbock to Amarillo.
The distance from the sign S to Lubbock L is 5 miles, so $SL = 5$.
The distance from the sign to Amarillo A is 128 miles, so $SA = 128$.
By the Segment Addition Postulate $SL + LA = SA$.
Use the equation to find the distance from Lubbock to Amarillo, LA.

$$\begin{aligned} SL + LA &= SA \\ 5 + LA &= 128 \\ LA &= 128 - 5 = 123 \end{aligned}$$

The picnic area P is the midpoint between Lubbock and Amarillo, so $LP = (1/2)123 = 61.5$.

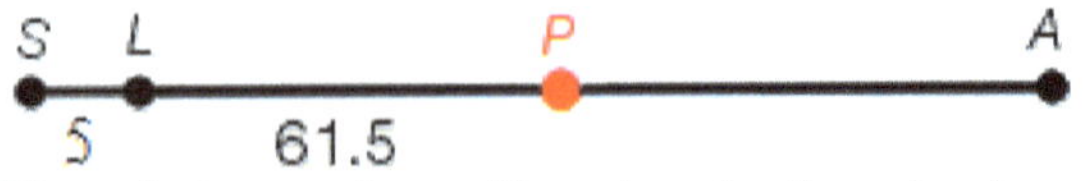

The distance from the sign to the picnic area is $SP = SL + LP = 5 + 61.5 = 66.5$ miles.

Measuring and Constructing Segments

7. **Step 1** Solve for x.

$$JL = 2JK$$
$$4x - 2 = 2(7)$$
$$4x - 2 = 14$$
$$\underline{\quad +2 \quad +2}$$
$$\frac{4x}{4} = \frac{16}{4}$$
$$x = 4$$

Step 2 Find KL and JL.

$KL = JK = 7$

$JL = 4x - 2$

$= 4(4) - 2 = 14$

8. **Step 1** Solve for y.

$$DE = EF$$
$$2y = 8y - 3$$
$$\underline{-8y \quad -8y}$$
$$-6y = -3$$
$$\frac{-6y}{-6} = \frac{-3}{-6}$$
$$y = 0.5$$

Step 2 Find DE, EF, and DF.

$DE = 2y$

$= 2(0.5) = 1$

$EF = 8y - 3$

$= 8(0.5) - 3 = 1$

$DF = DE + EF$

$= 1 + 1 = 2$

Fundamentals of Geometry

Points, Lines, Planes, and Angles

Measuring and Constructing Angles

1. $\angle AOB$, $\angle BOA$, or $\angle 1$; $\angle BOC$, $\angle COB$, or $\angle 2$; $\angle AOC$ or $\angle COA$

2. $m\angle VXW = 15°$
$\angle VXW$ is acute.

3. $m\angle TXW = 105°$
$\angle TXW$ is obtuse.

4. $m\angle RXU = 110°$
$\angle RXU$ is obtuse.

5. $m\angle JKM = m\angle JKL + m\angle LKM$
$= 42° + 28° = 70°$

6. $m\angle JKM = m\angle JKL + m\angle LKM$
$82.5° = 56.4° + m\angle LKM$
$m\angle LKM = 26.1°$

7. Step 1 Find x.
$m\angle ABD = m\angle DBC$
$(6x + 4)° = (8x - 4)°$
$4 = 2x - 4$
$8 = 2x$
$x = 4$

Step 2 Find $m\angle ABD$.
$m\angle ABD = 6x + 4$
$= 6(4) + 4$
$= 28°$

8. Step 1 Find y.
$m\angle ABD = m\angle DBC$
$(5y - 3)° = (3y + 15)°$
$2y - 3 = 15$
$2y = 18$
$y = 9$

Step 2 Find $m\angle ABC$.
$m\angle ABC = 2m\angle ABD$
$= 2(5y - 3)$
$= 2(5(9) - 3)$
$= 84°$

Fundamentals of Geometry

Points, Lines, Planes, and Angles

Pairs of Angles

1. $\angle 1$ and $\angle 2$ are adj. angles. Their noncommon sides, $\overrightarrow{EG}$ and $\overrightarrow{EJ}$, are opposite rays, so $\angle 1$ and $\angle 2$ also form a lin. pair.

2. $\angle 1$ and $\angle 3$ share vertex E but do not have a common side, so $\angle 1$ and $\angle 3$ are not adj. angles.

3. $\angle 2$ and $\angle 4$ share vertex E but do not have a common side, so $\angle 2$ and $\angle 4$ are not adj. angles.

4. $\angle 2$ and $\angle 3$ are adj. angles. Their noncommon sides, $\overrightarrow{EF}$ and $\overrightarrow{EH}$, are not opposite rays, so $\angle 1$ and $\angle 2$ are only adj. angles.

5. $(180 - x)°$
   $180° - 81.2° = 98.8°$

6. $(90 - x)°$
   $90° - 81.2° = 8.8°$

7. $(180 - y)°$
   $180° - (6x - 5)° = 180° - 6x + 5$
   $= (185 - 6x)°$

8. $(90 - y)°$
   $90° - (6x - 5)° = 90° - 6x + 5$
   $= (95 - 6x)°$

9. **Step 1** Let $m\angle A = x°$. Then $\angle B$, its comp., measures $(90 - x)°$.
   Step 2 Write and solve an equation.
   $x = 3(90 - x) + 6$
   $x = 270 - 3x + 6$
   $x = 276 - 3x$
   $4x = 276$
   $x = 69$
   $m\angle A = x = 69°$

10. **1 Understand the Problem**
    Answers are measures of $\angle 2$, $\angle 3$, and $\angle 4$.
    List **Important Information:**
    - $\angle 1 \cong \angle 2$
    - $\angle 1$ and $\angle 3$ are comp., and $\angle 2$ and $\angle 4$ are comp.
    - $m\angle 1 = 18.5°$

    2 Make a Plan
    If $\angle 1 \cong \angle 2$, then $m\angle 1 = m\angle 2$.
    If $\angle 1$ and $\angle 3$ are comp., then $m\angle 3 = (90 - 18.5)°$.
    If $\angle 2$ and $\angle 4$ are comp., then $m\angle 4 = (90 - 18.5)°$.

3 Solve
$m\angle 1 = m\angle 2 = 18.5°$
$m\angle 3 = m\angle 4 = (90 - 18.5)° = 71.5°$

4 Look Back
Answer makes sense because
$18.5° + 71.5° = 90°$, so $\angle 1$ and $\angle 3$ are comp., and
$\angle 2$ and $\angle 4$ are comp. Thus $m\angle 2 = 18.5°$,
$m\angle 3 = 71.5°$, and $m\angle 4 = 71.5°$.

11. $\angle ABE$ and $\angle CBD$ are vert. ∠s; $\angle ABC$ and $\angle EBD$ are vert. ∠s.

Fundamentals of Geometry

Review 1.1 / Points, Lines, Planes, and Angles

1. *B, E, A*

2. Possible answer: *B, C, D, E*

3. Possible answer: plane *ABC*

4. 

5. 

6. Possible answer: *G, J*, and ℓ

7. Possible answer: planes $\mathcal{T}$ and $\mathcal{S}$

8. 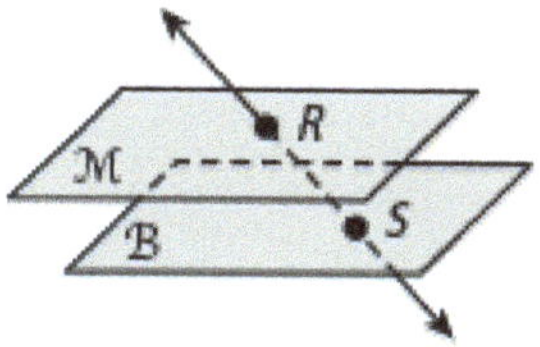

9. 

10. $$DB = \left|\frac{2}{3} - \left(-5\frac{1}{4}\right)\right| = \left|\frac{8}{12} + 5\frac{3}{12}\right| = \left|5\frac{11}{12}\right| = 5\frac{11}{12}$$

11. $$CD = \left|-5\frac{1}{4} - (-2)\right| = \left|-3\frac{1}{4}\right| = 3\frac{1}{4}$$

12. 

13. $$\begin{aligned} CE &= CD + DE \\ 17.1 &= CD + 8 \\ -8 & \quad\; -8 \\ 9.1 &= CD \end{aligned}$$

14. $$\begin{aligned} MR &= MN + NR \\ 5x - 3 &= 2.5x + x \\ 5x - 3 &= 3.5x \\ -5x \quad & \quad -5x \\ -3 &= -1.5x \\ \frac{-3}{-1.5} &= \frac{-1.5x}{-1.5} \\ 2 &= x \end{aligned}$$

$MN = 2.5x$
$= 2.5(2) = 5$

15. Total yards = pass + run
$= (24 - 9) + \frac{1}{2}(50 - 24) = 28$ yd

16. **Step 1** Solve for *x*.

$$\begin{aligned} DE &= EF \\ 2x + 4 &= 3x - 1 \\ -2x \quad & \quad -2x \\ 4 &= x - 1 \\ +1 \quad & \quad +1 \\ 5 &= x \end{aligned}$$

Step 2 Find *DE, EF*, and *DF*.

$DE = 2x + 4$
$= 2(5) + 4 = 14$
$EF = 3x - 1$
$= 3(5) - 1 = 14$
$DF = DE + EF$
$= 14 + 14 = 28$

17. $PQ = \frac{1}{2}PR$

$3y = \frac{1}{2}(42)$

$3y = 21$

$\frac{3y}{3} = \frac{21}{3}$

$y = 7$

$QR = \frac{1}{2}PR$

$= \frac{1}{2}(42) = 21$

18. $\angle 1$ or $\angle JMK$; $\angle 2$ or $\angle LMK$; $\angle M$ or $\angle JML$

19. $m\angle CGE = |110 - 20| = 90°$ $\angle CGE$ is a rt. $\angle$.

20. $m\angle BGD = |113 - 20| = 93°$ $\angle BGD$ is obtuse.

21. $m\angle AGB = 20°$

$\angle AGB$ is acute.

22. $m\angle RSU = m\angle RST + m\angle TSU$

$= 38° + 28.6° = 66.6°$

23. $m\angle RSU = m\angle RST + m\angle TSU$

$83.5° = m\angle RST + 46.7°$

$m\angle RST = 36.8°$

24. **Step 1** Find x.

$m\angle RSP = m\angle PST$

$(3x - 2)° = (9x - 26)°$

$-2 = 6x - 26$

$24 = 6x$

$x = 4$

Step 2 Find $m\angle RST$.

$m\angle RST = 2m\angle RSP$

$= 2(3x - 2)$

$= 2(3(4) - 2)$

$= 20°$

25. **Step 1** Find y.

$m\angle PST = \frac{1}{2}m\angle RST$

$(y + 5)° = \frac{1}{2}\left(\frac{5}{2}y\right)°$

$y + 5 = \frac{5}{4}y$

$5 = \frac{1}{4}y$

$y = 20$

Step 2 Find $m\angle RSP$.

$m\angle RSP = m\angle PST$

$= y + 5$

$= 20 + 5$

$= 25°$

26. $\angle 1$ and $\angle 4$ are adj. angles. Their noncommon sides are opposite rays, so $\angle 1$ and $\angle 4$ also form a lin. pair.

27. $\angle 2$ and $\angle 3$ are adj. angles. Their noncommon sides are opposite rays, so $\angle 2$ and $\angle 3$ also form a lin. pair.

28. $\angle 3$ and $\angle 4$ are adj. angles. Their noncommon sides are not opposite rays, so $\angle 3$ and $\angle 4$ are only adj. angles.

29. $\angle 3$ and $\angle 1$ share a vertex but do not have a common side, so $\angle 3$ and $\angle 1$ are not adj. angles.

30. $(180 - x)°$

$180° - 56.4° = 123.6°$

31. $(90 - x)°$

$90° - 56.4° = 33.6°$

32. $(180 - y)°$

$180° - (2x - 4)° = 180° - 2x + 4$

$= (184 - 2x)°$

33. $(90 - y)°$

$90° - (2x - 4)° = 90° - 2x + 4$

$= (94 - 2x)°$

34. **Step 1** Let $m\angle A = x°$, so $m\angle B = (90 - x)°$.

Step 2 Write and solve an equation.

$x = 3(90 - x)$

$x = 270 - 3x$

$4x = 270$

$x = 67.5$

$m\angle A = x = 67.5°$

$m\angle B = 90 - 67.5° = 22.5°$

35. $\angle 1 \cong \angle 2$

$m\angle 1 = m\angle 2 = 22.3°$

$m\angle 4 = m\angle 3 = 90° - 22.3° = 67.7°$

36. $\angle PTU, \angle VTR$; $\angle UTQ, \angle STV$; $\angle QTR, \angle PTS$; $\angle PTQ, \angle STR$; $\angle UTR, \angle PTV$; $\angle QTV, \angle UTS$

Using Formulas in Geometry

1. $P = 2\ell + 2w = 2(11) + 2(4) = 22 + 8 = 30$ mm
   $A = \ell w = (11)(4) = 44$ mm^2

2. $P = 4s = 4(y - 3) = 4y - 12$
   $A = s^2 = (y - 3)^2 = y^2 - 6y + 9$

3. $P = a + b + c = 5 + (x + 3) + 13 = (x + 21)$ m
   $A = \frac{1}{2}bh = \frac{1}{2}(x + 3)(4) = (2x + 6)$ m^2

4. Area of one △ is $A = \frac{1}{2}bh = \frac{1}{2}(3)(4) = 6$ in^2.
   There are $80(20) = 1600$ ▲ in total. The total area of the 1600 ▲ is $1600(6) = 9600$ in^2.

5. $C = 2\pi r = 2\pi(2.1) = 4.2\pi \approx 13.2$ m
   $A = \pi r^2 = \pi(2.1)^2 = 4.41\pi \approx 13.9$ m^2

6. $C = 2\pi r = 2\pi(7) = 14\pi \approx 44.0$ in.
   $A = \pi r^2 = \pi(7)^2 = 49\pi \approx 153.9$ in^2

7. $r = \frac{d}{2} = \frac{16}{2} = 8$ cm
   $C = 2\pi r = 2\pi(8) = 16\pi \approx 50.3$ cm
   $A = \pi r^2 = \pi(8)^2 = 64\pi \approx 201.1$ cm^2

Midpoint and Distance in the Coordinate Plane

1. $M\left(\frac{x_1 + x_2}{2}, \frac{y_1 + y_2}{2}\right)$

$\left(\frac{4 + (-4)}{2}, \frac{-6 + 2}{2}\right) = \left(\frac{0}{2}, \frac{-4}{2}\right)$

$= (0, -2)$

2. $M\left(\frac{x_1 + x_2}{2}, \frac{y_1 + y_2}{2}\right)$

$\left(\frac{0 + 3}{2}, \frac{-8 + 0}{2}\right) = \left(\frac{3}{2}, \frac{-8}{2}\right)$

$= \left(1\frac{1}{2}, -4\right)$

3. **Step 1** Let coords. of N equal (x, y).

Step 2 Use Mdpt. Formula:

$(0, 1) = \left(\frac{-3 + x}{2}, \frac{-1 + y}{2}\right)$

Step 3 Find x-coord.

$0 = \frac{-3 + x}{2}$

$2(0) = 2\left(\frac{-3 + x}{2}\right)$

$0 = -3 + x$

$x = 3$

Find y-coord.

$1 = \frac{-1 + y}{2}$

$2(1) = 2\left(\frac{-1 + y}{2}\right)$

$2 = -1 + y$

$y = 3$

The coordinates of N are (3, 3).

4. **Step 1** Let coords. of C equal (x, y).

Step 2 Use Mdpt. Formula.

$\left(-1\frac{1}{2}, 1\right) = \left(\frac{-3 + x}{2}, \frac{4 + y}{2}\right)$

Step 3 Find x-coord.

$-1\frac{1}{2} = \frac{-3 + x}{2}$

$2\left(\frac{3}{2}\right) = 2\left(\frac{-3 + x}{2}\right)$

$-3 = -3 + x$

$x = 0$

Find y-coord.

$1 = \frac{4 + y}{2}$

$2(1) = 2\left(\frac{4 + y}{2}\right)$

$2 = 4 + y$

$y = -2$

The coordinates of C are (0, −2).

5. **Step 1** Find coords. of each point.

$F(5, 4)$, $G(3, -1)$, $J(-4, 0)$, and $K(1, -2)$

Step 2 Use Dist. Formula.

$d = \sqrt{(x_2 - x_1)^2 + (y_2 - y_1)^2}$

$JK = \sqrt{(1 - (-4))^2 + (-2 - 0)^2}$

$= \sqrt{5^2 + (-2)^2}$

$= \sqrt{25 + 4} = \sqrt{29}$

$FG = \sqrt{(3 - 5)^2 + (-1 - 4)^2}$

$= \sqrt{(-2)^2 + (-5)^2}$

$= \sqrt{4 + 25} = \sqrt{29}$

Since $JK = FG$, $\overline{JK} \cong \overline{FG}$.

6. **Step 1** Find coords. of each point.

$J(-4, 0)$, $K(1, -2)$, $R(-3, -2)$, and $S(3, -5)$

Step 2 Use Dist. Formula.

$d = \sqrt{(x_2 - x_1)^2 + (y_2 - y_1)^2}$

$JK = \sqrt{(1 - (-4))^2 + (-2 - 0)^2}$

$= \sqrt{5^2 + (-2)^2}$

$= \sqrt{25 + 4} = \sqrt{29}$

$RS = \sqrt{(3 - (-3))^2 + (-5 - (-2))^2}$

$= \sqrt{6^2 + (-3)^2}$

$= \sqrt{36 + 9}$

$= \sqrt{45} = 3\sqrt{5}$

Since $JK \neq RS$, $\overline{JK} \not\cong \overline{RS}$.

7. **Method 1** Use Dist. Formula. Subst. values for coords. of A and B into Dist. Formula.

$AB = \sqrt{(x_2 - x_1)^2 + (y_2 - y_1)^2}$

$= \sqrt{(-4 - 1)^2 + (-4 - (-2))^2}$

$= \sqrt{(-5)^2 + (-2)^2}$

$= \sqrt{25 + 4} = \sqrt{29} \approx 5.4$

Method 2 Use Pyth. Thm. Count the units for the legs of the rt. △ formed by A and B.

$a = 5$ and $b = 2$

$c^2 = a^2 + b^2$

$= 5^2 + 2^2$

$= 25 + 4$

$= 29$

$c = \sqrt{29} \approx 5.4$

Midpoint and Distance in the Coordinate Plane

8. **Method 1** Use Dist. Formula. Subst. values for coords. of X and Y into Dist. Formula.

$$XY = \sqrt{(x_2 - x_1)^2 + (y_2 - y_1)^2}$$
$$= \sqrt{(-2 - (-2))^2 + (-8 - 7)^2}$$
$$= \sqrt{(0)^2 + (-15)^2}$$
$$= \sqrt{0 + 225} = 15.0$$

Method 2 Use Pyth. Thm. Count the units for the legs of the rt. △ formed by X and Y.

$a = 0$ and $b = 15$

$$c^2 = a^2 + b^2$$
$$= 0^2 + 15^2$$
$$= 225$$
$$c = \sqrt{225} = 15.0$$

9. **Method 1** Use Dist. Formula. Subst. values for coords. of V and W into Dist. Formula.

$$VW = \sqrt{(x_2 - x_1)^2 + (y_2 - y_1)^2}$$
$$= \sqrt{(-4 - 2)^2 + (8 - (-1))^2}$$
$$= \sqrt{(-6)^2 + 9^2}$$
$$= \sqrt{36 + 81} = \sqrt{117} \approx 10.8$$

Method 2 Use Pyth. Thm. Count the units for the legs of the rt. △ formed by V and W.

$a = 6$ and $b = 9$

$$c^2 = a^2 + b^2$$
$$= 6^2 + 9^2$$
$$= 36 + 81$$
$$= 117$$
$$c = \sqrt{117} \approx 10.8$$

10. Use Pyth. Thm. From the plan, $a = 22$ and $b = 16$.

$$c^2 = a^2 + b^2$$
$$= 22^2 + 16^2$$
$$= 484 + 256 = 740$$
$$c = \sqrt{740} \approx 27.2 \text{ ft}$$

1. transformation is a reflection; $\triangle ABC \rightarrow \triangle A'B'C'$
2. transformation is a translation; $PQRS \rightarrow P'Q'R'S'$
3. 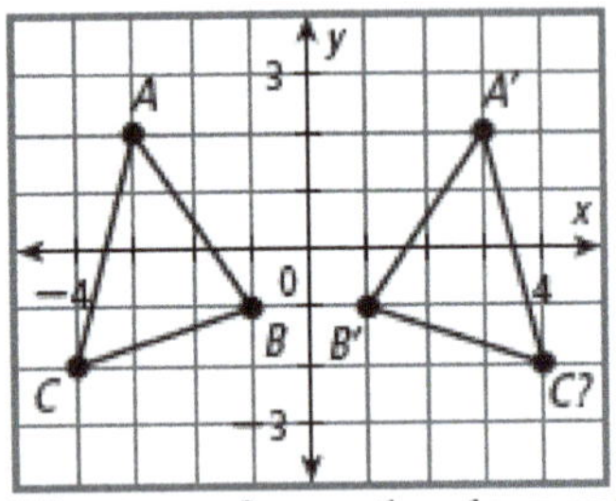

The transformation is a reflection across the *y*-axis because each pt. and its image are the same dist. from the *y*-axis.
4. **Step 1** State the coordinates of $\triangle DEF$.
The vertices of $\triangle DEF$ are $D(2, 3)$, $E(1, 1)$, and $F(4, 0)$.
Step 2 Apply the rule to find the vertices of the image.
$D'(2 - 3, 3 - 2) = D'(-1, 1)$
$E'(1 - 3, 1 - 2) = E'(-2, -1)$
$F'(4 - 3, 0 - 2) = F'(1, -2)$
Step 3 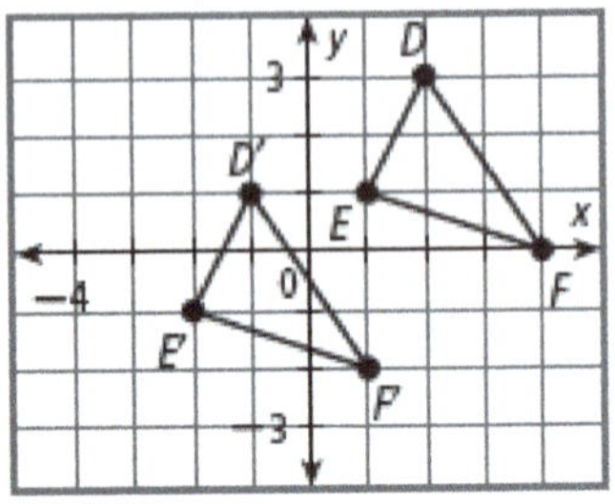

5. **Step 1** Choose 2 pts.
Choose a pt. A on the preimage and a corr. pt. A' on the image. A has coords. (0, 0) and A' has coords. (4, 4).
Step 2 Translate.
To translate A to A', 4 units are added to both the *x*-coord. and the *y*-coord. Therefore, the translation rule is $(x, y) \rightarrow (x + 4, y + 4)$.

Fundamentals of Geometry

Review 1.2 / Coordinate and Transformation Tools

1. $P = 4s$
$= 4(7.4) = 29.6$ m

$A = s^2$
$= (7.4)^2 = 54.76$ m^2

2. $P = 2\ell + 2w$
$= 2(x + 6) + 2x$
$= 4x + 12$

$A = \ell w$
$= (x + 6)x$
$= x^2 + 6x$

3. $P = a + b + c$
$= 5x + 8 + 4x$
$= 9x + 8$

$A = \frac{1}{2}bh$
$= \frac{1}{2}(8)(3x)$
$= 12x$

4. Area of one △ is $\frac{1}{2}bh = \frac{1}{2}(3)(1.5) = 2.25$ in^2.
Area of 32 ⚠ is $32(2.25) = 72$ in^2.

5. $C = 2\pi r$
$= 2\pi(12) = 24\pi$
≈ 75.4 m

$A = \pi r^2$
$= \pi(12)^2 = 144\pi$
≈ 452.4 m^2

6. $r = \frac{d}{2} = 6.25$ ft
$C = 2\pi r$
$= 2\pi(6.25) = 12.5\pi$
≈ 39.3 ft
$A = \pi r^2$
$= \pi(6.25)^2 = 39.0625\pi$
≈ 122.7 ft^2

7. $r = \frac{d}{2} = \frac{1}{4}$ mi
$C = 2\pi r$
$= 2\pi\left(\frac{1}{4}\right) = \frac{1}{4}\pi$
≈ 1.6 mi

$A = \pi r^2$
$= \pi\left(\frac{1}{4}\right)^2 = \frac{1}{16}\pi$
≈ 0.2 mi^2

8. $M\left(\frac{x_1 + x_2}{2}, \frac{y_1 + y_2}{2}\right)$
$\left(\frac{-3 + (-1)}{2}, \frac{-7 + 1}{2}\right) = \left(\frac{-4}{2}, \frac{-6}{2}\right)$
$= (-2, -3)$

9. $M\left(\frac{x_1 + x_2}{2}, \frac{y_1 + y_2}{2}\right)$
$\left(\frac{12 + (-5)}{2}, \frac{-7 + (-2)}{2}\right) = \left(\frac{7}{2}, \frac{-9}{2}\right)$
$= \left(3\frac{1}{2}, -4\frac{1}{2}\right)$

10. **Step 1** Let coords. of R equal (x, y).
Step 2 Use Mdpt. Formula.
$(7, -9) = \left(\frac{-3 + x}{2}, \frac{5 + y}{2}\right)$
Step 3 Find x-coord.
$7 = \frac{-3 + x}{2}$
$14 = -3 + x$
$x = 17$

Find y-coord.
$-9 = \frac{5 + y}{2}$
$-18 = 5 + y$
$y = -23$

The coordinates of R are $(17, -23)$.

11. **Step 1** Let coords. of C equal (x, y).
Step 2 Use Mdpt. Formula.
$\left(2\frac{1}{2}, 1\right) = \left(\frac{x + (-3)}{2}, \frac{y + (-2)}{2}\right)$
Step 3 Find x-coord.
$2\frac{1}{2} = \frac{x + (-3)}{2}$
$5 = x - 3$
$x = 8$

Find y-coord.
$1 = \frac{y + (-2)}{2}$
$2 = y - 2$
$y = 4$

The coordinates of C are $(8, 4)$.

12. **Step 1** Find coords. of each point.
$D(-4, 0)$, $E(0, -2)$, $F(2, 3)$, and $G(4, -1)$.
Step 2 Use Dist. Formula.
$DE = \sqrt{(0 - (-4))^2 + (-2 - 0)^2}$
$= \sqrt{4^2 + (-2)^2}$
$= \sqrt{16 + 4} = \sqrt{20} = 2\sqrt{5}$
$FG = \sqrt{(4 - 2)^2 + (-1 - 3)^2}$
$= \sqrt{2^2 + (-4)^2}$
$= \sqrt{4 + 16} = \sqrt{20} = 2\sqrt{5}$
Since $DE = FG$, $\overline{DE} \cong \overline{FG}$.

Fundamentals of Geometry

Review 1.2 / Coordinate and Transformation Tools

13. Step 1 Find coords. of each point.
$D(-4, 0)$, $E(0, -2)$, $R(-3, -4)$, and $S(2, -2)$.
Step 2 Use Dist. Formula.

$$DE = \sqrt{(0-(-4))^2 + (-2-0)^2}$$
$$= \sqrt{4^2 + (-2)^2}$$
$$= \sqrt{16+4} = \sqrt{20} = 2\sqrt{5}$$
$$RS = \sqrt{(2-(-3))^2 + (-2-(-4))^2}$$
$$= \sqrt{5^2 + 2^2}$$
$$= \sqrt{25+4} = \sqrt{29}$$

Since $DE \neq RS$, $\overline{DE} \not\cong \overline{RS}$.

14. Method 1 Dist. Formula.

$$UV = \sqrt{(-3-0)^2 + (-9-1)^2}$$
$$= \sqrt{(-3)^2 + (-10)^2}$$
$$= \sqrt{9+100} = \sqrt{109} \approx 10.4$$

th. Thm. Count the units for the legs of
ed by U and V.
: 10

$= 100 + 9 = 109$
$c = \sqrt{109} \approx 10.4$

15. Method 1 Dist. Formula.

$$MN = \sqrt{(2-10)^2 + (-5-(-1))^2}$$
$$= \sqrt{(-8)^2 + (-4)^2}$$
$$= \sqrt{64+16} = \sqrt{80} \approx 8.9$$

Method 2 Pyth. Thm. Count the units for the legs of the rt. △ formed by M and N.
$a = 8$ and $b = 4$
$c^2 = a^2 + b^2$
$= 8^2 + 4^2$
$= 64 + 16 = 80$
$c = \sqrt{80} \approx 8.9$

16. Method 1 Dist. Formula.

$$PQ = \sqrt{(5-(-10))^2 + (5-1)^2}$$
$$= \sqrt{15^2 + 4^2}$$
$$= \sqrt{225+16} = \sqrt{241} \approx 15.5$$

Method 2 Pyth. Thm. Count the units for the legs of the rt. △ formed by P and Q.
$a = 15$ and $b = 4$
$c^2 = a^2 + b^2$
$= 15^2 + 4^2$
$= 225 + 16 = 241$
$c = \sqrt{241} \approx 15.5$

17. Use Pyth. Thm.
$a = 11$ and $b = 14$
$c^2 = a^2 + b^2$
$= 11^2 + 14^2$
$= 121 + 196 = 317$
$c = \sqrt{317} \approx 18$ in.

18. rotation: $DEFG \rightarrow D'E'F'G'$

19. reflection: $WXYZ \rightarrow W'X'Y'Z'$

20.

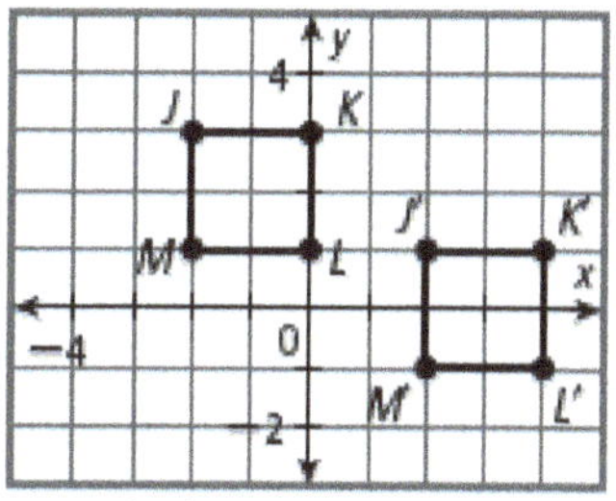

translation; each pt. moves the same dist. right and the same dist. down.

21. **Step 1** Apply the rule to find the vertices of the image.

$A'(-4 + 3, 1 - 2) = A'(-1, -1)$
$B'(1 + 3, 1 - 2) = B'(4, -1)$
$C'(1 + 3, -2 - 2) = C'(4, -4)$
$D'(-4 + 3, -2 - 2) = D'(-1, -4)$

Step 2

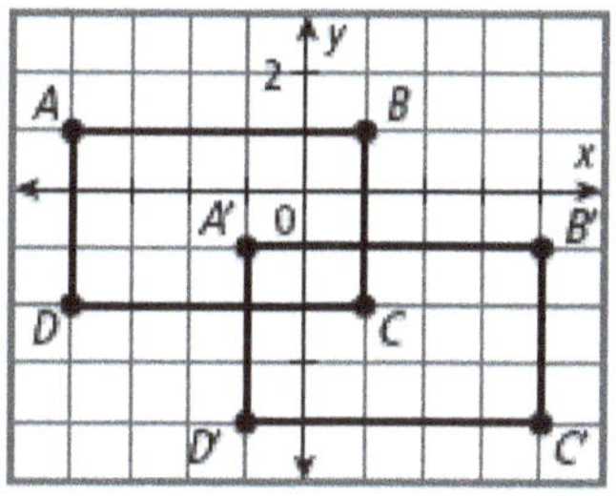

22. **Step 1** Choose 2 pts.
Choose a pt. A on the preimage and a corr. pt. A' on the image. A has coords. $(-5, 1)$ and A' has coords. $(6, -3)$.
Step 2 Translate.
To translate A to A', 11 units are added to the x-coord. and 4 units are subtracted from the y-coord. Therefore, the translation rule is $(x, y) \rightarrow (x + 11, y - 4)$.

Reasoning and Writing Geometric Proofs

Inductive and Deductive Reasoning

Using Inductive Reasoning to Make Conjectures

1. September
2. $\frac{4}{6}$
3. 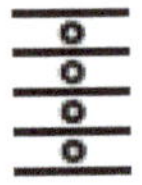
4. even
5. $1 = 1$
   $1 + 3 = 4$
   $1 + 3 + 5 = 9$
   $1 + 3 + 5 + 7 = 16$
   Rule is n^2.
6. The number of bacteria doubles every 20 min.
7. Roosevelt was inaugurated at age 42.
8. The 3 pts. are collinear.
9. Possible answer: $x = -3$

Reasoning and Writing Geometric Proofs

Inductive and Deductive Reasoning

Conditional Statements

1. Hypothesis: A person is at least 16 years old.
   Conclusion: The person can drive a car.
2. Hypothesis: A figure is a rectangle.
   Conclusion: A figure is a parallelogram.
3. Hypothesis: $a - b < a$.
   Conclusion: b is a positive number.
4. If a person is 18 years old, then that person is eligible to vote.
5. If $0 < a < b$, then $\left(\frac{a}{b}\right)^2 < \frac{a}{b}$.
6. T
7. F; possible answer: $x = 2$ and $y = -4$
8. F; possible answer: April
9. Converse: If Brielle travels 10 mi in 20 min, then she drives at exactly 30 mi/h; F.
   Inverse: If Brielle does not drive at exactly 30 mi/h, then she does not travel 10 mi in 20 min; F.
   Contrapositive: If Brielle does not travel 10 mi in 20 min, then she does not drive at exactly 30 mi/h; T.

Using Deductive Reasoning to Verify Conjectures

1. The conclusion is based on logical reasoning. It is a result of deductive reasoning.

2. The conclusion is based on logical reasoning. It is a result of deductive reasoning.

3. Hypothesis: You want to go on a field trip.
   Conclusion: You must have a signed permission slip.
   The given statement "Zola has a signed permission slip" matches the conclusion of a true conditional. But this does not mean the hypothesis is true. Zola could have a permission slip for another reason. The conclusion is not valid.

4. Hypothesis: The side lengths of a rect. are 3 ft and 4 ft.
   Conclusion: The rect.'s area is 12 ft^2.
   The given statement "A rect. has side lengths 3 ft and 4 ft" matches the hypothesis of the given conditional. By the Law of Detachment, the rect. has area 12 ft^2. The conjecture is valid.

5. Let p, q, and r represent the following:
   p: You fly from Texas to California.
   q: You travel from the central to the Pacific time zone.
   r: You gain two hours.
   You are given that $p \rightarrow q$ and $q \rightarrow r$. Since q is the conclusion of the 1st conditional and the hypothesis of the 2nd conditional, you can conclude that $p \rightarrow r$. The conjecture is valid by the Law of Syllogism.

6. Let p, q, and r represent the following:
   p: A figure is a square.
   q: A figure is a rectangle.
   r: A figure is a parallelogram.
   You are given that $p \rightarrow q$ and $p \rightarrow r$. The Law of Syllogism cannot be used to draw a conclusion since p is the hypothesis of both conditionals. Even though the conjecture $r \rightarrow q$ is true, the logic used to draw the conclusion is not valid.

7. Conclusion: Alex's car might not start.

Reasoning and Writing Geometric Proofs

Inductive and Deductive Reasoning

Biconditional Statements and Definitions

1. Let p and q represent the following:
   p: Perry can paint the entire living room.
   q: Perry has enough paint.
   2 parts of biconditional $p \leftrightarrow q$ are $p \rightarrow q$ and $q \rightarrow p$.
   Conditional: If Perry can paint the entire living room, then he has enough paint.
   Converse: If Perry has enough paint, then he can paint the entire living room.

2. Let p and q represent the following:
   p: Your medicine will be ready by 5 P.M.
   q: You drop your prescription off by 8 A.M.
   2 parts of biconditional $p \leftrightarrow q$ are $p \rightarrow q$ and $q \rightarrow p$.
   Conditional: If your medicine is ready by 5 P.M., then you dropped your prescription off by 8 A.M.
   Converse: If you drop your prescription off by 8 A.M., then your medicine will be ready by 5 P.M.

3. Converse: If a student is in the tenth grade, then the student is a sophomore.
   Biconditional: A student is a sophomore if and only if the student is in the tenth grade.

4. Converse: If 2 segs. are $\cong$, then they have the same length.
   Biconditional: 2 segs. have the same length if and only if they are $\cong$.

5. Conditional: $xy = 0 \rightarrow x = 0$ or $y = 0$; T
   Converse: $x = 0$ or $y = 0 \rightarrow xy = 0$; T
   Since conditional and converse are true, biconditional is true.

6. Conditional: If a figure is a quad., then it is a polygon; T
   Converse: If a figure is a polygon, then it is a quad.; F
   A $\triangle$ is a polygon but not a quad., so converse is false.
   Therefore, biconditional is false.

7. 2 lines are $\parallel$ if and only if they are coplanar and never intersect.

8. An animal is a hummingbird if and only if it is a tiny, brightly colored bird with narrow wings, a slender bill, and a long tongue.

Reasoning and Writing Geometric Proofs

Review 2.1 / Inductive and Deductive Reasoning

1. 5 P.M.

2. 42

3. 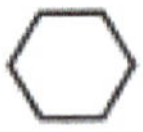

4. $2 = 1(2)$
$2 + 4 = 6 = 2(3)$
$2 + 4 + 6 = 12 = 3(4)$
Rule is $n(n + 1)$.

5. $n - 1$

6. About 5%(526) $\approx$ 26 students will participate.

7. Possible answer: $y = -1$

8. Possible Answer: $x = -1$

9. $m\angle 1 = m\angle 2 = 90°$

10. Hypothesis: An animal is a tabby.
Conclusion: The animal is a cat.

11. Hypothesis: Two lines intersect.
Conclusion: Four angles are formed.

12. Hypothesis: 8 oz of cereal cost $2.99.
Conclusion: 16 oz of cereal cost $5.98.

13. If a patient is ill, then you should monitor the patient's heart rate.

14. If the batter makes 3 strikes, then the batter is out.

15. If segs. are $\cong$, then they have equal measures.

16. T

17. F; there is a postulate that states that if 2 planes intersect, they do so in exactly 1 line.

18. T

19. Converse: If an event is unlikely to occur, then the probability of the event is 0.1; F.
Inverse: If the probability of an event is not 0.1, then the event is likely to occur; F.
Contrapositive: If an event is likely to occur, then the probability of the event is not 0.1; T.

20. Converse: If the air temperature is 32°F or less, then freezing rain is falling; F.
Inverse: If freezing rain is not falling, then the air temperature is greater than 32°F; F.
Contrapositive: If the air temperature is greater than 32°F, then freezing rain is not falling; T

21. The conclusion is based on mathematical calculation. So it is the result of deductive reasoning.

22. Since the conclusion is based on a pattern of observation, it is a result of inductive reasoning.

23. Hypothesis: One integer is odd and another integer is even.
Conclusion: The product of 2 integers is even.
Statement "The product of 2 integers is 24" matches the conclusion of a true conditional. However, hypothesis is not necessarily true. For example, $6(4) = 24$. Conclusion is not valid.

Reasoning and Writing Geometric Proofs

Review 2.1 / Inductive and Deductive Reasoning

24. Let p, q, and r represent the following:
p: An element is an alkali metal.
q: An element reacts with water.
r: An element is in the 1st column of the periodic table.
You are given that $p \rightarrow q$ and $r \rightarrow p$. Since p is the conclusion of the 2nd conditional and the hypothesis of the 1st conditional, you can conclude that $r \rightarrow q$. The conjecture is valid by the Law of Syllogism.

25. Conclusion: Dakota gets better grades in Social Studies.

26. Conditional: If 3 pts. are coplanar, then they lie in the same plane.
Converse: If 3 pts. lie in the same plane, then they are coplanar.

27. Conditional: If a ▱ is a rect., then it has 4 rt. ∠s.
Converse: If a ▱ has 4 rt. ∠s, then it is a rect.

28. Conditional: If a lunar eclipse occurs, then Earth is between the Sun and the Moon.
Converse: If Earth is between the Sun and the Moon, then a lunar eclipse occurs.

29. Converse: If it is the weekend, then today is Saturday or Sunday.
Biconditional: Today is Saturday or Sunday if and only if it is the weekend.

30. Converse: If Greg wins the race, then he has the fastest time.
Biconditional: Greg has the fastest time if and only if he wins the race.

31. Converse: If a △ is a rt. △, then it contains a rt. ∠.
Biconditional: A △ contains a rt. ∠ if and only if it is a rt. △

32. Conditional is true because a swimmer is an athlete. Converse is false because an athlete might not be a swimmer. Therefore, biconditional is false. Possible counterexample: Felipe could be a runner.

33. Conditional is true because if $2n$ is even, it is divisible by 2, so $\frac{2n}{2} = n$ is an integer. Converse is true because if n is an integer, then $2n$ is an integer divisible by 2, and so is even. Therefore, biconditional is true.

34. A figure is a ○ if and only if it is the set of all pts. that are a fixed dist. from a given pt.

35. A player is a catcher if and only if the player is positioned behind home plate and catches throws from the pitcher.

Reasoning and Writing Geometric Proofs

Mathematical Proof

Algebraic Proof

1.

$y + 1 = 5$	Given equation
$-1 \quad -1$	Subtr. Prop. of =
$y = 4$	Simplify.

2.

$t - 3.2 = -8.3$	Given equation
$+3.2 \quad +3.2$	Add. Prop. of =
$t = -5.1$	Simplify.

3.

$2p - 30 = -4p + 6$	Given equation
$+4p \quad +4p$	Add. Prop. of =
$6p - 30 = 6$	Simplify.
$+30 \quad +30$	Add. Prop. of =
$6p = 36$	Simplify.
$\frac{6p}{6} = \frac{36}{6}$	Div. Prop. of =
$p = 6$	Simplify.

4.

$\frac{x+3}{-2} = 8$	Given equation
$(-2)\left(\frac{x+3}{-2}\right) = -2(8)$	Mult. Prop. of =
$x + 3 = -16$	Simplify.
$-3 \quad -3$	Subtr. Prop. of =
$x = -19$	Simplify.

5.

$\frac{1}{2}n = \frac{3}{4}$	Given equation
$2\left(\frac{1}{2}n\right) = 2\left(\frac{3}{4}\right)$	Mult. Prop. of =
$n = \frac{3}{2}$	Simplify.

6.

$0 = 2(r - 3) + 4$	Given equation
$0 = 2r - 2$	Distribute.
$+2 \quad +2$	Add. Prop. of =
$2 = 2r$	Simplify.
$\frac{2}{2} = \frac{2r}{2}$	Div. Prop. of =
$1 = r$	Simplify.
$r = 1$	Sym. Prop. of =

7. 1 Understand the Problem

Answer will be amount of fat in g.

Important Information:

- $C = 9f + 90$
- $C = 102$ calories

2 Make a Plan

Subst. given information into formula and solve.

3 Solve

$C = 9f + 90$	Given equation
$102 = 9f + 90$	Subst.
$-90 \quad -90$	Subtr. Prop. of =
$12 = 9f$	Simplify.
$\frac{12}{9} = \frac{9f}{9}$	Div. Prop. of =
$\frac{4}{3} = f$	Simplify.

Cereal contains $\frac{4}{3}$ g of fat.

4 Look Back

Check answer by substituting it back into original formula.

$C = 9f + 90$

$102 \stackrel{?}{=} 9\left(\frac{4}{3}\right) + 90$

$90 \stackrel{?}{=} 12 + 90$

$102 = 102$ ✓

8. 1 Understand the Problem

Answer will be number of movie rentals.

Important Information:

- $C = \$5.75 + \$0.89m$
- $C = \$11.98$

2 Make a Plan

Subst. given information into formula and solve.

3 Solve

$C = 5.75 + 0.89m$	Given equation
$11.98 = 5.75 + 0.89m$	Subst.
$-5.75 \quad -5.75$	Subtr. Prop. of =
$6.23 = 0.89m$	Simplify.
$\frac{6.23}{0.89} = m$	Div. Prop. of =
$7 = m$	Simplify.

Elias rented 7 movies.

4 Look Back

Check answer by substituting it back into original formula.

$C = 5.75 + 0.89m$

$11.98 \stackrel{?}{=} 5.75 + 0.89(7)$

$11.98 \stackrel{?}{=} 5.75 + 6.23$

$11.98 = 11.98$ ✓

9. Def. of ≅ segs.
Subst.
Subtr. Prop. of =
Subtr. Prop. of =
Div. Prop. of =

10. Seg. Add. Post.
Subst.
Subtr. Prop. of =
Add. Prop. of =
Div. Prop. of =

11. Reflex. Prop. of ≅

12. Trans. Prop. of =

13. Sym. Prop. of =

14. Trans. Prop. of ≅

Reasoning and Writing Geometric Proofs

Mathematical Proof

Geometric Proof

1. 1. Given
2. Subst.
3. Simplify.
4. Add. Prop. of =
5. Simplify.
6. Def. of supp. ∠s

2. a. Def. of ≅ ∠s

b. ∠1 and ∠2 are supp.

c. Subst.

d. ∠1 and ∠3 are supp.

3.

Statements	Reasons
1. X is the mdpt. of $\overline{AY}$, Y is the mdpt. of $\overline{XB}$.	1. Given
2. $\overline{AX} \cong \overline{XY}$, $\overline{XY} \cong \overline{YB}$	2. Def. of mdpt.
3. $\overline{AX} \cong \overline{YB}$	3. Trans. Prop. of ≅

Reasoning and Writing Geometric Proofs

Mathematical Proof

Flowchart and Paragraph Proofs

1.

Statements	Reasons
1. $\angle 1 \cong \angle 2$	1. Given
2. $\angle 1$ and $\angle 2$ are supp.	2. Lin. Pair Thm.
3. $\angle 1$ and $\angle 2$ are rt. $\angle$s	3. $\cong$ $\angle$s supp. → rt. $\angle$s

2.

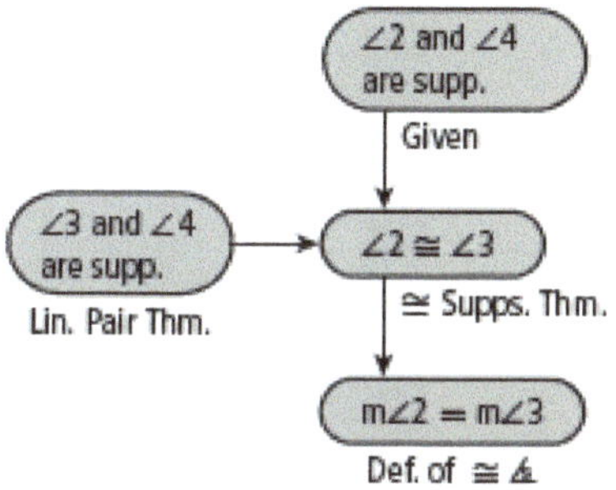

3.

Statements	Reasons
1. $\angle 2 \cong \angle 4$	1. Given
2. $\angle 1 \cong \angle 2$, $\angle 3 \cong \angle 4$	2. Vert. $\angle$s Thm.
3. $\angle 1 \cong \angle 4$	3. Trans. Prop. of $\cong$
4. $\angle 1 \cong \angle 3$	4. Trans. Prop. of $\cong$

4. It is given that $\overrightarrow{BD}$ bisects $\angle ABC$, so $\angle 1 \cong \angle 2$ by def. of $\angle$ bis. By Vert. $\angle$s Thm., $\angle 1 \cong \angle 4$ and $\angle 2 \cong \angle 3$. By Trans. Prop. of $\cong$, $\angle 4 \cong \angle 2$, and thus $\angle 4 \cong \angle 3$. Therefore $\overrightarrow{BG}$ bisects $\angle FBH$ by def. of $\angle$ bis.

Reasoning and Writing Geometric Proofs

Review 2.2 / Mathematical Proof

1. $5x - 3 = 4(x + 2)$ Given equation
$5x - 3 = 4x + 8$ Distrib. Prop.
$x - 3 = 8$ Subtr. Prop. of =
$x = 11$ Add. Prop. of =

2. $1.6 = 2n$ Given equation
$0.5 = n$ Div. Prop. of =

3. $\frac{z}{3} - 2 = -10$ Given equation
$\frac{z}{3} = -8$ Add. Prop. of =
$z = -24$ Mult. Prop. of =

4. $-(h + 3) = 72$ Given equation
$-h - 3 = 72$ Distrib. Prop.
$-h = 75$ Add. Prop. of =
$h = -75$ Mult. Prop. of =

5. $9y + 17 = -19$ Given equation
$9y = -36$ Subtr. Prop. of =
$y = -4$ Div. Prop. of =

6. $\frac{1}{2}(p - 16) = 13$ Given equation
$p - 16 = 26$ Mult. Prop. of =
$p = 42$ Add. Prop. of =

7. $T = 0.03c + 0.05b$ Given equation
$147 = 0.03c + 0.05(150)$ Subst.
$147 = 0.03c + 7.5$ Simplify.
$139.5 = 0.03c$ Subtr. Prop. of =
$4{,}650 = c$ Div. Prop. of =
4,650 bottles were collected.

8. ∠ Add. Post.
Subst.
Simplify.
Subtr. Prop. of =
Add. Prop. of =
Div. Prop. of =

9. ∠ Add. Post.
Subst.
Distrib. Prop.
Simplify.
Subtr. Prop. of =
Div. Prop. of =

10. Sym. Prop. of ≅

11. Reflex. Prop. of =

12. Trans. Prop. of =

13. Reflex. Prop. of ≅

14.
1. Given
2. Def. of ∠ bisector
3. Def. of ≅ ∡
4. Given
5. Subst.
6. ∠ Add. Post.
7. Subst.
8. Simplify.
9. Def. of rt. ∠

15. **a.** m∠1 + m∠2 = 180°, m∠3 + m∠4 = 180°
b. Subst.
c. m∠1 = m∠4
d. Def. of ≅ ∡

16. **a.** Def. of rt. ∠
b. m∠1 + m∠2 = m∠*BAC*
c. m∠2 = m∠3
d. Subst.
e. ∠1 and ∠3 are comp.

17.

Statements	Reasons
1. $\overline{BE} \cong \overline{CE}$, $\overline{DE} \cong \overline{AE}$	1. Given
2. $BE = CE$, $DE = AE$	2. Def. of ≅ segs.
3. $AE + BE = AB$, $CE + DE = CD$	3. Seg. Add. Post.
4. $DE + CE = AB$	4. Subst.
5. $AB = CD$	5. Subst.
6. $\overline{AB} \cong \overline{CD}$	6. Def. of ≅ segs.

18.

Statements	Reasons
1. $\angle 1$ and $\angle 3$ are comp., $\angle 2$ and $\angle 4$ are comp.	1. Given
2. $m\angle 1 + m\angle 3 = 90°$, $m\angle 2 + m\angle 4 = 90°$	2. Def. of comp. ∠s
3. $m\angle 1 + m\angle 3 = m\angle 2 + m\angle 4$	3. Subst.
4. $\angle 3 \cong \angle 4$	4. Given
5. $m\angle 3 = m\angle 4$	5. Def. of ≅ ∠s
6. $m\angle 1 = m\angle 2$	6. Subtr. Prop. of =
7. $\angle 1 \cong \angle 2$	7. Def. of ≅ ∠s

19.

Statements	Reasons
1. B is mdpt. of $\overline{AC}$.	1. Given
2. $\overline{AB} \cong \overline{BC}$	2. Def. of mdpt.
3. $AB = BC$	3. Def. of ≅ segs.
4. $AD + DB = AB$, $BE + EC = BC$	4. Seg. Add. Post.
5. $AD + DB = BE + EC$	5. Subst.
6. $AD = EC$	6. Given
7. $DB = BE$	7. Subtr. Prop. of =

20.

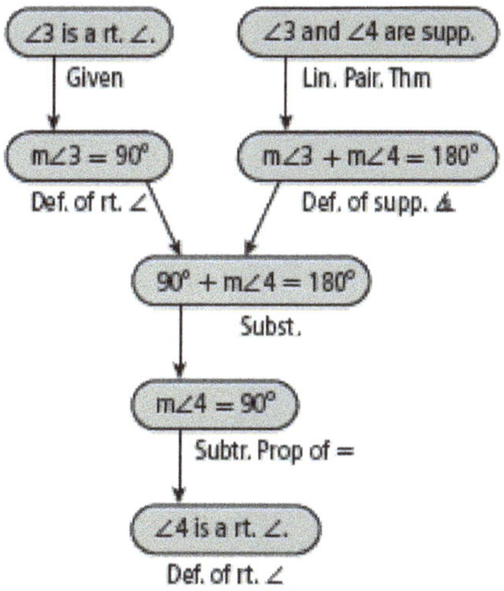

Reasoning and Writing Geometric Proofs

Review 2.2 / Mathematical Proof

21.

Statements	Reasons
1. $\angle 1 \cong \angle 4$	1. Given
2. $\angle 1 \cong \angle 2$	2. Vert. ∡ Thm.
3. $\angle 4 \cong \angle 2$	3. Trans. Prop. of $\cong$
4. $m\angle 4 = m\angle 2$	4. Def. of $\cong$ ∡
5. $\angle 3$ and $\angle 4$ are supp.	5. Lin. Pair Thm.
6. $m\angle 3 + m\angle 4 = 180°$	6. Def. of supp. ∡
7. $m\angle 3 + m\angle 2 = 180°$	7. Subst.
8. $\angle 2$ and $\angle 3$ are supp.	8. Def. of supp. ∡

22. Since $\angle 1$ and $\angle 2$ are comp., $m\angle 1 + m\angle 2 = 90°$. $\angle 1 \cong \angle 3$ by Vert. ∡ Thm. Thus $m\angle 1 = m\angle 3$. By subst., $m\angle 2 + m\angle 3 = 90°$, so $\angle 2$ and $\angle 3$ are comp.

Parallel and Perpendicular Lines

Lines with Transversals

Planes, Lines, and Angles

1. Possible answer: $\overline{EH} \perp \overline{DH}$
2. Possible answer: $\overline{AB}$ and $\overline{DH}$ are skew.
3. Possible answer: $\overline{AB} \parallel \overline{CD}$
4. Possible answer: plane *ABC* ∥ plane *EFG*
5. Possible answer: ∠2 and ∠4
6. Possible answer: ∠6 and ∠8
7. Possible answer: ∠6 and ∠3
8. Possible answer: ∠2 and ∠3
9. transv.: *n*; corr. ∡
10. transv.: *m*; alt. ext. ∡
11. transv.: *n*; alt. int. ∡
12. transv.: *p*; same-side. int. ∡

Parallel and Perpendicular Lines

Lines with Transversals

Angles, Parallel Lines, and Transversals

1. $x = 127$

$m\angle JKL = x = 127°$

2. $(7x - 14)° = (4x + 19)°$

$3x - 14 = 19$

$3x = 33$

$x = 11$

$m\angle BEF = 4x + 19$

$= 4(11) + 19$

$= 63°$

3. $m\angle 1 = 90°$

4. $(6x)° + (3x + 9)° = 180°$

$9x + 9 = 180$

$9x = 171$

$x = 19$

$m\angle CBY = 3x + 9$

$= 3(19) + 9$

$= 66°$

5. $5x + 6y = 94$

$\underline{4x + 6y = 86}$

$x = 8$

$4x + 6y = 86$

$4(8) + 6y = 86$

$32 + 6y = 86$

$6y = 54$

$y = 9$

Parallel and Perpendicular Lines

Lines with Transversals

Proving that Lines are Parallel

1. $\angle 4 \cong \angle 5$, so $p \parallel q$ by the Conv. of the Corr. ∡ Post.
2. $m\angle 1 = (4x + 16)° = 4(28) + 16 = 128°$
   $m\angle 8 = (5x - 12)° = 5(28) - 12 = 128°$
   So $\angle 1 \cong \angle 8$. $p \parallel q$ by the Conv. of the Corr. ∡ Post.
3. $m\angle 4 = (6x - 19)° = 6(11) - 19 = 47°$
   $m\angle 5 = (3x + 14)° = 3(11) + 14 = 47°$
   So $\angle 4 \cong \angle 5$. $p \parallel q$ by the Conv. of the Corr. ∡ Post.
4. $\angle 1 \cong \angle 5$, so $r \parallel s$ by the Conv. of the Alt. Ext. ∡ Thm.
5. $\angle 3 \cong \angle 7$, so $r \parallel s$ by the Conv. of the Alt. Int. ∡ Thm.
6. $m\angle 4 = (13x - 4)° = 13(5) - 4 = 61°$
   $m\angle 8 = (9x + 16)° = 9(5) + 16 = 61°$
   So $\angle 4 \cong \angle 8$. $r \parallel s$ by the Conv. of the Alt. Int. ∡ Thm.
7. $m\angle 2 = (25x + 7)° = 25(5) + 7 = 132°$
   $m\angle 6 = (24x + 12)° = 24(5) + 12 = 132°$
   So $\angle 2 \cong \angle 6$. $r \parallel s$ by the Conv. of the Alt. Ext. ∡ Thm.
8. $\angle 3$ and $\angle 4$ are supp., so $r \parallel s$ by the Conv. of the Same-Side Int. ∡ Thm.
9. $m\angle 8 = (17x + 37)° = 17(6) + 37 = 139°$
   $m\angle 7 = (9x - 13)° = 9(6) - 13 = 41°$
   $m\angle 8 + m\angle 7 = 139° + 41° = 180°$
   So $m\angle 8$ and $m\angle 7$ are supp. $r \parallel s$ by the Conv. of the Same-Side Int. ∡ Thm.

10a. Trans. Prop. of ≅ **b.** $\overline{XY} \parallel \overline{WV}$

10c. Conv. of the Alt. Int. ∡ Thm.

11. $m\angle 1 = (17x + 9)° = 17(3) + 9 = 60°$
    $m\angle 2 = (14x + 18)° = 14(3) + 18 = 60°$
    So $\angle 1 \cong \angle 2$. By the Conv. of the Alt. Int. ∡ Thm., landings are ∥.

Parallel and Perpendicular Lines

Lines with Transversals

Properties of Perpendicular Lines

1. $\overline{EB}$

2.
$$\begin{aligned} ED &> EB \\ x + 12 &> 7 \\ -12 & \quad -12 \\ \hline x &> -5 \end{aligned}$$

3. a. 2 intersecting lines form a lin. pair of ≅ ∡ → lines ⊥.

   b. $\overleftrightarrow{DE} \perp \overleftrightarrow{AF}$

   c. 2 lines ⊥ to same line → 2 lines ||.

4. The service lines are coplanar lines that are ⊥ to the same line (the center line), so they must be || to each other.

Parallel and Perpendicular Lines

Review 3.1 / Lines with Transversals

1. Possible answer: $\overline{AB} \parallel \overline{DE}$
2. Possible answer: $\overline{AB}$ and $\overline{CF}$ are skew.
3. Possible answer: $\overline{BD} \perp \overline{DF}$
4. Possible answer: plane $ABC \parallel$ plane DEF
5. Possible answer: $\angle 2$ and $\angle 6$
6. Possible answer: $\angle 1$ and $\angle 8$
7. Possible answer: $\angle 1$ and $\angle 6$
8. Possible answer: $\angle 2$ and $\angle 5$
9. transv.: p; corr. ∡
10. transv.: q; alt. int. ∡
11. transv.: ℓ; alt. ext. ∡
12. transv.: p; same-side. int. ∡
13. $m\angle KLM = y = 115°$
14. $4a = 2a + 50$
    $2a = 50$
    $a = 25$
    $m\angle VYX = 4(25) = 100°$
15. $m\angle ABC = x = 116°$
16. $13x + 17x = 180$
    $30x = 180$
    $x = 6$
    $m\angle EFG = 17(6) = 102°$
17. $3n - 45 = 2n + 15$
    $n - 45 = 15$
    $n = 60$
    $m\angle PQR + (2n + 15) = 180$
    $m\angle PQR + 2(60) + 15 = 180$
    $m\angle PQR + 135 = 180$
    $m\angle PQR = 45°$
18. $4x - 14 = 3x + 12$
    $x - 14 = 12$
    $x = 26$
    $m\angle STU + (3x + 12) = 180$
    $m\angle STU + 3(26) + 12 = 180$
    $m\angle STU + 90 = 180$
    $m\angle STU = 90°$
19. $m\angle 1 = 60$ and $m\angle 2 + 60 = 180$
    $2x - 3y = 60$
    $\underline{x + 3y + 60 = 180}$
    $3x + 60 = 240$
    $3x = 180$
    $x = 60$
    $2x - 3y = 60$
    $2(60) - 3y = 60$
    $120 - 3y = 60$
    $-3y = -60$
    $y = 20$
20. $\angle 3 \cong \angle 7$, so $\ell \parallel m$ by the Conv. of the Corr. ∡ Post.
21. $m\angle 4 = 54°$
    $m\angle 8 = (7x + 5)° = 7(7) + 5 = 54°$
    So $\angle 4 \cong \angle 8$. $\ell \parallel m$ by the Conv. of the Corr. ∡ Post.
22. $m\angle 2 = (8x + 4)° = 8(15) + 4 = 124°$
    $m\angle 6 = (11x - 41)° = 11(15) - 41 = 124°$
    So $\angle 2 \cong \angle 6$. $\ell \parallel m$ by the Conv. of the Corr. ∡ Post.
23. $m\angle 1 = (3x + 19)° = 3(12) + 19 = 55°$
    $m\angle 5 = (4x + 7)° = 4(12) + 7 = 55°$
    So $\angle 1 \cong \angle 5$. $\ell \parallel m$ by the Conv. of the Corr. ∡ Post.
24. $\angle 3 \cong \angle 6$, so $n \parallel p$ by the Conv. of the Alt. Int. ∡ Thm.

Parallel and Perpendicular Lines

Review 3.1 / Lines with Transversals

25. $\angle 2 \cong \angle 7$, so $n \parallel p$ by the Conv. of the Alt. Ext. ∡ Thm.

26. $\angle 4$ and $\angle 6$ are supp., so $n \parallel p$ by the Conv. of the Same-Side Int. ∡ Thm.

27. $m\angle 1 = (8x - 7)° = 8(14) - 7 = 105°$
$m\angle 8 = (6x + 21)° = 6(14) + 21 = 105°$
So $\angle 1 \cong \angle 8$. $n \parallel p$ by the Conv. of the Alt. Ext. ∡ Thm.

28. $m\angle 4 = (4x + 3)° = 4(25) + 3 = 103°$
$m\angle 5 = (5x - 22)° = 5(25) - 22 = 103°$
So $\angle 4 \cong \angle 5$. $n \parallel p$ by the Conv. of the Alt. Int. ∡ Thm.

29. $m\angle 3 = (2x + 15)° = 2(30) + 15 = 75°$
$m\angle 5 = (3x + 15)° = 3(30) + 15 = 105°$
$m\angle 3 + m\angle 5 = 75° + 105° = 180°$
So $m\angle 3$ and $m\angle 5$ are supp. $n \parallel p$ by the Conv. of the Same-Side Int. ∡ Thm.

30. **a.** Corr. ∡ Post. **b.** Given
c. Trans. Prop. of $\cong$ **d.** $\overline{BC} \parallel \overline{DE}$
e. Conv. of the Corr. ∡ Post.

31. $m\angle 1 = (3x + 2)° = 3(6) + 2 = 20°$
$m\angle 2 = (5x - 10)° = 5(6) - 10 = 20°$
So $\angle 1 \cong \angle 2$. $\overline{DJ} \parallel \overline{EK}$ by the Conv. of the Corr. ∡ Post.

32. $\overline{WY}$

33.
$$\begin{aligned} WY &< WZ \\ x + 8 &< 19 \\ -8 & \quad -8 \\ x &< 11 \end{aligned}$$

34. **a.** Given **b.** $\overleftrightarrow{AB} \parallel \overleftrightarrow{CD}$
c. $\perp$ Transv. Thm.

35. The frets are lines that are $\perp$ to the same line (the string), so the frets must be $\parallel$ to each other.

Parallel and Perpendicular Lines

Slope and the Equation of a Line

Finding the Slope Given Two Points

1. Substitute (5, 7) for (x_1, y_1) and (−2, 1) for (x_2, y_2) in the slope formula and then simplify.
$$m = \frac{y_2 - y_1}{x_2 - x_1} = \frac{1 - 7}{-2 - 5} = \frac{-6}{-7} = \frac{6}{7}$$

2. Substitute (−5, 3) for (x_1, y_1) and (4, −2) for (x_2, y_2) in the slope formula and then simplify.
$$m = \frac{y_2 - y_1}{x_2 - x_1} = \frac{-2 - 3}{4 - (-5)} = \frac{-5}{9} = -\frac{5}{9}$$

3. Substitute (0, 1) for (x_1, y_1) and (−5, 1) for (x_2, y_2) in the slope formula and then simplify.
$$m = \frac{y_2 - y_1}{x_2 - x_1} = \frac{1 - 1}{-5 - 0} = \frac{0}{-5} = 0$$

4. Substitute (4, −2) for (x_1, y_1) and (6, 3) for (x_2, y_2) in the slope formula and then simplify.
$$m = \frac{y_2 - y_1}{x_2 - x_1} = \frac{3 - (-2)}{6 - 4} = \frac{5}{2}$$

5. Use the points (8, 80) and (11, 200) to graph the line and find the slope.
$$m = \frac{y_2 - y_1}{x_2 - x_1} = \frac{200 - 80}{11 - 8} = \frac{120}{3} = 40$$
The slope is 40, which means the bird is flying at an average speed of 40 mi/h.

6.

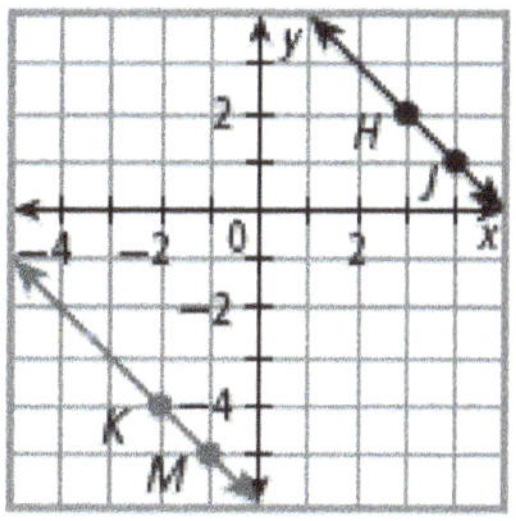

$$\text{slope of } \overleftrightarrow{HJ} = \frac{1 - 2}{4 - 3} = \frac{-1}{1} = -1$$
$$\text{slope of } \overleftrightarrow{KM} = \frac{-5 - (-4)}{-1 - (-2)} = \frac{-1}{1} = -1$$
The slopes are the same, so the lines are parallel.

7.

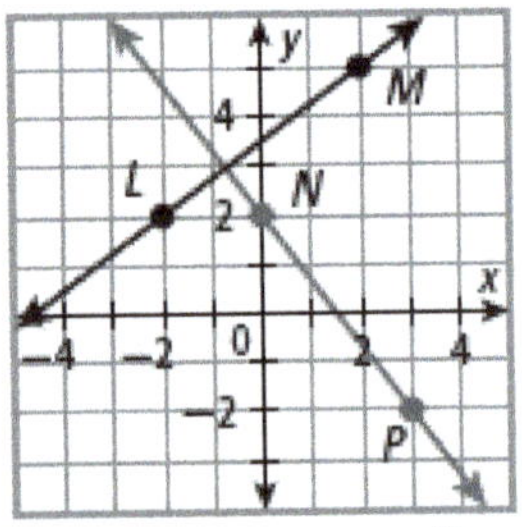

slope of $\overleftrightarrow{LM} = \dfrac{5-2}{2-(-2)} = \dfrac{3}{4}$

slope of $\overleftrightarrow{NP} = \dfrac{-2-2}{3-0} = \dfrac{-4}{3} = -\dfrac{4}{3}$

The product of the slopes is $\left(\dfrac{3}{4}\right)\left(-\dfrac{4}{3}\right) = -1$, so lines are perpendicular.

8.

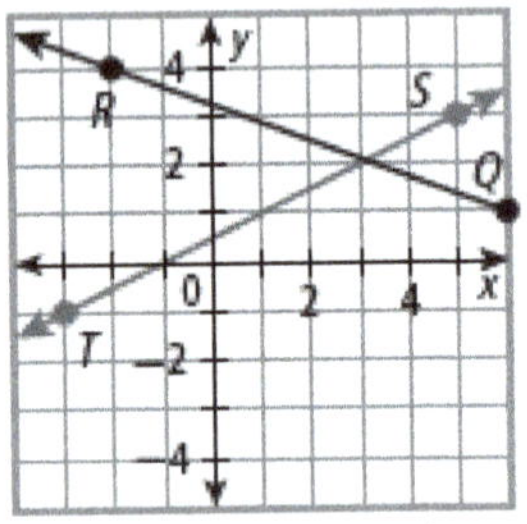

slope of $\overleftrightarrow{QR} = \dfrac{4-1}{-2-6} = \dfrac{3}{-8} = -\dfrac{3}{8}$

slope of $\overleftrightarrow{ST} = \dfrac{-1-3}{-3-5} = \dfrac{-4}{-8} = \dfrac{1}{2}$

The slopes are not the same, so the lines are not parallel. The product of the slopes is not −1, so lines are not perpendicular.

Parallel and Perpendicular Lines

Slope and the Equation of a Line

Slope-Intercept Form

1.

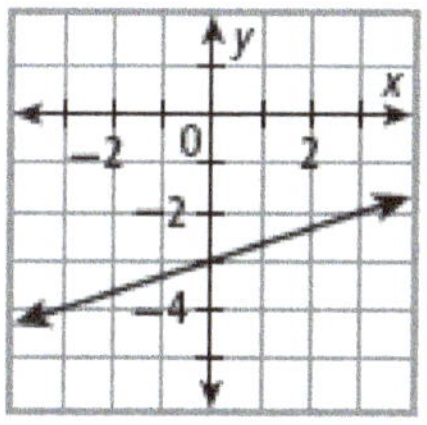

Plot (0, −3). Count 1 unit up and 3 units right and plot another point. Draw the line connecting the two points.

2.

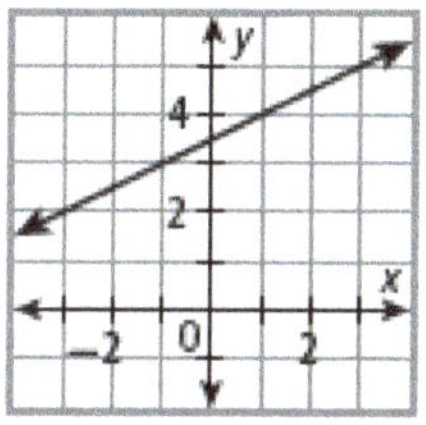

Plot (0, 3.5). Count 0.5 units up and 1 unit right and plot another point. Draw the line connecting the two points.

3.

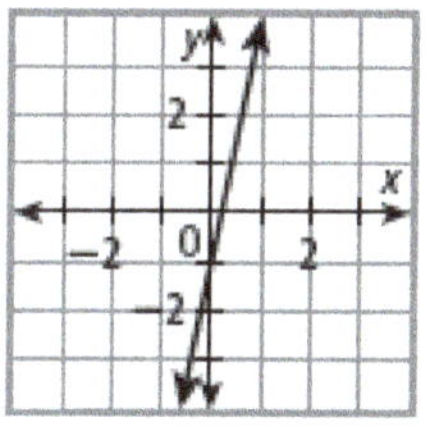

Plot (0, −1). Count 5 units up and 1 unit right and plot another point. Draw the line connecting the two points.

4.

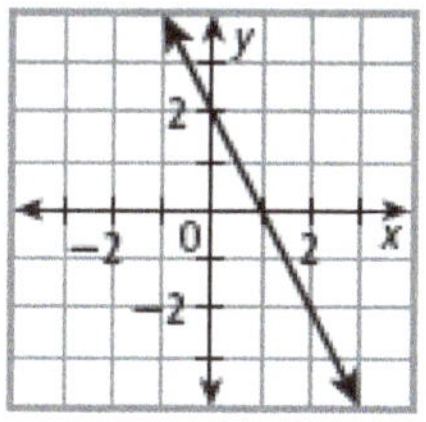

Plot (0, 2). Count 2 units down and 1 unit right and plot another point. Draw the line connecting the two points.

5. $y = mx + b$

$y = 8x + 2$

6. $y = mx + b$

$y = \frac{1}{2}x + (-6)$

$y = \frac{1}{2}x - 6$

7. $y = mx + b$

$y = 0x + (-3)$

$y = -3$

8.
$$\begin{aligned} y &= mx + b \\ 7 &= 5(2) + b \\ 7 &= 10 + b \\ -10 & \quad -10 \\ -3 &= b \end{aligned}$$

$y = mx + b$

$y = 5x + (-3)$

$y = 5x - 3$

Slope-Intercept Form

9. $y = \frac{2}{5}x - 6$ is in the form $y = mx + b$.

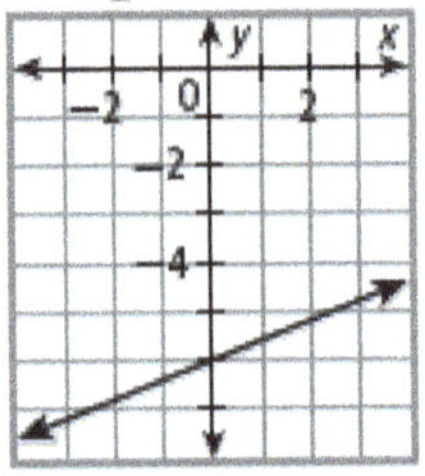

Plot (0, −6). Count 2 units up and 5 units right and plot another point. Draw the line connecting the two points.

10.

$$3x - y = -1$$
$$\underline{-3x \qquad -3x}$$
$$-y = -3x - 1$$
$$-1(-y) = -1(-3x - 1)$$
$$y = 3x + 1$$

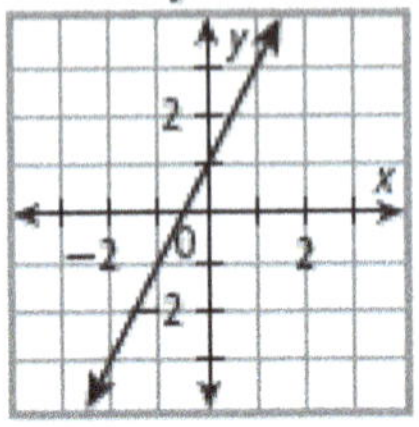

Plot (0, 1). Count 3 units up and 1 unit right and plot another point. Draw the line connecting the two points.

11.

$$2x + y = 4$$
$$\underline{-2x \qquad -2x}$$
$$y = -2x + 4$$

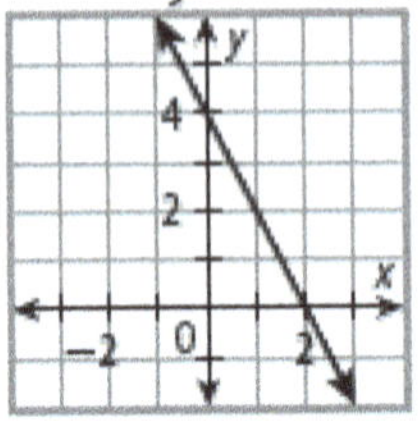

Plot (0, 4). Count 2 units down and 1 unit right and plot another point. Draw the line connecting the two points.

12a. An equation is $y = 18x + 10$.

b. The y-intercept is 10. This is the distance she has already biked.
The slope is 18. This is Helen's speed.

c. $y = 18x + 10$
$= 18(2) + 10 = 46$
Helen will have biked 46 mi after 2 hours.

Parallel and Perpendicular Lines

Slope and the Equation of a Line

Point-Slope Form

1. 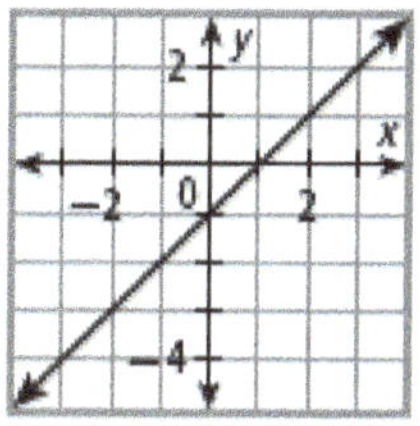

Plot (1, 0). Move 1 unit up and 1 unit right and plot another point. Draw the line connecting the two points.

2. 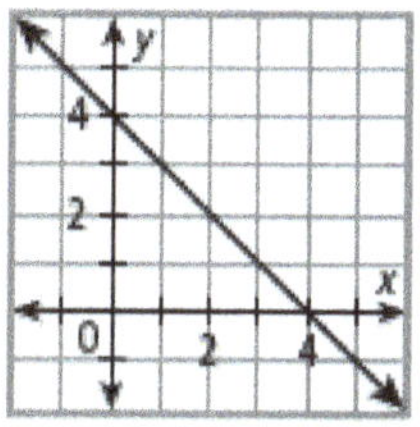

Plot (3, 1). Move 1 unit down and 1 unit right and plot another point. Draw the line connecting the two points.

3. 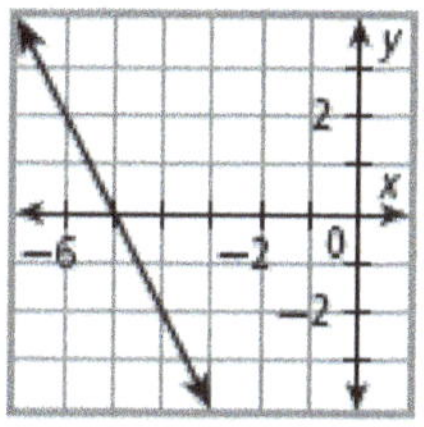

Plot (−4, −2). Move 2 units down and 1 unit right and plot another point. Draw the line connecting the two points.

4. 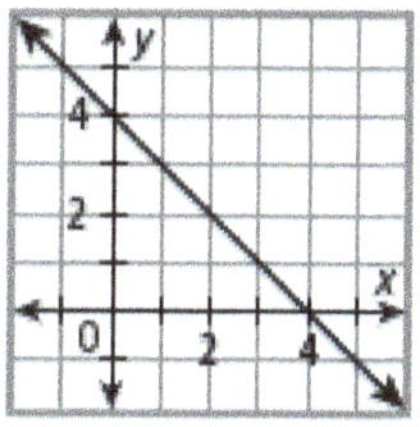

5. 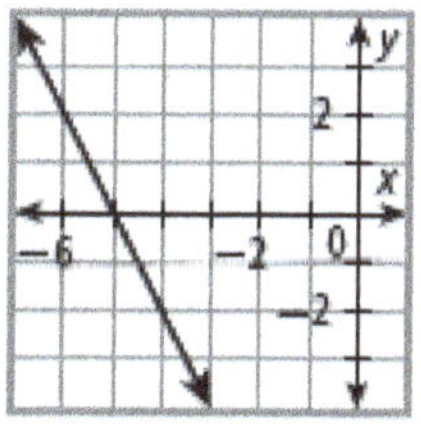

6. 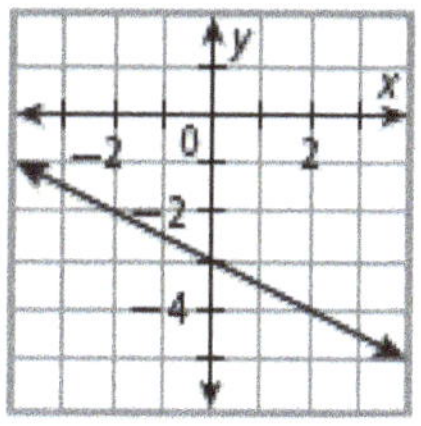

7. $y - 8 = -\frac{1}{3}(x - (-3))$

$y - 8 = -\frac{1}{3}x - 1$

$\underline{+8} \qquad \underline{+8}$

$y = -\frac{1}{3}x + 7$

8. $y - 1 = 2(x - 1)$

$y - 1 = 2x - 2$

$\underline{+1} \qquad \underline{+1}$

$y = 2x - 1$

9. $m = \frac{y_2 - y_1}{x_2 - x_1} = \frac{-2 - 2}{2 - (-2)} = \frac{-4}{4} = -1$

$y - 2 = -1(x - (-2))$

$y - 2 = -x - 2$

$\underline{+2} \qquad \underline{+2}$

$y = -x$

10. $m = \frac{y_2 - y_1}{x_2 - x_1} = \frac{3 - 1}{-5 - 1} = \frac{2}{-6} = -\frac{1}{3}$

$y - 1 = -\frac{1}{3}(x - 1)$

$y - 1 = -\frac{1}{3}x + \frac{1}{3}$

$\underline{+1} \qquad \underline{+1}$

$y = -\frac{1}{3}x + \frac{4}{3}$

11. $m = \frac{y_2 - y_1}{x_2 - x_1} = \frac{4 - 0}{0 - 8} = \frac{4}{-8} = -\frac{1}{2}$

$y - 4 = -\frac{1}{2}(x - 0)$

$y - 4 = -\frac{1}{2}x$

$\underline{+4} \qquad \underline{+4}$

$y = -\frac{1}{2}x + 4$

12. $m = \frac{y_2 - y_1}{x_2 - x_1} = \frac{3 - 0}{0 - (-2)} = \frac{3}{2}$

$y - 3 = \frac{3}{2}(x - 0)$

Parallel and Perpendicular Lines

Slope and the Equation of a Line

Point-Slope Form

13. $m = \frac{y_2 - y_1}{x_2 - x_1} = \frac{4-2}{7-5} = \frac{2}{2} = 1$

$y - 2 = 1(x - 5)$

$y - 2 = x - 5$

$\underline{+2} \quad \underline{+2}$

$y = x - 3$

x-intercept:

$0 = x - 3$

$\underline{+3} \quad \underline{+3}$

$x = 3$

y-intercept:

$y = 0 - 3$

$y = -3$

14. $m = \frac{y_2 - y_1}{x_2 - x_1} = \frac{-5-5}{-3-(-1)} = \frac{-10}{-2} = 5$

$y - 5 = 5(x - (-1))$

$y - 5 = 5x + 5$

$\underline{+5} \quad \underline{+5}$

$y = 5x + 10$

x-intercept:

$0 = 5x + 10$

$-10 \quad -10$

$\frac{-10}{5} = \frac{5x}{5}$

$x = -2$

y-intercept:

$y = 5(0) + 10$

$y = 10$

15. $m = \frac{y_2 - y_1}{x_2 - x_1} = \frac{-9-9}{-4-2} = \frac{-18}{-6} = 3$

$y - 9 = 3(x - 2)$

$y - 9 = 3x - 6$

$\underline{+9} \quad \underline{+9}$

$y = 3x + 3$

x-intercept:

$0 = 3x + 3$

$\underline{-3} \quad \underline{-3}$

$\frac{-3}{3} = \frac{3x}{3}$

$x = -1$

y-intercept:

$y = 3(0) + 3$

$y = 3$

16. $m = \frac{y_2 - y_1}{x_2 - x_1} = \frac{5-3}{10-0} = \frac{2}{10} = \frac{1}{5}$

$y - y_1 = m(x - x_1)$

$y - 6 = \frac{1}{5}(x - 15)$

$y - 6 = \frac{1}{5}x - 3$

$\underline{+6} \quad \underline{+6}$

$y = \frac{1}{5}x + 3$

$y = \frac{1}{5}x + 3$

$= \frac{1}{5}(30) + 3 = 9$

The depth of the oil after half an hour is 9 ft.

Parallel and Perpendicular Lines

Slope and the Equation of a Line

Slopes of Parallel and Perpendicular Lines

1. The lines described by $y = 6$ and $y = -8$ both have slope 0. These lines are parallel.
   The lines described by $y = 6x + 5$ and $y = 6x - 7$ both have slope 6. These lines are parallel.

2. $$y - 3 = \frac{3}{4}(x - 5)$$
   $$y - 3 = \frac{3}{4}x - \frac{15}{4}$$
   $$\underline{+3} \qquad \underline{+3}$$
   $$y = \frac{3}{4}x - \frac{3}{4}$$

   $$y - 4 = -2(x + 2)$$
   $$y - 4 = -2x - 4$$
   $$\underline{+4} \qquad \underline{+4}$$
   $$y = -2x$$

   The lines described by $y = \frac{3}{4}x - 1$ and $y - 3 = \frac{3}{4}(x - 5)$ represent parallel lines. They each have slope $\frac{3}{4}$.
   The lines described by $y = -2x$ and $y - 4 = -2(x + 2)$ have the same slope, but they are not parallel lines. They are the same line.

3. slope of $\overline{AB} = \frac{5 - 5}{4 - (-2)} = \frac{0}{6} = 0$
   slope of $\overline{BC} = \frac{2 - 5}{6 - 4} = \frac{-3}{2} = -\frac{3}{2}$
   slope of $\overline{CD} = \frac{-1 - 2}{2 - 6} = \frac{-3}{-4} = \frac{3}{4}$
   slope of $\overline{AD} = \frac{-1 - 5}{2 - (-2)} = \frac{-6}{4} = -\frac{3}{2}$
   $\overline{AD}$ and $\overline{BC}$ are parallel because they have the same slope. Therefore, $ABCD$ is a trapezoid.

4. The slope of the line described by $y = \frac{2}{3}x - 4$ is $\frac{2}{3}$. The slope of the line described by $y = -\frac{3}{2}x + 2$ is $-\frac{3}{2}$.
   $$\frac{2}{3}\left(-\frac{3}{2}\right) = -1$$
   These lines are perpendicular because the product of their slopes is -1.
   The graph described by $y = -1$ is a horizontal line, and the graph described by $x = 3$ is a vertical line. These lines are perpendicular.

5. The slope of the line described by $y = -\frac{3}{7}x - 4$ is $-\frac{3}{7}$. The slope of the line described by $y - 7 = \frac{7}{3}(x - 3)$ is $\frac{7}{3}$.

   $-\frac{3}{7}\left(\frac{7}{3}\right) = -1$

   These lines are perpendicular because the product of their slopes is -1.

   The slope of the line described by $y - 4 = -7(x + 2)$ is -7. The slope of the line described by $y - 1 = \frac{1}{7}(x - 4)$ is $\frac{1}{7}$.

   $-7\left(\frac{1}{7}\right) = -1$

   These lines are perpendicular because the product of their slopes is -1.

6. slope of $\overline{PQ} = \frac{6 - 4}{2 - 1} = \frac{2}{1} = 2$

   slope of $\overline{QR} = \frac{3 - 6}{8 - 2} = \frac{-3}{6} = -\frac{1}{2}$

   slope of $\overline{RS} = \frac{1 - 3}{7 - 8} = \frac{-2}{-1} = 2$

   slope of $\overline{PS} = \frac{1 - 4}{7 - 1} = \frac{-3}{6} = -\frac{1}{2}$

   $\overline{PQ}$ is perpendicular to $\overline{QR}$ because the product of their slopes is -1. $\overline{QR}$ is perpendicular to $\overline{RS}$ because the product of their slopes is -1. $\overline{RS}$ is perpendicular to $\overline{PS}$ because the product of their slopes is -1. $\overline{PS}$ is perpendicular to $\overline{PQ}$ because the product of their slopes is -1. Therefore, all the angles are right angles, and $PQRS$ is a rectangle.

7. The perpendicular line has a slope of $\frac{2}{5}$, because $-\frac{5}{2}\left(\frac{2}{5}\right) = -1$.

   $y - y_1 = m(x - x_1)$

   $y - 0 = \frac{2}{5}(x - 5)$

   $y = \frac{2}{5}(x - 5)$

   $y = \frac{2}{5}x - 2$

Parallel and Perpendicular Lines

Review 3.2 / Slope and the Equation of a Line

1. Substitute (0, 7) for (x_1, y_1) and (0, 3) for (x_2, y_2) in the slope formula and then simplify.
$$m = \frac{y_2 - y_1}{x_2 - x_1} = \frac{3 - 7}{0 - 0} = \frac{-4}{0}$$
The slope is undefined.

2. Substitute (5, −2) for (x_1, y_1) and (3, −2) for (x_2, y_2) in the slope formula and then simplify.
$$m = \frac{y_2 - y_1}{x_2 - x_1} = \frac{-2 - (-2)}{3 - 5} = \frac{0}{-2} = 0$$

3. Substitute (3, 4) for (x_1, y_1) and (4, 3) for (x_2, y_2) in the slope formula and then simplify.
$$m = \frac{y_2 - y_1}{x_2 - x_1} = \frac{3 - 4}{4 - 3} = \frac{-1}{1} = -1$$

4. Substitute (0, 4) for (x_1, y_1) and (3, −3) for (x_2, y_2) in the slope formula and then simplify.
$$m = \frac{y_2 - y_1}{x_2 - x_1} = \frac{-3 - 4}{3 - 0} = \frac{-7}{3} = -\frac{7}{3}$$

5. Use the points (2.5, 100) and (5, 475) to graph the line and find the slope.
$$m = \frac{y_2 - y_1}{x_2 - x_1} = \frac{475 - 100}{5 - 2.5} = \frac{375}{2.5} = 150$$
The slope is 150, which means that the plane is flying at an average speed of 150 mi/h.

6. $y = mx + b$
$$y = -\frac{2}{3}x + 2$$

7. $$\begin{aligned} y &= mx + b \\ -8 &= 0(6) + b \\ -8 &= b \end{aligned}$$
$$\begin{aligned} y &= mx + b \\ y &= 0x + (-8) \\ y &= -8 \end{aligned}$$

8. $$m = \frac{2 - 2}{-2 - 5} = 0$$
$$\begin{aligned} y &= mx + b \\ 2 &= 0(5) + b \\ 2 &= b \\ y &= 0x + 2 \\ y &= 2 \end{aligned}$$

9. The equation is given in slope-intercept form, with a slope of $\frac{1}{2}$ and a y-intercept of −2. Plot the point (0, −2) and then rise 1 and run 2 to find another point. Draw the line containing the two points.

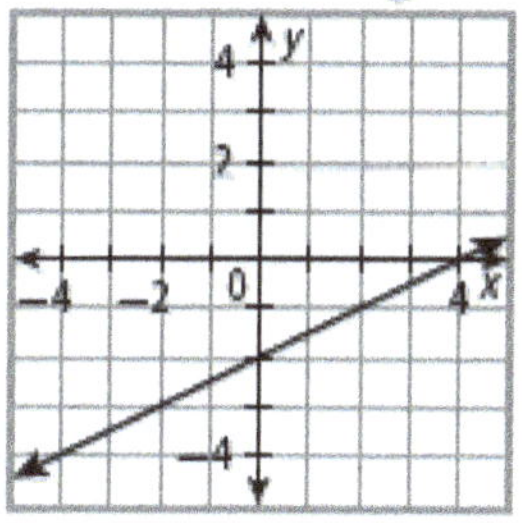

10. The equation is given in the form for a horizontal line with a y-intercept of 2. The equation tells you that the y-coordinate of every point on the line is 2. Draw the horizontal line through (0, 2).

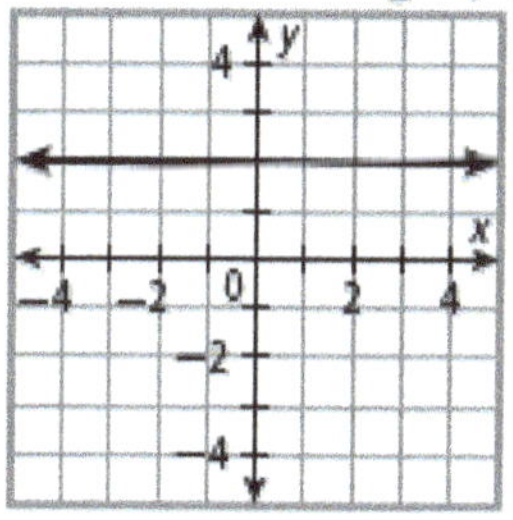

11. $$m = \frac{6 - (-2)}{4 - 0} = \frac{8}{4} = 2$$
$$\begin{aligned} y - y_1 &= m(x - x_1) \\ y - (-2) &= 2(x - 0) \\ y + 2 &= 2x \end{aligned}$$

12. $$\begin{aligned} y - y_1 &= m(x - x_1) \\ y - (-4) &= \frac{2}{3}(x - 6) \\ y + 4 &= \frac{2}{3}(x - 6) \end{aligned}$$

Parallel and Perpendicular Lines

Review 3.2 / Slope and the Equation of a Line

13. The equation is given in point-slope form, with a slope of 1 through the point $(-4, 7)$. Plot the point $(-4, 7)$ and then rise 1 and run 1 to find another point. Draw the line containing the two points.

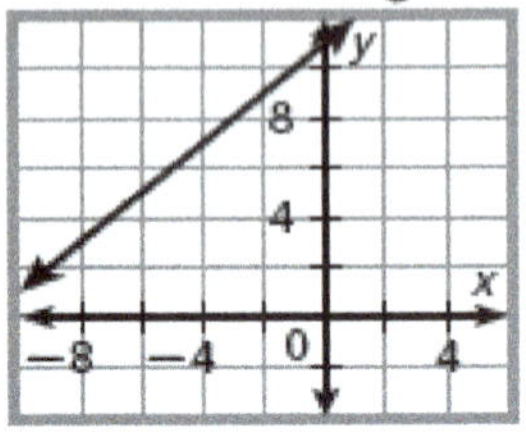

14.

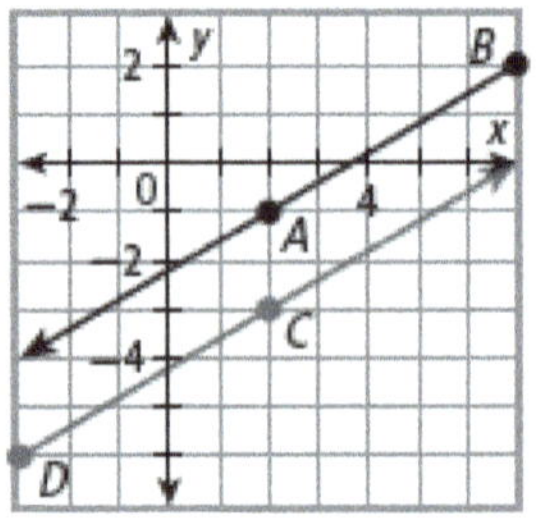

slope of $\overleftrightarrow{AB} = \frac{2-(-1)}{7-2} = \frac{3}{5}$

slope of $\overleftrightarrow{CD} = \frac{-6-(-3)}{-3-2} = \frac{-3}{-5} = \frac{3}{5}$

The slopes are the same, so the lines are parallel.

15.

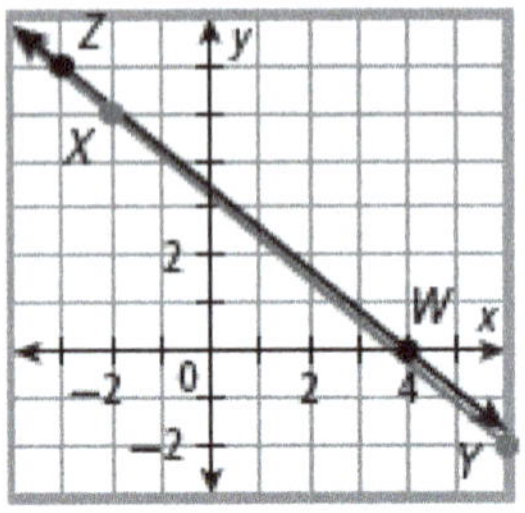

slope of $\overleftrightarrow{XY} = \frac{-2-5}{6-(-2)} = \frac{-7}{8} = -\frac{7}{8}$

slope of $\overleftrightarrow{ZW} = \frac{0-6}{4-(-3)} = \frac{-6}{7} = -\frac{6}{7}$

The slopes are not the same, so the lines are not parallel. The product of the slopes is not -1, so lines are not perpendicular.

16.

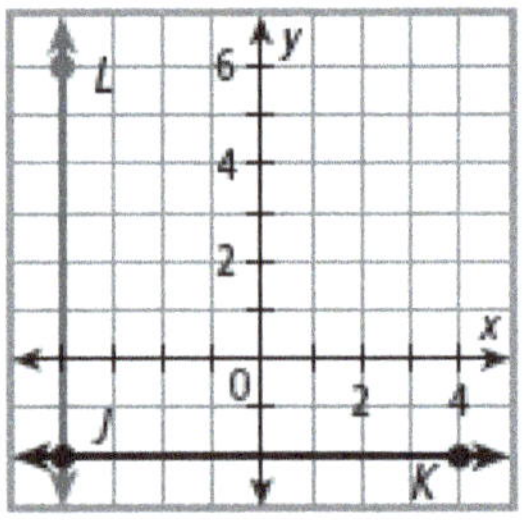

$\overleftrightarrow{JK}$ is horiz. and $\overleftrightarrow{JL}$ is vert, so the lines are perpendicular.

17.

$$\begin{aligned} -x + 2y &= 17 \\ +x \quad\quad &\;\; +x \\ 2y &= x + 17 \\ \frac{2y}{2} &= \frac{x+17}{2} \\ y &= \frac{1}{2}x + \frac{17}{2} \end{aligned}$$

$$\begin{aligned} 3x + y &= 27 \\ -3x \quad\quad &\;\; -3x \\ y &= -3x + 27 \end{aligned}$$

The lines described by $y = -3x + 2$ and $3x + y = 27$ represent parallel lines. They each have slope -3.

The lines described by $y = \frac{1}{2}x - 1$ and $-x + 2y = 17$ represent parallel lines. They each have slope $\frac{1}{2}$.

18. The slope of the line described by $y - 9 = 3(x + 1)$ is 3. The slope of the line described by $y = -\frac{1}{3}x + 5$ is $-\frac{1}{3}$.

$3\left(-\frac{1}{3}\right) = -1$

These lines are perpendicular because the product of their slopes is -1.

The graph described by $y = 0$ is a horizontal line, and the graph described by $x = 6$ is a vertical line. These lines are perpendicular.

Parallel and Perpendicular Lines

Review 3.2 / Slope and the Equation of a Line

19. slope of $\overline{LM} = \dfrac{4-0}{0-(-3)} = \dfrac{4}{3}$

slope of $\overline{MN} = \dfrac{4-0}{0-3} = \dfrac{4}{-3} = -\dfrac{4}{3}$

slope of $\overline{NP} = \dfrac{-4-0}{0-3} = \dfrac{-4}{-3} = \dfrac{4}{3}$

slope of $\overline{LP} = \dfrac{-4-0}{0-(-3)} = \dfrac{-4}{3} = -\dfrac{4}{3}$

$\overline{LM}$ is parallel to $\overline{NP}$ because they have the same slope. $\overline{MN}$ is parallel to $\overline{LP}$ because they have the same slope. Since opposite sides are parallel, *LMNP* is a parallelogram.

20. slope of $\overline{AB} = \dfrac{-3-(-2)}{-3-(-7)} = \dfrac{-1}{4} = -\dfrac{1}{4}$

slope of $\overline{BC} = \dfrac{-7-(-3)}{(-4-(-3))} = \dfrac{-4}{-1} = 4$

$\overline{AB}$ is perpendicualr to $\overline{BC}$ because the product of their slopes is -1. Since *ABC* contains a right angle, *ABC* is a right triangle.

Triangle Congruence

Triangles and Congruence

Classifying Triangles

1. $\angle DBC$ is a rt. $\angle$.
   So $\triangle DBC$ is a rt. $\triangle$.
2. $\angle ABD$ and $\angle DBC$ are supp.
   $\angle ABD + \angle DBC = 180°$
   $\angle ABD + 90 = 180$
   $\angle ABD = 90°$
   $\angle ABD$ is a rt. $\angle$. So $\triangle ABD$ is a rt. $\triangle$.
3. $m\angle ADC = m\angle ADB + m\angle BDC$
   $= 31 + 70 = 101°$
   $\angle ADC$ is obtuse. So $\triangle ADC$ is an obtuse $\triangle$.
4. $EG = 3 + 3 = 6$,
   $EH = 8$, $GH = 8$
   $\overline{EH} \cong \overline{GH}$
   Exactly two sides are $\cong$, so $\triangle EGH$ is isosc.
5. $EF = 3$, $EH = 8$,
   $FH = 7.4$
   No sides are congruent, so $\triangle EFH$ is scalene.
6. $GF = 3$, $GH = 8$, $FH = 7.4$
   No sides are congruent, so $\triangle HFG$ is scalene.
7. **Step 1** Find y.
   $6y = 4y + 12$
   $2y = 12$
   $y = 6$
   Step 2 Find side lengths.
   $\triangle$ is equilateral, so all three side lengths $= 6y = 36$.
8. **Step 1** Find x.
   $2x + 1.7 = x + 2.4$
   $2x = x + 0.7$
   $x = 0.7$
   Step 2 Find side lengths.
   $x + 2.4 = 0.7 + 2.4 = 3.1$
   $2x + 1.7 = 2(0.7) + 1.7 = 1.4 + 1.7 = 3.1$
   $4x + 0.5 = 4(0.7) + 0.5 = 2.8 + 0.5 = 3.3$
9. Perimeter is
   $P = 3 + 3 + 1.5$
   $= 7.5$ cm
   $50 \div 7.5 = 6\frac{2}{3}$ earrings
   The jeweler can make 6 earrings.

Triangle Congruence

Triangles and Congruence

Angle Relationships in Triangles

1. Think: Use △ ∠ Sum Thm.
   $180 = 3y + 13 + 2y + 2 + 5y - 5$
   $180 = 10y + 10$
   $170 = 10y$
   $y = 17$

2. Deneb: $3y + 13 = 3(17) + 13 = 64°$
   Altair: $2y + 2 = 2(17) + 2 = 36°$
   Vega: $5y - 5 = 5(17) - 5 = 80°$

3. $20.8 + m\angle = 90$
   $m\angle = 69.2°$

4. $y + m\angle = 90$
   $m\angle = (90 - y)°$

5. $24\frac{2}{3} + m\angle = 90$
   $m\angle = 65\frac{1}{3}°$

6. $m\angle M + m\angle N = m\angle NPQ$
   $3y + 1 + 2y + 2 = 48$
   $5y + 3 = 48$
   $5y = 45$
   $y = 9$
   $m\angle M = 3y + 1 = 3(9) + 1 = 28°$

7. $m\angle K + m\angle L = m\angle HJL$
   $7x + 6x - 1 = 90$
   $13x = 91$
   $x = 7$
   $m\angle L = 6x - 1 = 6(7) - 1 = 41°$

8. $m\angle A + m\angle B = 117$
   $65 + m\angle B = 117$
   $m\angle B = 52°$
   $m\angle A + m\angle B + m\angle BCA = 180$
   $117 + m\angle BCA = 180$
   $m\angle BCA = 63°$

9. $\angle C \cong \angle F$
   $m\angle C = m\angle F$
   $4x^2 = 3x^2 + 25$
   $x^2 = 25$
   $m\angle C = 4x^2 = 100°$
   $m\angle F = m\angle C = 100°$

10. $\angle S \cong \angle U$
    $m\angle S = m\angle U$
    $5x - 11 = 4x + 9$
    $x = 20$
    $m\angle S = 5x - 11$
    $= 5(20) - 11$
    $= 89°$
    $m\angle U = m\angle S = 89°$

11. $\angle C \cong \angle Z$
    $m\angle C = m\angle Z$
    $4x + 7 = 3(x + 5)$
    $4x + 7 = 3x + 15$
    $x = 8$
    $m\angle C = 4x + 7 = 4(8) + 7 = 39°$
    $m\angle Z = m\angle C = 39°$

Congruent Triangles

1. $\overline{LM}$
2. $\overline{RT}$
3. $\angle M$
4. $\overline{NM}$
5. $\angle R$
6. $\angle T$
7.
$$\begin{aligned} \overline{JK} &\cong \overline{FG} \\ JK &= FG \\ 3y - 15 &= 12 \\ 3y &= 27 \\ y &= 9 \\ KL &= y = 9 \end{aligned}$$
8.
$$\begin{aligned} \angle G &\cong \angle K \\ m\angle G &= m\angle K \\ 4x - 20 &= 108 \\ 4x &= 128 \\ x &= 32 \end{aligned}$$
9.

Statements	Reasons
1. $\overline{AB} \parallel \overline{CD}$	1. Given
2. $\angle ABE \cong \angle CDE$, $\angle BAE \cong \angle DCE$	2. Alt. Int. $\angle$ Thm.
3. $\overline{AB} \cong \overline{CD}$	3. Given
4. E is the mdpt. of $\overline{AC}$ and $\overline{BD}$	4. Given
5. $\overline{AE} \cong \overline{CE}$, $\overline{BE} \cong \overline{DE}$	5. Def. of mdpt.
6. $\angle AEB \cong \angle CED$	6. Vert. $\angle$ Thm
7. $\triangle ABE \cong \triangle CDE$	7. Def. of $\cong$ ⚠

10.

Statements	Reasons
1. $\angle UST \cong \angle RST$, $\angle U \cong \angle R$	1. Given
2. $\angle STU \cong \angle STR$	2. Third ⚠ Thm.
3. $\overline{SU} \cong \overline{SR}$	3. Given
4. $\overline{ST} \cong \overline{ST}$	4. Reflex. Prop. of $\cong$
5. $\overline{TU} \cong \overline{TR}$	5. Given
6. $\triangle RTS \cong \triangle UTS$	6. Def. of $\cong$ ⚠

Triangle Congruence

Review 4.1 / Triangles and Congruence

1. $m\angle BEA = 90°$; rt. $\triangle$

2. $m\angle BCD = 60 + 60 = 120°$; obtuse

3. $m\angle ABC = 30 + 30 = 60°$
$m\angle ABC = m\angle ACB = m\angle BAC$; equiangular

4. $\overline{PS} \cong \overline{ST} \cong \overline{PT}$; equilateral

5. $\overline{PS} \cong \overline{RS}$, so $PS = RS = 10$; $RP = 17$; isosc.

6. $RT = 10 + 10 = 20$, $RP = 17$, $PT = 10$; scalene

7. **Step 1** Find z.

$3z - 1 = z + 5$
$3z = z + 6$
$2z = 6$
$z = 3$

Step 2 Find side lengths.

$z + 5 = 3 + 5 = 8$
$3z - 1 = 3(3) - 1 = 8$
$4z - 4 = 4(3) - 4 = 8$

8. **Step 1** Find x.

$8x + 1.4 = 2x + 6.8$
$8x = 2x + 5.4$
$6x = 5.4$
$x = 0.9$

Step 2 Find side lengths.

$8x + 1.4 = 8(0.9) + 1.4$
$= 7.2 + 1.4$
$= 8.6$
$2x + 6.8 = 2(0.9) + 6.8$
$= 1.8 + 6.8$
$= 8.6$

9. a. Check students' drawings.
$\overline{XY}, \overline{YZ}, \overline{XZ}, \angle X, \angle Y, \angle Z$

b. Possible answer: scalene obtuse

10. $PQ + PR + QR = 60$

$PQ + PQ + \frac{4}{3}PQ = 60$

$\frac{10}{3}PQ = 60$

$PQ = \frac{3}{10}(60) = 18$ ft

$PR = PQ = 18$ ft

$QR = \frac{4}{3}PQ = \frac{4}{3}(18) = 24$ ft

11. $150 \div 60 = 2\frac{1}{2}$; 2 complete trusses

12. $m\angle A + m\angle B + m\angle P = 180$
$39 + 57 + m\angle P = 180$
$96 + m\angle P = 180$
$m\angle P = 84°$

13. $76\frac{1}{4} + m\angle = 90$
$m\angle = 13\frac{3}{4}°$

14. $2x + m\angle = 90$
$m\angle = (90 - 2x)°$

15. $56.8 + m\angle = 90$
$m\angle = 33.2°$

16. Think: Use Ext. $\angle$ Thm.
$m\angle W + m\angle X = m\angle XYZ$
$5x + 2 + 8x + 4 = 15x - 18$
$13x + 6 = 15x - 18$
$24 = 2x$
$x = 12$
$m\angle XYZ = 15x - 18$
$= 15(12) - 18 = 162°$

Triangle Congruence

Review 4.1 / Triangles and Congruence

17. Think: Use Ext. ∠ Thm and subst. $m\angle C = m\angle D$.

$m\angle C + m\angle D = m\angle ABD$
$2m\angle D = m\angle ABD$
$2(6x - 5) = 11x + 1$
$12x - 10 = 11x + 1$
$x = 11$
$m\angle C = m\angle D$
$= 6x - 5$
$= 6(11) - 5 = 61°$

18. Think: Use Third ∆ Thm.

$\angle N \cong \angle P$
$m\angle N = m\angle P$
$3y^2 = 12y^2 - 144$
$-9y^2 = -144$
$y^2 = 16$
$m\angle N = 3y^2 = 3(16) = 48°$
$m\angle P = m\angle N = 48°$

19. Think: Use Third ∆ Thm.

$\angle Q \cong \angle S$
$m\angle Q = m\angle S$
$2x^2 = 3x^2 - 64$
$64 = x^2$
$m\angle Q = 2x^2 = 2(64) = 128°$
$m\angle S = m\angle Q = 128°$

20. $\overline{LM}$

21. $\overline{CF}$

22. $\angle N$

23. $\angle D$

24. $\angle ADB \cong \angle CDB$

$m\angle ADB = m\angle CDB$
$4x + 10 = 90$
$4x = 80$
$x = 20$
$m\angle C = x + 11 = 31°$

25. $\overline{AB} \cong \overline{CB}$

$AB = CB$
$y - 7 = 12$
$y = 19$

26.

Statements	Reasons
1. $\angle N \cong \angle R$	1. Given
2. $\overline{MP}$ bisects $\angle NMR$	2. Given
3. $\angle NMP \cong \angle RMP$	3. Def. of ∠ bisector
4. $\angle NPM \cong \angle RPM$	4. Third ∆ Thm.
5. P is the mdpt. of $\overline{NR}$	5. Given
6. $\overline{PN} \cong \overline{PR}$	6. Def. of mdpt.
7. $\overline{MN} \cong \overline{MR}$	7. Given
8. $\overline{MP} \cong \overline{MP}$	8. Reflex. Prop. of ≅
9. $\triangle MNP \cong \triangle MRP$	9. Def. of ≅ ∆s

27.

Statements	Reasons
1. ∠ADC and ∠BCD are rt. ∠s	1. Given
2. ∠ADC ≅ ∠BCD	2. Rt. ∠ ≅ Thm.
3. ∠DAC ≅ ∠CBD	3. Given
4. ∠ACD ≅ ∠BDC	4. Third ∠s Thm.
5. $\overline{AC} \cong \overline{BD}$, $\overline{AD} \cong \overline{BC}$	5. Given
6. $\overline{DC} \cong \overline{DC}$	6. Reflex. Prop. of ≅
7. △ADC ≅ △BCD	7. Def. of ≅ △s

Triangle Congruence

Proving Triangle Congruence

Triangle Congruence: SSS and SAS

1. It is given that $\overline{DA} \cong \overline{BC}$ and $\overline{AB} \cong \overline{CD}$. $\overline{BD} \cong \overline{DB}$ by the Reflex. Prop. of $\cong$. Thus $\triangle ABD \cong \triangle CBD$ by SSS.
2. It is given that $\overline{MN} \cong \overline{MQ}$ and $\overline{NP} \cong \overline{QP}$. $\overline{MP} \cong \overline{MP}$ by the Relex. Prop. of $\cong$. Thus $\triangle MNP \cong \triangle MQP$ by SSS.
3. It is given that $\overline{JG} \cong \overline{LG}$, and $\overline{GK} \cong \overline{GH}$. $\angle JGK \cong \angle LGH$ by the Vert. ∡ Thm. So $\triangle JGK \cong \triangle LGH$ by SAS.
4. When $x = 4$, $HI = GH = 3$, and $IJ = GJ = 5$. $\overline{HJ} \cong \overline{HJ}$ by the Reflex. Prop. of $\cong$. Therefore, $\triangle GHJ \cong \triangle IHJ$ by SSS.
5. When $x = 18$, $RS = UT = 61$, and $m\angle SRT = m\angle UTR = 36°$. $\overline{RT} \cong \overline{TR}$ by the Reflex. Prop. of $\cong$. So $\triangle RST \cong \triangle TUR$ by SAS.
6.

Statements	Reasons
1. $\overline{JK} \cong \overline{ML}$	1. Given
2. $\angle JKL \cong \angle MLK$	2. Given
3. $\overline{KL} \cong \overline{LK}$	3. Reflex. Prop. of $\cong$
4. $\triangle JKL \cong \triangle MLK$	4. SAS Steps 1, 2, 3

Triangle Congruence

Proving Triangle Congruence

Triangle Congruence: ASA, AAS, and HL

1. Yes; by the Def. of $\angle$ bisector, $\angle TSV \cong \angle RSV$ and $\angle TVS \cong \angle RVS$. $\overline{SV} \cong \overline{SV}$ by the Reflex. Prop. of $\cong$. So $\triangle VRS \cong \triangle VTS$ by ASA.
2. No; you need to know that a pair of corr. sides are $\cong$.
3. **a.** $\overline{QS} \cong \overline{SQ}$ **b.** $\angle RQS \cong \angle PSQ$

   c. Rt. $\angle \cong$ Thm. **d.** AAS
4. Yes; it is given that $\angle D$ and $\angle B$ are rt. $\angle$s and $\overline{AD} \cong \overline{BC}$. $\triangle ABC$ and $\triangle CDA$ are rt. $\triangle$s by def. $\overline{AC} \cong \overline{CA}$ by the Reflex. Prop. of $\cong$. So $\triangle ABC \cong \triangle CDA$ by HL.
5. No; you need to know that $\overline{VX} \cong \overline{VZ}$.

Triangle Congruence

Proving Triangle Congruence

Triangle Congruence: CPCTC

1. $\angle BCA \cong \angle DCE$ by Vert. $\measuredangle$ Thm, $\angle CBA \cong \angle CDE$ by Rt. $\angle \cong$ Thm., and $\overline{BC} \cong \overline{DC}$ (given). Therefore $\triangle ABC \cong \triangle EDC$ by ASA. By CPCTC, $\overline{AB} \cong \overline{DE}$, so $AB = DE = 6.3$ m.

2. **a.** Def. of $\perp$ **b.** Rt. $\angle \cong$ Thm. **c.** Reflex. Prop. of $\cong$ **d.** Def. of mdpt. **e.** $\triangle RXS \cong \triangle RXT$ **f.** CPCTC

3.

Statements	Reasons
1. $\overline{AC} \cong \overline{AD}$, $\overline{CB} \cong \overline{DB}$	1. Given
2. $\overline{AB} \cong \overline{AB}$	2. Reflex. Prop. of $\cong$
3. $\triangle ACB \cong \triangle ADB$	3. SSS Steps 1, 2
4. $\angle CAB \cong \angle DAB$	4. CPCTC
5. $\overline{AB}$ bisects $\angle CAD$.	5. Def. of $\angle$ bisector

4. Use Distance Formula to find side lengths.

$EF = \sqrt{((-1) - (-3))^2 + (3 - 3)^2}$
$= \sqrt{4 + 0} = 2$

$FG = \sqrt{((-2) - (-3))^2 + (0 - 3)^2}$
$= \sqrt{1 + 9} = \sqrt{10}$

$EG = \sqrt{((-2) - (-1))^2 + (0 - 3)^2}$
$= \sqrt{1 + 9} = \sqrt{10}$

$JK = \sqrt{(0 - 2)^2 + ((-1) - (-1))^2}$
$= \sqrt{4 + 0} = 2$

$KL = \sqrt{(1 - 2)^2 + (2 - (-1))^2}$
$= \sqrt{1 + 9} = \sqrt{10}$

$JL = \sqrt{(1 - 0)^2 + (2 - (-1))^2}$
$= \sqrt{1 + 9} = \sqrt{10}$

So $\overline{EF} \cong \overline{JK}$, $\overline{FG} \cong \overline{KL}$, and $\overline{EG} \cong \overline{JL}$. Therefore $\triangle EFG \cong \triangle JKL$ by SSS, and $\angle EFG \cong \angle JKL$ by CPCTC.

5. Use Distance Formula to find side lengths.

$AB = \sqrt{(4 - 2)^2 + (1 - 3)^2}$
$= \sqrt{4 + 4} = 2\sqrt{2}$

$BC = \sqrt{(1 - 4)^2 + ((-1) - 1)^2}$
$= \sqrt{9 + 4} = \sqrt{13}$

$AC = \sqrt{(1 - 2)^2 + ((-1) - 3)^2}$
$= \sqrt{1 + 16} = \sqrt{17}$

$RS = \sqrt{((-3) - (-1))^2 + ((-2) - 0)^2}$
$= \sqrt{4 + 4} = 2\sqrt{2}$

$ST = \sqrt{(0 - (-3))^2 + ((-4) - (-2))^2}$
$= \sqrt{9 + 4} = \sqrt{13}$

$RT = \sqrt{(0 - (-1))^2 + ((-4) - 0)^2}$
$= \sqrt{1 + 16} = \sqrt{17}$

So $\overline{AB} \cong \overline{RS}$, $\overline{BC} \cong \overline{ST}$, and $\overline{AC} \cong \overline{RT}$. Therefore $\triangle ABC \cong \triangle RST$ by SSS, and $\angle ACB \cong \angle RTS$ by CPCTC.

Triangle Congruence

Proving Triangle Congruence

Introduction to Coordinate Proof

1. 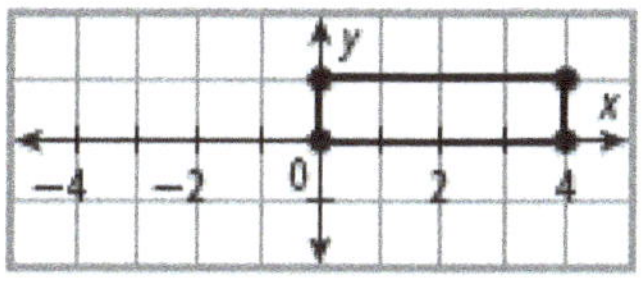

2. 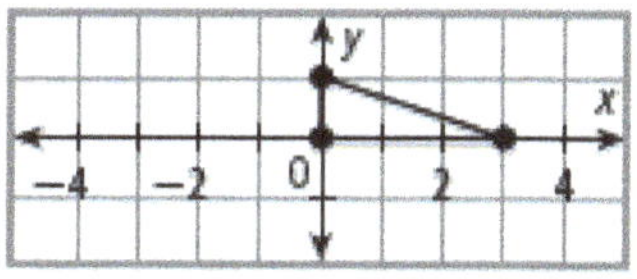

3. By the Mdpt. Forumla, the coords of *A* are (0, 3) and the coords. of *B* are (4, 0).
By the Dist. Formula,

$$PQ = \sqrt{(0-8)^2 + (6-0)^2}$$
$$= \sqrt{(-8)^2 + 6^2} = \sqrt{64 + 36}$$
$$= \sqrt{100} = 10 \text{ units.}$$
$$AB = \sqrt{(0-4)^2 + (3-0)^2}$$
$$= \sqrt{(-4)^2 + 3^2} = \sqrt{16 + 9}$$
$$= \sqrt{25} = 5 \text{ units.}$$

So $AB = \frac{1}{2}PQ$.

4. Possible answer:

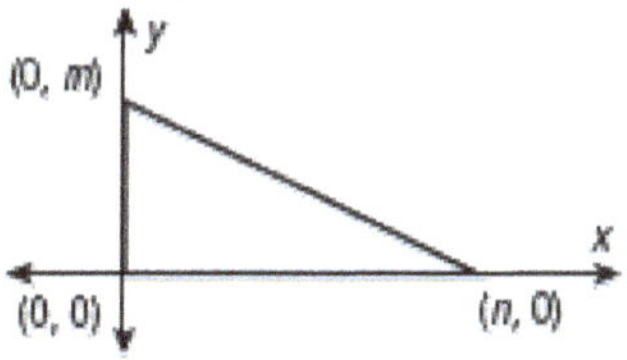

5. Possible answer:

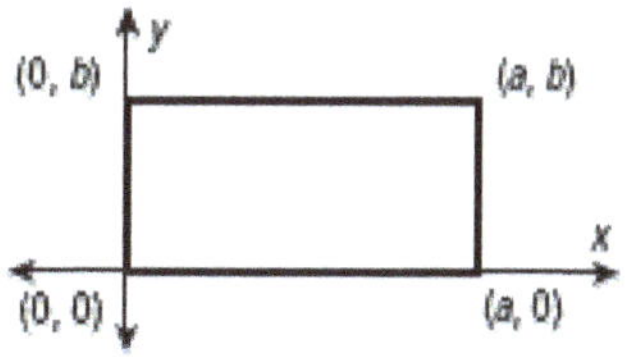

6. 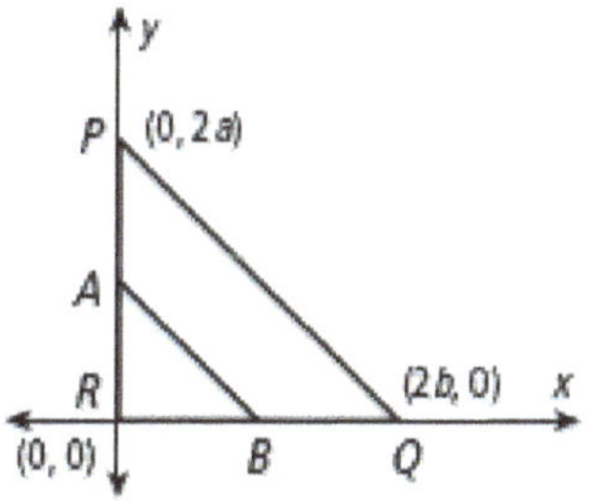

By the Mdpt. Formula, the coords. of *A* are (0, *a*) and the coords of *B* are (*b*, 0).
By the Dist. Formula,

$$PQ = \sqrt{(0-2b)^2 + (2a)^2} \quad AB = \sqrt{(0-b)^2 + (a-0)^2}$$
$$= \sqrt{(-2b)^2 + (2a)^2} \quad = \sqrt{(-b)^2 + a^2}$$
$$= \sqrt{4b^2 + 4a^2} \quad = \sqrt{b^2 + a^2} \text{ units}$$
$$= 2\sqrt{b^2 + a^2} \text{ units}$$

So $AB = \frac{1}{2}PQ$.

Triangle Congruence

Proving Triangle Congruence

Isosceles and Equilateral Triangles

1.

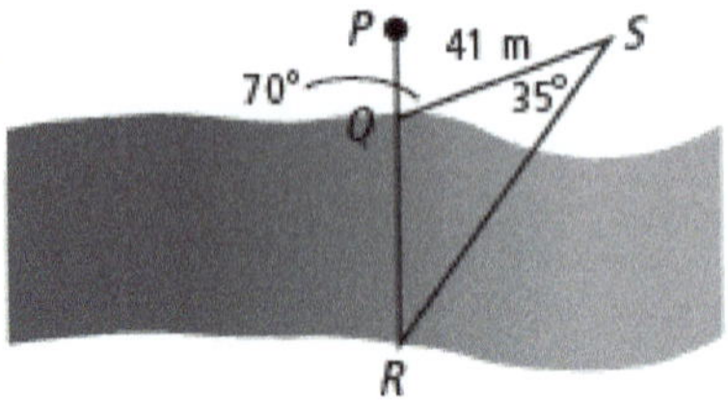

By the Ext. ∠ Thm., $m\angle R = 35°$. Since $m\angle R = m\angle S$ by the Conv. of the Isosc. △ Thm., $QR = QS = 41$ m.

2. Think: Use Isosc. △ Thm., △ ∠ Sum Thm., and Vert. ∠ Thm.

$m\angle B = m\angle A = 31°$

$m\angle A + m\angle B + m\angle ABC = 180$

$31 + 31 + m\angle ABC = 180$

$m\angle ABC = 118°$

$m\angle ECD = m\angle ABC = 118°$

3. Think: Use Isosc. △ Thm. and △ ∠ Sum Thm.

$m\angle J = m\angle K$

$m\angle J + m\angle K + m\angle L = 180$

$2m\angle K + 82 = 180$

$2m\angle K = 98$

$m\angle K = 49°$

4. Think: Use Isosc. △ Thm.

$m\angle X = m\angle Y$

$5t - 13 = 3t + 3$

$2t = 16$

$t = 8$

$m\angle X = 5t - 13 = 27°$

5. Think: Use Isosc. △ Thm. and △ ∠ Sum Thm.

$m\angle B = m\angle C = 2x$

$m\angle A + m\angle B + m\angle C = 180$

$4x + 2x + 2x = 180$

$8x = 180$

$x = 22.5$

$m\angle A = 4x = 90°$

6. Think: Use Equilat. △ Thm. and △ ∠ Sum Thm.

$\angle R \cong \angle S \cong \angle T$

$m\angle R + m\angle S + m\angle T = 180$

$12y + 12y + 12y = 180$

$36y = 180$

$y = 5$

7. Think: Use Equilat. △ Thm. and △ ∠ Sum Thm.

$\angle L \cong \angle M \cong \angle N$

$m\angle L + m\angle M + m\angle N = 180$

$3(10x + 20) = 180$

$30x = 120$

$x = 4$

8. Think: Use Equiang. △ Thm.

$\overline{AB} \cong \overline{BC} \cong \overline{AC}$

$BC = AC$

$6y + 2 = -y + 23$

$7y = 21$

$y = 3$

$BC = 6y + 2$

$= 6(3) + 2 = 20$

9. Think: Use Equiang. △ Thm.

$\overline{HJ} \cong \overline{JK} \cong \overline{HK}$

$HJ = JK$

$7t + 15 = 10t$

$15 = 3t$

$t = 5$

$JK = 10t$

$= 10(5) = 50$

10. Proof:

It is given that $\triangle ABC$ is rt. isosc., $\overline{AB} \cong \overline{BC}$, and X is the mdpt. of $\overline{AC}$. By Mdpt. Formula, coords. of X are $\left(\frac{0 + 2a}{2}, \frac{2a + 0}{2}\right) = (a, a)$. By Dist. Formula, $AX = \sqrt{(a-0)^2 + (a-2a)^2} = a\sqrt{2}$ and $BX = \sqrt{(a-0)^2 + (a-0)^2} = a\sqrt{2} = AX$. So $\triangle AXB$ is isosc. by def. of an isosc. △.

Triangle Congruence

Review 4.2 / Proving Triangle Congruence

1. It is given that $BC = ED = 4$ in. and $BD = EC = 3$ in. So by the def. of ≅, $\overline{BC} \cong \overline{ED}$, and $\overline{BD} \cong \overline{EC}$. $\overline{DC} \cong \overline{CD}$ by the Reflex. Prop. of ≅. Thus $\triangle BCD \cong \triangle EDC$ by SSS.

2. It is given that $\overline{KJ} \cong \overline{LJ}$ and $\overline{GK} \cong \overline{GL}$. $\overline{GJ} \cong \overline{GJ}$ by the Reflex. Prop. of ≅. So $\triangle GJK \cong \triangle GJL$ by SSS.

3. It is given that $\angle C$ and $\angle B$ are rt. ∠s and $\overline{EC} \cong \overline{DB}$. $\angle C \cong \angle B$ by the Rt. ∠ ≅ Thm. $\overline{CB} \cong \overline{BC}$ by the Reflex. Prop. of ≅. So $\triangle ECB \cong \triangle DBC$ by SAS.

4. When $y = 3$, $NQ = NM = 3$, and $QP = MP = 4$. So by the def. of ≅, $\overline{NQ} \cong \overline{NM}$, and $\overline{QP} \cong \overline{MP}$. $m\angle M = m\angle Q = 90^\circ$, so $\angle M \cong \angle Q$ by the def. of ≅. Thus $\triangle MNP \cong \triangle QNP$ by SAS.

5. When $t = 5$, $YZ = 24$, $ST = 20$, and $SU = 22$. So by the def. of ≅, $\overline{XY} \cong \overline{ST}$, $\overline{YZ} \cong \overline{TU}$, and $\overline{XZ} \cong \overline{SU}$. This $\triangle XYZ \cong \triangle STU$ by SSS.

6.

Statements	Reasons
1. B is mdpt. of $\overline{DC}$	1. Given
2. $\overline{DB} \cong \overline{CB}$	2. Def. of mdpt.
3. $\overline{AB} \perp \overline{DC}$	3. Given
4. $\angle ABD$ and $\angle ABC$ are rt. ∠s	4. Def. of ⊥
5. $\angle ABD \cong \angle ABC$	5. Rt. ∠ ≅ Thm.
6. $\overline{AB} \cong \overline{AB}$	6. Reflex. Prop. of ≅
7. $\triangle ABD \cong \triangle ABC$	7. SAS Steps 2, 5, 6

7. No; you need to know that $\angle MKJ \cong \angle MKL$.

8. Yes; by the Alt. Int. ∠s Thm., $\angle SRT \cong \angle UTR$, and $\angle STR \cong \angle URT$. $\overline{RT} \cong \overline{TR}$ by the Reflex. Prop. of ≅. So $\triangle RST \cong \triangle TUR$ by ASA.

9. a. $\angle A \cong \angle D$ b. Given
   c. $\angle C \cong \angle F$ d. AAS

10. No; you need to know that $\angle K$ and $\angle H$ are rt. ∠s.

11. Yes; E is a mdpt. So by def., $\overline{BE} \cong \overline{CE}$, and $\overline{AE} \cong \overline{DE}$. $\angle A$ and $\angle D$ are ≅ by the Rt. ∠ ≅ Thm. By def., $\triangle ABE$ and $\triangle DCE$ are rt. △s. So $\triangle ABE \cong \triangle DCE$ by HL.

12. $\angle ABC \cong \angle EDC$ by Rt. ∠ ≅ Thm., $\angle ACB \cong \angle ECD$ by Vert. ∠s Thm., and $\overline{BC} \cong \overline{DC}$. So $\triangle ABC \cong \triangle EDC$ by ASA. By CPCTC, $AB = DE = 420$ ft.

13.

Statements	Reasons
1. M is mdpt. of $\overline{PQ}$ and $\overline{RS}$.	1. Given
2. $\overline{PM} \cong \overline{QM}$, $\overline{RM} \cong \overline{SM}$	2. Def. of mdpt.
3. $\angle PMS \cong \angle QMR$	3. Vert. ∠s Thm.
4. $\triangle PMS \cong \triangle QMR$	4. SAS Steps 2, 3
5. $\overline{QR} \cong \overline{PS}$	5. CPCTC

14.

Statements	Reasons
1. $\overline{WX} \cong \overline{XY} \cong \overline{YZ} \cong \overline{ZW}$	1. Given
2. $\overline{ZX} \cong \overline{XZ}$	2. Reflex. Prop. of ≅
3. $\triangle WXZ \cong \triangle YZX$	3. SSS, steps 1, 2
4. $\angle W \cong \angle Y$	4. CPCTC

15.

Statements	Reasons
1. G is mdpt. of $\overline{FH}$.	1. Given
2. $FG = HG$	2. Def. of mdpt.
3. $\overline{FG} \cong \overline{HG}$	3. Def. of ≅.
4. Draw $\overline{EG}$.	4. Exactly 1 line through any 2 pts.
5. $\overline{EG} \cong \overline{EG}$	5. Reflex. Prop. of ≅
6. $\overline{EF} \cong \overline{EH}$	6. Given
7. $\triangle EGF \cong \triangle EGH$	7. SSS Steps 3, 5, 6
8. $\angle EFG \cong \angle EHG$	8. CPCTC
9. $\angle 1 \cong \angle 2$	9. ≅ Supp. Thm.

16.

Statements	Reasons
1. $\overline{LM}$ bisects $\angle JLK$.	1. Given
2. $\angle JLM \cong \angle KLM$	2. Def. of ∠ bisector
3. $\overline{JL} \cong \overline{KL}$	3. Given
4. $\overline{LM} \cong \overline{LM}$	4. Reflex. Prop. of ≅
5. $\triangle JLM \cong \triangle KLM$	5. SAS Steps 3, 2, 4
6. $\overline{JM} \cong \overline{KM}$	6. CPCTC
7. M is mdpt. of $\overline{JK}$.	7. Def. of mdpt.

17. $RS = \sqrt{(2-0)^2 + (4-0)^2}$
$= \sqrt{4+16} = 2\sqrt{5}$
$ST = \sqrt{((-1)-2)^2 + (4-3)^2}$
$= \sqrt{9+1} = \sqrt{10}$
$RT = \sqrt{((-1)-0)^2 + (3-0)^2}$
$= \sqrt{1+9} = \sqrt{10}$
$UV = \sqrt{((-3)-(-1))^2 + ((-4)-0)^2}$
$= \sqrt{4+16} = 2\sqrt{5}$
$VW = \sqrt{((-4)-(-3))^2 + ((-1)-(-4))^2}$
$= \sqrt{1+9} = \sqrt{10}$
$UW = \sqrt{((-4)-(-1))^2 + ((-1)-0)^2}$
$= \sqrt{9+1} = \sqrt{10}$
So $\overline{RS} \cong \overline{UV}$, $\overline{ST} \cong \overline{VW}$, and $\overline{RT} \cong \overline{UW}$. Therefore, $\triangle RST \cong \triangle UVW$ by SSS, and $\angle RST \cong \angle UVW$ by CPCTC.

18. $AB = \sqrt{(2-(-1))^2 + (3-1)^2}$
$= \sqrt{9+4} = \sqrt{13}$
$BC = \sqrt{(2-2)^2 + ((-2)-3)^2}$
$= \sqrt{0+25} = 5$
$AC = \sqrt{(2-(-1))^2 + ((-2)-1)^2}$
$= \sqrt{9+9} = 3\sqrt{2}$
$DE = \sqrt{((-1)-2)^2 + ((-5)-(-3))^2}$
$= \sqrt{9+4} = \sqrt{13}$
$EF = \sqrt{((-1)-(-1))^2 + (0-(-5))^2}$
$= \sqrt{0+25} = 5$
$DF = \sqrt{((-1)-2)^2 + (0-(-3))^2}$
$= \sqrt{9+9} = 3\sqrt{2}$
So $\overline{AB} \cong \overline{DE}$, $\overline{BC} \cong \overline{EF}$, and $\overline{CA} \cong \overline{DF}$. Therefore, $\triangle ABC \cong \triangle DEF$ by SSS, and $\angle BAC \cong \angle EDF$ by CPCTC.

19. Possible answer:

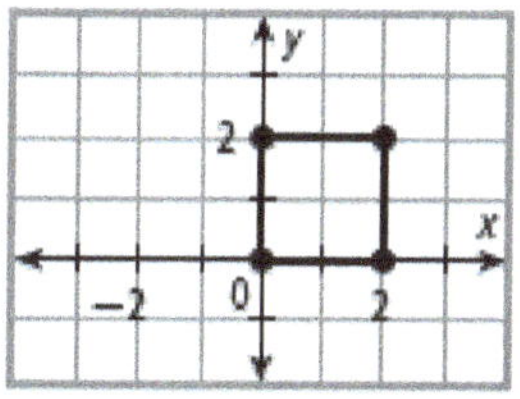

20. Possible answer:

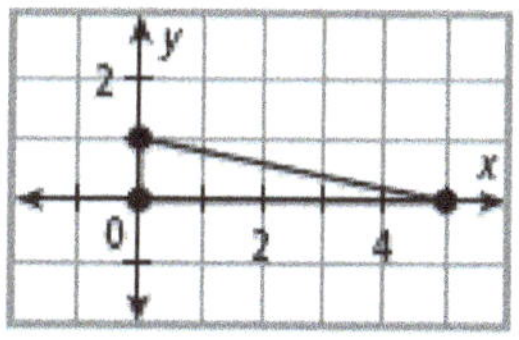

21.

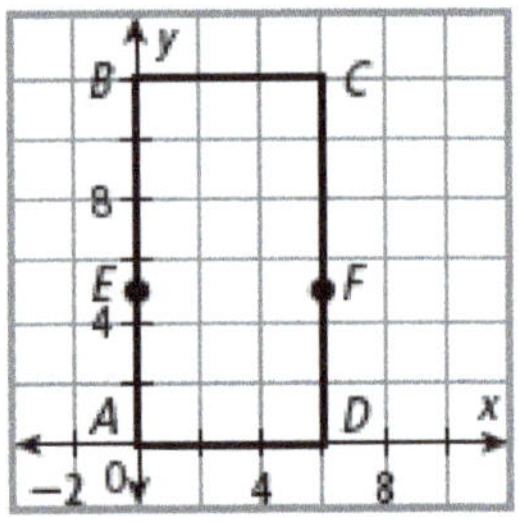

$E = \left(\frac{0+0}{2}, \frac{0+10}{2}\right) = (0, 5)$

$F = \left(\frac{6+6}{2}, \frac{0+10}{2}\right) = (6, 5)$

$BC = \sqrt{(6-0)^2 + (10-10)^2} = 6$ units.

$EF = \sqrt{(6-0)^2 + (5-5)^2} = 6$ units.

So $EF = BC$.

22. Possible answer:

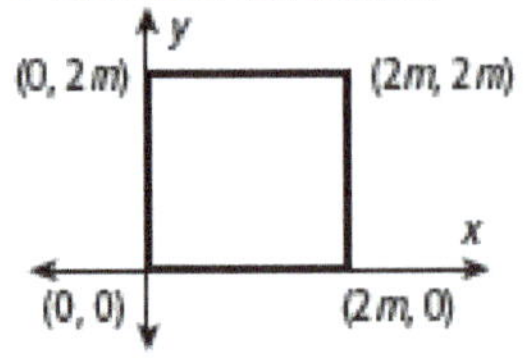

23. Possible answer:

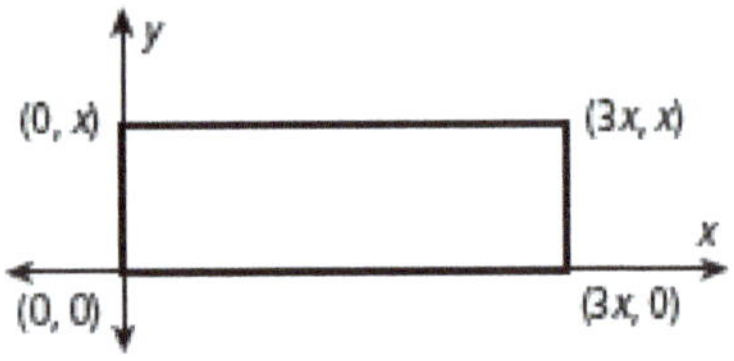

24.

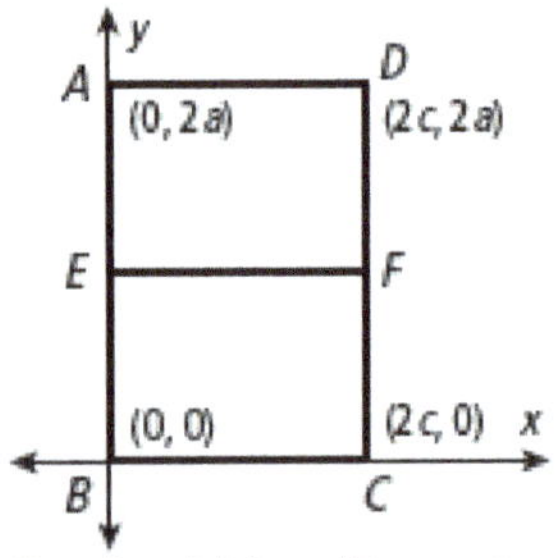

By the Mdpt. Formula, the coords. of E are $(0, a)$ and the coords of F are $(2c, a)$.
By the Dist. Formula,

$AD = \sqrt{(2c-0)^2 + (2a-2a)^2}$
$= \sqrt{(2c)^2} = 2c$ units.

Similarly,

$EF = \sqrt{(2c-0)^2 + (a-a)^2}$
$= \sqrt{(2c)^2} = 2c$ units.

So $EF = AD$.

25. By ∠ Add. Post., $m\angle ATB = 80 - 40 = 40°$. $m\angle BAT = 40°$ by Alt. Int. ∠s Thm. $\angle ATB \cong \angle BAT$ by def. of ≅. Since $\triangle ABT$ is isosc. by Conv. of Isosc. △ Thm., $BT = BA = 2.4$ mi.

26. Think: use Isosc. △ Thm., △ ∠ Sum Thm., and Vert. ∠s Thm.

$m\angle B = m\angle ACB$

$m\angle A + m\angle B + m\angle ACB = 180$

$96 + 2m\angle ACB = 180$

$m\angle ACB = 42°$

$m\angle DCE = m\angle ACB = 42°$

$m\angle D = m\angle E$

$m\angle D + m\angle E + m\angle DCE = 180$

$2m\angle E + 42 = 180$

$m\angle E = 69°$

27. Think: Use Isosc. △ Thm. and △ ∠ Sum Thm.

$m\angle U = m\angle S = 57°$

$m\angle SRU + m\angle S + m\angle U = 180$

$m\angle SRT + m\angle TRU + 57 + 57 = 180$

$2m\angle TRU = 66$

$m\angle TRU = 33°$

28. $m\angle D = m\angle E$

$x^2 = 3x + 10$

$x^2 - 3x - 10 = 0$

$(x - 5)(x + 2) = 0$

$x = 5 \text{ or } -2$

$m\angle D + m\angle E + m\angle F = 180$

$x^2 + 3x + 10 + m\angle F = 180$

$m\angle F = 180 - 50 \text{ or } 180 - 8$

$= 130° \text{ or } 172°$

29. Think: Use Isosc. △ Thm. and △ ∠ Sum Thm.

$m\angle A = m\angle B = (6y + 1)°$

$m\angle A + m\angle B + m\angle C = 180$

$2(6y + 1) + 21y + 3 = 180$

$33y = 165$

$y = 5°$

$m\angle A = 6y + 1 = 31°$

30. Think: Use Equilat. △ Thm. and △ ∠ Sum Thm.

$\angle F \cong \angle G \cong \angle H$

$m\angle F + m\angle G + m\angle H = 180$

$3\left(\frac{z}{2} + 14\right) = 180$

$z + 28 = 120$

$z = 92$

31. Think: Use Equilat. △ Thm. and △ ∠ Sum Thm.

$\angle L \cong \angle M \cong \angle N$

$m\angle L + m\angle M + m\angle N = 180$

$3(1.5y - 12) = 180$

$y - 8 = 40$

$y = 48$

32. Think:
use Equiang. △ Thm.

$\overline{BC} \cong \overline{CD} \cong \overline{BD}$

$BC = CD$

$\frac{3}{2}x + 2 = \frac{5}{4}x + 6$

$6x + 8 = 5x + 24$

$x = 16$

$BC = \frac{3}{2}x + 2$

$= \frac{3}{2}(16) + 2 = 26$

33. Think:
use Equiang. △ Thm.

$\overline{XY} \cong \overline{YZ} \cong \overline{XZ}$

$XY = XZ$

$2x = \frac{5}{2}x - 5$

$5 = \frac{1}{2}x$

$x = 10$

$XZ = XY$

$= 2x$

$= 2(10) = 20$

34. **Proof:** It is given that $\triangle ABC$ is isosc., $\overline{AB} \cong \overline{AC}$, P is mdpt. of $\overline{AB}$, and Q is mdpt. of $\overline{AC}$. By Mdpt. Formula, coords. of P are (a, b), and coords. of Q are $(3a, b)$. By Dist. Formula, $PC = QB = \sqrt{9a^2 + b^2}$, so $\overline{PC} \cong \overline{QB}$ by def. of ≅.

Properties and Attributes of Triangles

Segments in Triangles

Perpendicular and Angle Bisector Theorems

1. Since $PS = QS$ and $m \perp \overline{PQ}$, m is the $\perp$ bisector of $\overline{PQ}$ by the Conv. of the $\perp$ Bisector Thm.
   $PQ = 2QT$
   $PQ = 2(47.7) = 95.4$

2. $SP = SQ$
   $SP = 25.9$

3. $PS = QS$
   $4a = 2a + 26$
   $2a = 26$
   $a = 13$
   So $QS = 2(13) + 26 = 52$.

4. $AD = CD$
   $AD = 21.9$

5. Since $AD = CD$, $\overline{AC} \perp \overline{AB}$, and $\overline{CD} \perp \overline{BC}$, $\overrightarrow{BD}$ bisects $\angle ABC$ by the Conv. of the $\angle$ Bisector Thm.
   $m\angle CBD = \frac{1}{2}m\angle ABC$
   $m\angle CBD = \frac{1}{2}(48°) = 24°$

6. Since $DA = DC$, $\overline{AD} \perp \overline{AB}$, and $\overline{CD} \perp \overline{BC}$, $\overrightarrow{BD}$ bisects $\angle ABC$ by the Conv. of the $\angle$ Bisector Thm.
   $m\angle DBC = m\angle DBA$
   $10y + 3 = 8y + 10$
   $2y + 3 = 10$
   $2y = 7$
   $y = \frac{7}{2}$
   So $m\angle DBC = [10\left(\frac{7}{2}\right) + 3]° = 38°$

7. The braces can be installed so that $\overline{PK} \perp \overline{JL}$, $\overline{PM} \perp \overline{NL}$, and $PK = PM$. Then by the Conv. of the $\angle$ Bisector Thm., P will be on the bisector of $\angle JLN$.

8. 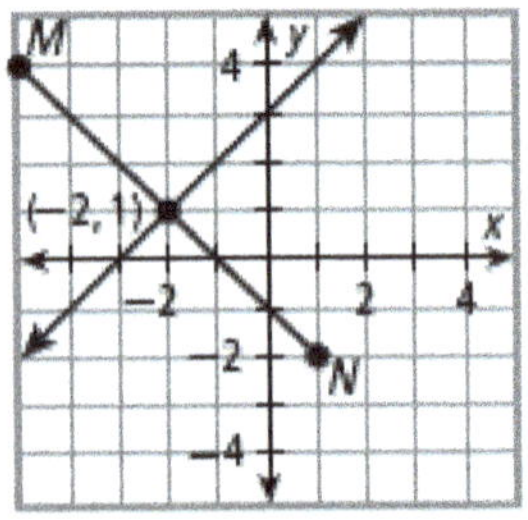

Step 1 Graph $\overline{MN}$.
The $\perp$ bisector of $\overline{MN}$ is $\perp$ to $\overline{MN}$ at its midpoint.

Step 2 Find the midpoint of $\overline{MN}$.

midpoint of $\overline{MN} = \left(\frac{-5+1}{2}, \frac{4+(-2)}{2}\right) = (-2, 1)$

Step 3 Find the slope of the perpendicular bisector.

slope of $\overline{MN} = \frac{-2-4}{1-(-5)} = \frac{-6}{6} = -1$

Since the slopes of $\perp$ lines are opposite reciprocals, the slope of the $\perp$ bisector is 1.

Step 4 Use point-slope form to write an equation.
The $\perp$ bisector of $\overline{MN}$ has slope 1 and passes through $(-2, 1)$.

$y - y_1 = m(x - x_1)$
$y - 1 = 1[x - (-2)]$
$y - 1 = x + 2$

Perpendicular and Angle Bisector Theorems

9.

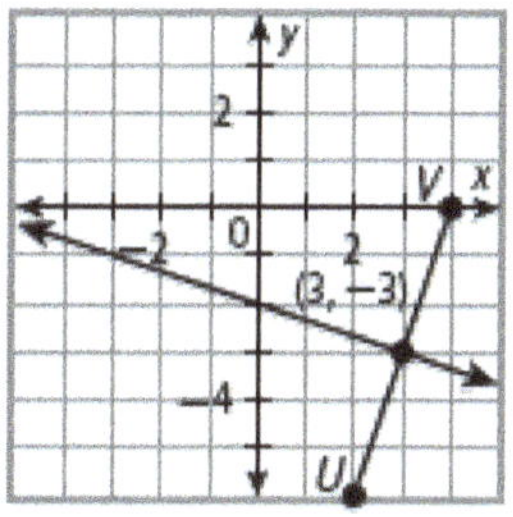

Step 1 Graph $\overline{UV}$.

The ⊥ bisector of $\overline{UV}$ is ⊥ to $\overline{UV}$ at its midpoint

Step 2 Find the midpoint of $\overline{UV}$.

midpoint of $\overline{UV} = \left(\frac{2+4}{2}, \frac{-6+0}{2}\right) = (3, -3)$

Step 3 Find the slope of the perpendicular bisector.

slope of $\overline{UV} = \frac{0-(-6)}{4-2} = \frac{6}{2} = 3$

Since the slopes of ⊥ lines are opposite reciprocals, the slope of the ⊥ bisector is $-\frac{1}{3}$.

Step 4 Use point-slope form to write an equation.

The ⊥ bisector of $\overline{UV}$ has slope $-\frac{1}{3}$ and passes through (3, −3).

$$y - y_1 = m(x - x_1)$$
$$y - (-3) = -\frac{1}{3}(x - 3)$$
$$y + 3 = -\frac{1}{3}(x - 3)$$

10.

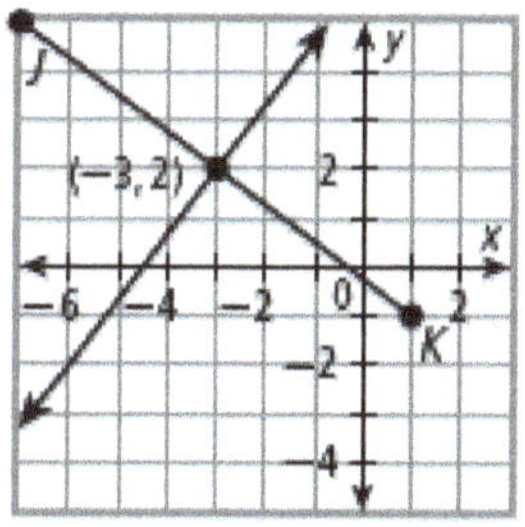

Step 1 Graph $\overline{JK}$.

The ⊥ bisector of $\overline{JK}$ is ⊥ to $\overline{JK}$ at its midpoint

Step 2 Find the midpoint of $\overline{JK}$.

midpoint of $\overline{JK} = \left(\frac{-7+1}{2}, \frac{5+(-1)}{2}\right) = (-3, 2)$

Step 3 Find the slope of the perpendicular bisector.

slope of $\overline{JK} = \frac{-1-5}{1-(-7)} = \frac{-6}{8} = -\frac{3}{4}$

Since the slopes of ⊥ lines are opposite reciprocals, the slope of the ⊥ bisector is $\frac{4}{3}$.

Step 4 Use point-slope form to write an equation.

The ⊥ bisector of $\overline{JK}$ has slope $\frac{4}{3}$ and passes through (−3, 2).

$$y - y_1 = m(x - x_1)$$
$$y - 2 = \frac{4}{3}[x - (-3)]$$
$$y - 2 = \frac{4}{3}(x + 3)$$

Properties and Attributes of Triangles

Segments in Triangles

Medians, Altitudes, and Midsegments in Triangles

1. Since $\overline{BD}$ is a median, B is the midpoint of $\overline{AC}$. Therefore, $AB = BC$ and $AB = AC/2$.
   Since $AC = 10$,
   $AB = 10/2 = 5$.

2. Since $x = 6$,
   $AB = 2(6) = 12$ and
   $BC = 3(6) - 6 = 12$.
   So, $AB = BC$. Therefore, B is the midpoint of $\overline{AC}$ and so $\overline{BD}$ is a median.

3. Since $\overline{BD}$ is a median, B is the midpoint of $\overline{AC}$. Therefore, $AB = BC$ and $AB = AC/2$. If $AB = AC/2$, then $2AB = AC$. Since $AB = x + 2$ and $AC = 3x - 1$, $2(x + 2) = 3x - 1$. Solve.
   $$2(x+2) = 3x - 1$$
   $$2x + 4 = 3x - 1$$
   $$5 = x$$

4. Plot the vertices of ΔDEF. The median through E passes through the midpoint of the opposite side, $\overline{DF}$. So, find the midpoint of $\overline{DF}$, which is the midpoint of the line through $D(•4, •3)$ and $F(2, 1)$.
   $$M = \left(\frac{-4+2}{2}, \frac{-3+1}{2}\right) = \left(\frac{-2}{2}, \frac{-2}{2}\right) = (-1, -1)$$
   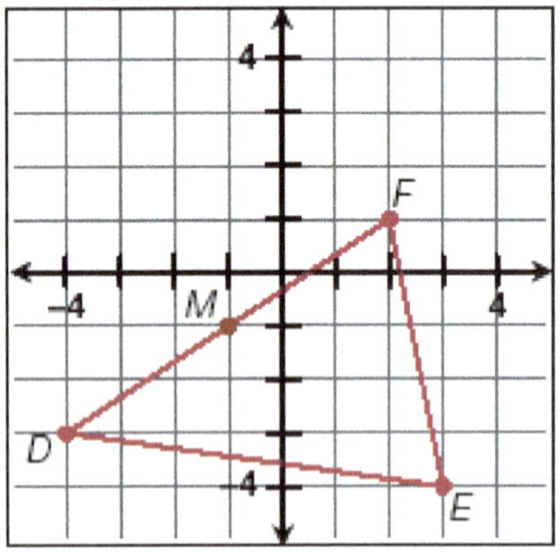

   Therefore, the median through E passes through $E(3, -4)$ and $M(-1, -1)$. Use these two points to find the equation of the line. First, find the slope.
   $$m = \frac{-1-(-4)}{-1-3} = \frac{3}{-4} = -\frac{3}{4}$$
   Substitute the slope and one of the points into point-slope form and solve for y to write the equation of the line in slope-intercept form.
   $$y - (-4) = -\frac{3}{4}(x-3)$$
   $$y + 4 = -\frac{3}{4}x + \frac{9}{4}$$
   $$y = -\frac{3}{4}x - \frac{7}{4}$$

5. Plot the vertices of ΔDEF. The altitude through D must be perpendicular to the opposite side, $\overline{EF}$. Find the slope of the line perpendicular to $\overline{EF}$ by first finding the slope of $\overline{EF}$ and then finding the opposite reciprocal of that slope.
   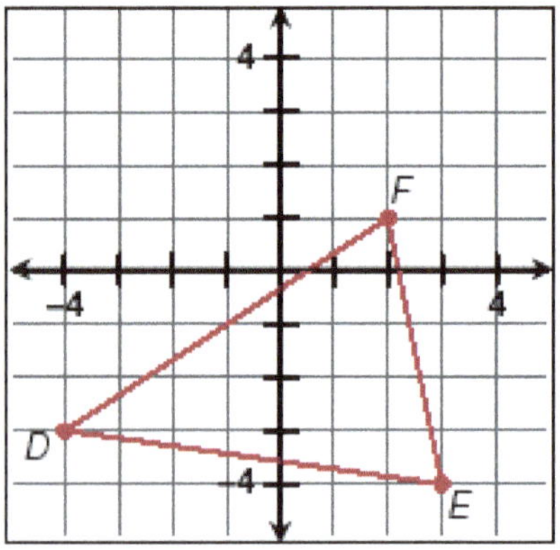

   The slope of $\overline{EF}$ is –5. So, the slope of the line perpendicular to $\overline{EF}$ is 1/5. Substitute the slope and the coordinates of D into point-slope form and solve for y to write the equation of the line in slope-intercept form.
   $$y - (-3) = \frac{1}{5}(x - (-4))$$
   $$y + 3 = \frac{1}{5}(x+4)$$
   $$y + 3 = \frac{1}{5}x + \frac{4}{5}$$
   $$y = \frac{1}{5}x - \frac{11}{5}$$

Medians, Altitudes, and Midsegments in Triangles

6. $NM = \frac{1}{2}XY$
$= \frac{1}{2}(10.2) = 5.1$

7. $XZ = 2LM$
$= 2(5.6) = 11.2$

8. $NZ = \frac{1}{2}XZ$
$= \frac{1}{2}(11.2) = 5.6$

9. $m\angle LMN = m\angle MNZ = 29°$

10. $m\angle YXZ = m\angle MNZ = 29°$

11. $m\angle XLM + m\angle LMN = 180$
$m\angle XLM + 29 = 180$
$m\angle XLM = 151°$

Properties and Attributes of Triangles

Review 5.1 / Segments in Triangles

1. $GJ = GK$
   $GJ = 8.25$

2. $JG = KG$
   $x + 12 = 3x - 17$
   $12 = 2x - 17$
   $29 = 2x$
   $14.5 = x$
   So $KG = 3(14.5) - 17 = 26.5$.

3. Since $GJ = GK$ and $t \perp \overline{JK}$, t is the $\perp$ bisector of $\overline{JK}$ by the Conv. of the $\perp$ Bisector Thm.
   $JK = 2JH$
   $JK = 2(26.5) = 53$

4. $RQ = TQ$
   $RQ = 1.3$

5. Since $RQ = TQ$, $\overline{RQ} \perp \overline{RS}$, and $\overline{TQ} \perp \overline{TS}$, $\overrightarrow{SQ}$ bisects $\angle RST$ by the Conv. of the Bisector Thm.
   $m\angle RST = 2m\angle RSQ$
   $m\angle RST = 2(58°) = 116°$

6. $m\angle QSR = m\angle QST$
   $9a + 48 = 6a + 50$
   $3a + 48 = 50$
   $3a = 2$
   $m\angle QST = 6\left(\frac{2}{3}\right) + 50 = 54°$

7. They can position Main St. so that the $\angle$ formed by Elm St. and Main St. is $\cong$ to the $\angle$ formed by Grove St. and Main St. Then by the $\angle$ Bisector Thm., every point on Main St. will be equidistant from Elm St. and Grove St.

8. **Step 1** Graph $\overline{EF}$.
   The $\perp$ bisector of $\overline{EF}$ is $\perp$ to $\overline{EF}$ at its midpoint
   Step 2 Find the midpoint of $\overline{EF}$.
   $\left(\frac{x_1 + x_2}{2}, \frac{y_1 + y_2}{2}\right)$
   midpoint of $\overline{EF} = \left(\frac{-4 + 0}{2}, \frac{-7 + 1}{2}\right) = (-2, -3)$
   Step 3 Find the slope of the perpendicular bisector.
   slope $= \frac{y_2 - y_1}{x_2 - x_1}$
   slope of $\overline{EF} = \frac{1 - (-7)}{0 - (-4)} = \frac{8}{4} = 2$
   Since the slopes of $\perp$ lines are opposite reciprocals, the slope of the $\perp$ bisector is $-\frac{1}{2}$.
   Step 4 Use point-slope form to write an equation.
   The $\perp$ bisector of $\overline{EF}$ has slope $-\frac{1}{2}$ and passes through (–2, –3).
   $y - y_1 = m(x - x_1)$
   $y - (-3) = -\frac{1}{2}[x - (-2)]$
   $y + 3 = -\frac{1}{2}(x + 2)$

9. **Step 1** Graph $\overline{XY}$.
The ⊥ bisector of $\overline{XY}$ is ⊥ to $\overline{XY}$ at its midpoint
Step 2 Find the midpoint of $\overline{XY}$.
$$\left(\frac{x_1 + x_2}{2}, \frac{y_1 + y_2}{2}\right)$$
$$\text{midpoint of } \overline{XY} = \left(\frac{-7 + (-1)}{2}, \frac{5 + (-1)}{2}\right) = (-4, 2)$$
Step 3 Find the slope of the perpendicular bisector.
$$\text{slope} = \frac{y_2 - y_1}{x_2 - x_1}$$
$$\text{slope of } \overline{XY} = \frac{-1 - 5}{-1 - (-7)} = \frac{-6}{6} = -1$$
Since the slopes of ⊥ lines are opposite reciprocals, the slope of the ⊥ bisector is 1.
Step 4 Use point-slope form to write an equation.
The ⊥ bisector of $\overline{XY}$ has slope $-\frac{1}{2}$ and passes through (−2, −3).
$$y - y_1 = m(x - x_1)$$
$$y - 2 = 1[x - (-4)]$$
$$y - 2 = x + 4$$

10. **Step 1** Graph $\overline{MN}$.
The ⊥ bisector of $\overline{MN}$ is ⊥ to $\overline{MN}$ at its midpoint
Step 2 Find the midpoint of $\overline{MN}$.
$$\left(\frac{x_1 + x_2}{2}, \frac{y_1 + y_2}{2}\right)$$
$$\text{midpoint of } \overline{MN} = \left(\frac{-3 + 7}{2}, \frac{1 + (-5)}{2}\right) = (2, -3)$$
Step 3 Find the slope of the ⊥ bisector.
$$\text{slope} = \frac{y_2 - y_1}{x_2 - x_1}$$
$$\text{slope of } \overline{MN} = \frac{-5 - (-1)}{7 - (-3)} = \frac{-4}{10} = -\frac{2}{5}$$
Since the slopes of ⊥ lines are opposite reciprocals, the slope of the bisector is $\frac{5}{2}$.
Step 4 Use point-slope form to write an equation.
The bisector of ⊥ has slope $-\frac{1}{2}$ and passes through (−2, −3).
$$y - y_1 = m(x - x_1)$$
$$y - (-3) = \frac{5}{2}(x - 2)$$
$$y + 3 = \frac{5}{2}(x - 2)$$

11. $GJ = 2PQ$
$= 2(19) = 38$

12. $RQ = \frac{1}{2}GH$
$= \frac{1}{2}(27) = 13.5$

13. $RJ = \frac{1}{2}GJ$
$= \frac{1}{2}(38) = 19$

14. $m\angle PQR = m\angle QRJ$
$= 55°$

15. $m\angle HGJ = m\angle QRJ$
$= 55°$

16. $m\angle GPQ + m\angle HGJ = 180°$
$m\angle GPQ + 55 = 180°$
$m\angle GPQ = 125°$

17. Yes; $\overline{DE}$ is a midsegment of $\triangle ABC$, so its length is half of $4\frac{1}{2}$ ft, or $2\frac{1}{4}$ ft, which is 27 in. This is less than 30 in. So the carpenter can use the 30 in. timber to make the crossbar.

Properties and Attributes of Triangles

Relationships in Triangles

Indirect Proof and Inequalities in One Triangle **Page [1 of 1]**

1. Possible answer:
Given: $\triangle ABC$ is a scalene triangle.
Prove: $\triangle ABC$ cannot have 2 ≅ ∡.
Proof: Assume that $\triangle ABC$ does have 2 ≅ ∡. Let $\angle A$ and $\angle C$ be the ≅ ∡. Then $\overline{AB} \cong \overline{CB}$ by the Converse of the Isosc. △ Theorem. However, a scalene △ by definition has no ≅ sides. So $\triangle ABC$ is not scalene, which contradicts the given information. This means the assumption is false, and therefore $\triangle ABC$ can not have 2 ≅ ∡.

2. Possible answer:
Given: $\triangle PQR$ is an isosc. △ with base $\overline{PR}$.
Prove: $\triangle PQR$ cannot have a base ∠ that is a rt. ∠.
Proof: Assume that $\triangle PQR$ has a base ∠ that is a rt. ∠. Let $\angle P$ be the rt. ∠. By the Isosc. △ Theorem, $\angle P \cong \angle R$, so $\angle R$ is also a rt. ∠. By the definition of rt. ∡, $m\angle P = m\angle R = 90°$. By the △ Sum Theorem, $m\angle P + m\angle Q + m\angle R = 180°$. By Subst. $m\angle Q = 0°$. However, a △ cannot have an ∠ with a measure of 0°. So there is no $\triangle PQR$, which contradicts the given information. This means the assumption is false, and therefore $\triangle PQR$ can not have a base ∠ that is rt.

3. The shortest side is $\overline{PQ}$, so the smallest ∠ is $\angle R$. The longest side is $\overline{PR}$, so the greatest ∠ is $\angle Q$. ∡ from smallest to greatest are $\angle R$, $\angle P$, $\angle Q$.

4. $m\angle Z = 180 - (39 + 46) = 95°$
The smallest ∠ is $\angle X$, so the shortest side is $\overline{YZ}$. The greatest ∠ is $\angle Z$, so the longest side is $\overline{XY}$. Sides from shortest to longest are $\overline{YZ}$, $\overline{XZ}$, $\overline{XY}$.

5. $4 + 7 \overset{?}{>} 10$ $\quad 4 + 10 \overset{?}{>} 7$ $\quad 7 + 10 \overset{?}{>} 4$
$11 > 10$ ✓ $\quad 14 > 7$ ✓ $\quad 17 > 4$ ✓
Yes; the sum of each pair of 2 lengths is greater than the third length.

6. $2 + 9 \overset{?}{>} 12$
$11 \not> 12$
No; $2 + 9 = 11$, which is not greater than the third side length.

7. $3.5 + 3.5 \overset{?}{>} 6$ $\quad 3.5 + 6 \overset{?}{>} 3.5$
$7 > 6$ ✓ $\quad 9.5 > 3.5$ ✓
Yes; the sum of each pair of 2 lengths is greater than the third length.

8. $1.1 + 1.7 \overset{?}{>} 3$
$2.8 \not> 3$
No; $1.1 + 1.7 = 2.8$, which is not greater than the third side length.

9. When $x = 5$, $3x = 15$, $2x - 1 = 9$, $x^2 = 25$.
$9 + 15 \overset{?}{>} 25$
$24 \not> 25$
No; when $x = 5$, the value of $3x$ is 15, the value of $2x - 1$ is 9 and the value of x^2 is 25. $15 + 9 = 24$, which is not greater than the third side length.

10. When $c = 2$, $7c + 6 = 20$, $10c - 7 = 13$, $3c^2 = 12$.
$12 + 13 \overset{?}{>} 20$ $\quad 12 + 20 \overset{?}{>} 13$ $\quad 13 + 20 \overset{?}{>} 12$
$25 > 20$ ✓ $\quad 32 > 13$ ✓ $\quad 33 > 12$ ✓
Yes; when $c = 2$, the value of $7c + 6$ is 20, the value of $10c - 7$ is 13, and the value of $3c^2$ is 12. The sum of each pair of 2 lengths is greater than the third length.

11. Let s be the length of the 3rd side. Apply the △ Inequal. Theorem.
$s + 8 > 12$ $\quad s + 12 > 8$ $\quad 8 + 12 > s$
$s > 4$ $\quad s > -4$ $\quad 20 > s$
Combine the inequals. So $4 < s < 20$. The length of the 3rd side is > 4 mm and < 20 mm.

12. Let s be the length of the 3rd side. Apply the △ Inequal. Theorem.
$s + 16 > 16$ $\quad 16 + 16 > s$
$s > 0$ $\quad 32 > s$
Combine the inequals. So $0 < s < 32$. The length of the 3rd side is > 0 ft and < 32 ft.

13. Let s be the length of the 3rd side. Apply the △ Inequal. Theorem.
$s + 11.4 > 12$ $\quad s + 12 > 11.4$ $\quad 11.4 + 12 > s$
$s > 0.6$ $\quad s > -0.6$ $\quad 23.4 > s$
Combine the inequals. So $0.6 < s < 23.4$. the length of the 3rd side is > 0.6 cm and < 23.4 cm.

14. a. The longest side is opposite the greatest ∠. So the longest side is between the refrigerator and the stove.

b. No; $4 + 5 = 9 \not> 9$. By the △ Inequal. Theorem, a △ cannot have these side lengths.

Properties and Attributes of Triangles

Relationships in Triangles

Inequalities in Two Triangles

1. Compare the sides and the ∡ in $\triangle ABC$ and $\triangle XYZ$.
$AB = YZ$ $BC = XY$ $m\angle B < m\angle Y$
By the Hinge Theorem, $AC < XZ$.

2. Compare the side lengths in $\triangle SRT$ and $\triangle QRT$.
$RT = RT$ $RS = RQ$ $ST > QT$
By the Converse of Hinge Theorem, $m\angle SRT > m\angle QRT$.

3. Compare the sides and ∡ in $\triangle KLM$ and $\triangle KNM$.
$KM = KM$ $LM = NM$ $m\angle KML > m\angle KMN$
By the Hinge Theorem, $KL > KN$.

4. **Step 1** Compare the side lengths in ∆s. By the Converse of the Hinge Theorem,
$2x + 8 < 25$
$2x < 17$
$x < 8.5$.
Step 2 Since $(2x + 8)°$ is an angle in a △,
$2x + 8 > 0$
$2x > -8$
$x > -4$.
Step 3 Combine the inequals.
The range of values is $-4 < x < 8.5$.

5. **Step 1** Compare the sides and the ∡ in ∆s. By the Hinge Theorem,
$5x - 6 < 9$
$5x < 15$
$x < 3$.
Step 2 Since $5x - 6$ is a length,
$5x - 6 > 0$
$5x > 6$
$x > 1.2$.
Step 3 Combine the inequals.
The range of values is $1.2 < x < 3$.

6. **Step 1** Compare the sides and the ∡ in ∆s. By the Hinge Theorem,
$2x - 5 < x + 7$
$x < 12$.
Step 2 Since $2x - 5$ is a length,
$2x - 5 > 0$
$2x > 5$
$x > 2.5$.
Step 3 Combine the inequals.
The range of values is $2.5 < x < 12$.

7. The 2nd position; the lengths of the upper and lower arm are the same in both positions, but the distance from the shoulder to the wrist is greater in the 2nd position. So the included ∠ measure is greater by the Converse of the Hinge Theorem.

8.

Statements	Reasons
1. $\overline{FH}$ is a median of $\triangle DFG$; $m\angle DHF > m\angle GHF$	1. Given
2. H is midpoint of $\overline{DG}$.	2. Def. of median
3. $\overline{DH} \cong \overline{GH}$	3. Def. of midpoint
4. $\overline{FH} \cong \overline{FH}$	4. Reflex. Prop. of ≅
5. $DF > GF$	5. Hinge Thm.

The Pythagorean Theorem

1. $a^2 + b^2 = c^2$
$3^2 + 9^2 = x^2$
$90 = x^2$
$\sqrt{90} = x$
$x = \sqrt{(9)(10)}$
$= 3\sqrt{10}$

2. $a^2 + b^2 = c^2$
$x^2 + 7^2 = 11^2$
$x^2 = 72$
$x = \sqrt{72}$
$x = \sqrt{(36)(2)}$
$= 6\sqrt{2}$

3. $a^2 + b^2 = c^2$
$(x - 2)^2 + 8^2 = x^2$
$x^2 - 4x + 4 + 64 = x^2$
$-4x + 68 = 0$
$68 = 4x$
$x = 17$

4. Let the width and the height of the monitor be $w = 5x$ and $h = 4x$, respectively.
$a^2 + b^2 = c^2$
$(4x)^2 + (5x)^2 = 19^2$
$41x^2 = 361$
$x^2 = \frac{361}{41}$
$x = \sqrt{\frac{361}{41}}$
$w = 5x = 5\sqrt{3\frac{61}{41}} \approx 14.8$ in.
$h = 4x = 4\sqrt{\frac{361}{41}} \approx 11.9$ in.

5. $a^2 + b^2 = c^2$
$4^2 + 5^2 = c^2$
$41 = c^2$
$c = \sqrt{41}$
The side lengths do not form a Pythagorean triple because $\sqrt{41}$ is not a whole number.

6. $a^2 + b^2 = c^2$
$12^2 + b^2 = 20^2$
$b^2 = 256$
$b = 16$
The side lengths are nonzero whole numbers that satisfy the equation $a^2 + b^2 = c^2$, so they form a Pythagorean triple.

7. $a^2 + b^2 = c^2$
$1.5^2 + b^2 = 1.7^2$
$b^2 = 0.64$
$b = 0.8$
The side lengths do not form a Pythagorean triple because they are not whole numbers.

8. **Step 1** Determine if the measures form a $\triangle$.
By the $\triangle$ Inequal. Theorem, 7, 10, and 12 can be the side lengths of a $\triangle$.
Step 2 Classify the $\triangle$.
$c^2 \stackrel{?}{=} a^2 + b^2$
$12^2 \stackrel{?}{=} 7^2 + 10^2$
$144 \stackrel{?}{=} 49 + 100$
$144 < 149$
Since $c^2 < a^2 + b^2$, $\triangle$ is acute.

The Pythagorean Theorem

9. **Step 1** Determine if the measures form a △.
By the △ Inequal. Theorem, 9, 11, and 15 can be the side lengths of a △.
Step 2 Classify the △.
$c^2 \overset{?}{=} a^2 + b^2$
$15^2 \overset{?}{=} 9^2 + 11^2$
$225 \overset{?}{=} 81 + 121$
$225 > 202$
Since $c^2 > a^2 + b^2$, △ is obtuse.

10. **Step 1** Determine if the measures form a △.
By the △ Inequal. Theorem, 9, 40, and 41 can be the side lengths of a △.
Step 2 Classify the △.
$c^2 \overset{?}{=} a^2 + b^2$
$41^2 \overset{?}{=} 9^2 + 40^2$
$1681 \overset{?}{=} 81 + 1600$
$1681 = 1681$
Since $c^2 = a^2 + b^2$, △ is a rt. △.

11. **Step 1** Determine if measures form a △.
Since $1\frac{1}{2} + 1\frac{3}{4} = 3\frac{1}{4} \not> 3\frac{1}{4}$, these cannot be the side lengths of a △.

12. **Step 1** Determine if the measures form a △.
By the △ Inequal. Theorem, 5.9, 6, and 8.4 can be the side lengths of a △.
Step 2 Classify the △.
$c^2 \overset{?}{=} a^2 + b^2$
$8.4^2 \overset{?}{=} 5.9^2 + 6^2$
$70.56 \overset{?}{=} 34.81 + 36$
$70.56 < 70.81$
Since $c^2 < a^2 + b^2$, △ is acute.

13. **Step 1** Determine if the measures form a △.
By the △ Inequal. Theorem, 11, 13, and $7\sqrt{6}$ can be the side lengths of a △.
Step 2 Classify the △.
$c^2 \overset{?}{=} a^2 + b^2$
$(7\sqrt{6})^2 \overset{?}{=} 11^2 + 13^2$
$294 \overset{?}{=} 121 + 169$
$294 > 290$
Since $c^2 > a^2 + b^2$, △ is obtuse.

Properties and Attributes of Triangles

Relationships in Triangles

Applying Special Right Triangles

1. The △ is an isosc. rt. △, which is a 45°-45°-90° △.
$x = 14\sqrt{2}$

2. The △ is an isosc. rt. △, which is a 45°-45°-90° △.
$12 = x\sqrt{2}$
$\frac{12}{\sqrt{2}} = x$
$\frac{12\sqrt{2}}{2} = x$
$x = 6\sqrt{2}$

3. The △ is an isosc. rt. △, which is a 45°-45°-90° △.
$9\sqrt{2} = x\sqrt{2}$
$x = 9$

4. The sign forms a right △. Using the Pyth. Thm., we get
$d = \sqrt{19.5^2 + 19.5^2}$
$d \approx 27.6$ in.

5. $6 = 2x$
$3 = x$
$y = x\sqrt{3}$
$y = 3\sqrt{3}$

6. $15 = x\sqrt{3}$
$\frac{15}{\sqrt{3}} = x$
$\frac{15\sqrt{3}}{3} = x$
$5\sqrt{3} = x$
$y = 2x$
$y = 10\sqrt{3}$

7. $x = (7\sqrt{3})\sqrt{3}$
$x = 21$
$y = 2(7\sqrt{3})$
$y = 14\sqrt{3}$

8. **Step 1** Divide the equil. △ into two 30°-60°-90° △s. The height of the frame is the length of the longer leg.

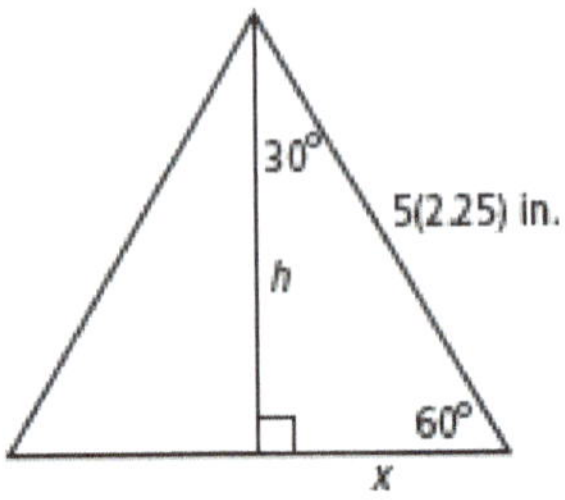

Step 2 Find the length x of the shorter leg.
$5(2.25) = 2x$
$5.625 = x$

Step 3 Find the length h of the longer leg.
$h = 5.625\sqrt{3} \approx 9.75$ in.

Properties and Attributes of Triangles

Review 5.2 / Relationships in Triangles

1. Possible answer:
Given: $\triangle ABC$ is scalene. $\overline{XZ}$ and $\overline{YZ}$ are midsegments of $\triangle ABC$.
Prove: $\triangle ABC$ cannot have 2 $\cong$ midsegments.
Proof: Assume that $\triangle ABC$ does have 2 $\cong$ midsegments. Let $\overline{XZ}$ and $\overline{YZ}$ be the $\cong$ midsegments. By the def. of $\cong$ segs., $XZ = YZ$. By the $\triangle$ Mid segment Thm., $XZ = \frac{1}{2}BC$ and $YZ = \frac{1}{2}BA$. So $\frac{1}{2}BC = \frac{1}{2}BA$ by subst. But then $BC = BA$, and by the def. of $\cong$ segs., $\overline{BC} \cong \overline{BA}$. However, a scalene $\triangle$ by def. has no $\cong$ sides. So $\triangle ABC$ is not scalene, which contradicts the given information. This means the assumption is false, and therefore
a scalene $\triangle$ cannot have 2 $\cong$ midsegments.

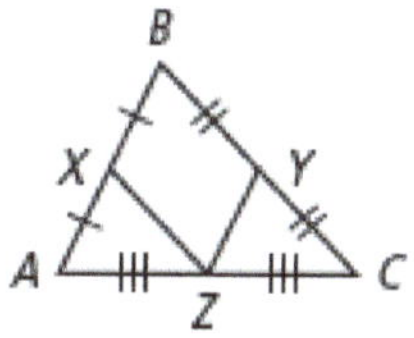

2. Possible answer:
Given: $\angle J$ and $\angle K$ are supp.
Prove: $\angle J$ and $\angle K$ cannot both be obtuse.
Proof: Assume that $\angle J$ and $\angle K$ are both obtuse. Then $m\angle J > 90°$ and $m\angle K > 90°$ by the definition of obtuse. Add the 2 inequals., $m\angle J + m\angle K > 180°$. However, by the definition of supp. ∡, $m\angle J + m\angle K = 180°$. So $m\angle J + m\angle K > 180°$ contradicts the given information. This means the assumption is false, and therefore a pair of supp. ∡ cannot both be obtuse.

3. The shortest side is $\overline{KL}$, so the smallest $\angle$ is $\angle J$. The longest side is $\overline{JL}$, so the greatest $\angle$ is $\angle K$. ∡ from smallest to greatest are $\angle J$, $\angle L$, $\angle K$.

4. $m\angle S = 90$, $m\angle T = 90 - m\angle R = 24°$
The smallest $\angle$ is $\angle T$, so the shortest side is $\overline{RS}$. The greatest $\angle$ is $\angle S$, so the longest side is $\overline{RT}$. Sides from shortest to longest are $\overline{RS}$, $\overline{ST}$, $\overline{RT}$.

5. $6 + 10 \overset{?}{>} 15$
$16 > 15$ ✓
Yes; the sum of each pair of 2 lengths is greater than the third length.

6. $14 + 18 \overset{?}{>} 32$
$32 \not> 32$
No; $14 + 18 = 32$, which is not greater than the third side length.

7. $5.8 + 5.8 \overset{?}{>} 11.9$
$11.6 \not> 11.9$
No; $5.8 + 5.8 = 11.6$, which is not greater than the third side length.

8. $41.9 + 62.5 \overset{?}{>} 103$
$104.4 > 103$ ✓
Yes; the sum of each pair of 2 lengths is greater than the third.

Properties and Attributes of Triangles

Review 5.2 / Relationships in Triangles

9. When $z = 6$, $z + 8 = 14$, $3z + 5 = 23$, $4z - 11 = 13$.

$13 + 14 \overset{?}{>} 23$

$27 > 23$ ✓

Yes; when $z = 6$, the value of $z + 8$ is 14, the value of $3z + 5$ is 23, and the value of $4z - 11$ is 13. The sum of each pair of 2 lengths is greater than the third side length.

10. When $m = 3$, $m + 11 = 14$, $8m = 24$, $m^2 + 1 = 10$.

$10 + 14 \overset{?}{>} 24$

$24 \not> 24$

No; when $m = 3$, the value of $m + 11 = 14$, the value of $8m$ is 24, and the valu e of $m^2 + 1$ is 10. the sum of 14 and 10 is 24, which is not greater than the third side length.

11. $b + c > s$

$s + 4 > 19$ $\quad$ $4 + 19 > s$

$s > 15$ $\quad$ $23 > s$

15 yd $< s <$ 23 yd

12. $s + 23 > 28$ $\quad$ $23 + 28 > s$

$s > 5$ $\quad$ $51 > s$

5 km $< s <$ 51 km

13. $s + 3.8 > 9.2$ $\quad$ $3.8 + 9.2 > s$

$s > 5.4$ $\quad$ $13.0 > s$

5.4 cm $< s <$ 13.0 cm

14. $s + 1.89 > 3.07$ $\quad$ $1.89 + 3.07 > s$

$s > 1.18$ $\quad$ $4.96 > s$

1.18 m $< s <$ 4.96 m

15. $s + 2\frac{1}{8} > 3\frac{5}{8}$ $\quad$ $2\frac{1}{8} + 3\frac{5}{8} > s$

$s > 1\frac{1}{2}$ $\quad$ $5\frac{3}{4} > s$

$1\frac{1}{2}$ in. $< s < 5\frac{3}{4}$ in.

16. $s + 3\frac{5}{6} > 6\frac{1}{2}$ $\quad$ $3\frac{5}{6} + 6\frac{1}{2} > s$

$s > 2\frac{2}{3}$ $\quad$ $10\frac{1}{3} > s$

$2\frac{2}{3}$ ft $< s < 10\frac{1}{3}$ ft

17. $\overline{AD}$, $\overline{BD}$, $\overline{AB}$, $\overline{BC}$, $\overline{CD}$; possible answer: in $\triangle ABD$, $m\angle ABD = 50°$. In $\triangle BCD$, $m\angle DBC = 74°$. In $\triangle ABD$, the order of the tubes from shortest to longest is $\overline{AD}$, $\overline{BD}$, $\overline{AB}$. In $\triangle BCD$, the order of the tubes from shortest to longest is $\overline{BD}$, $\overline{BC}$, $\overline{CD}$. So $AD < BD < AB$, and $BD < BC < CD$. Since $AB = 50.8$ and $BC = 54.1$, it is also true that $AB < BC$. So $\overline{AD} < \overline{BD} < \overline{AB} < \overline{BC} < \overline{CD}$.

18. $BC = CD$, $CA = CA$, $AD > AB$; by Converse of Hinge Theorem, $m\angle DCA > m\angle BCA$.

19. $GH = KL$, $HJ = LM$, $GJ < KM$; by Converse of Hinge Theorem, $m\angle GHJ < m\angle KLM$.

20. $ST = UV$, $SU = SU$, $m\angle UST > m\angle SUV$; by Hinge Theorem, $TU > SV$.

21. $4z - 12 < 16$ $\quad$ $4z - 12 > 0$

$4z < 28$ $\quad$ $4z > 12$

$z < 7$ $\quad$ $z > 3$

Combining, $3 < z < 7$.

22. $2z + 7 < 72$ $\quad$ $2z + 7 > 0$

$2z < 65$ $\quad$ $2z > -7$

$z < 32.5$ $\quad$ $z > -3.5$

Combining, $-3.5 < z < 32.5$.

23. $4z - 6 < z + 11$
$3z < 17$
$z < \frac{17}{3}$

$4z - 6 > 0$
$4z > 6$
$z > \frac{3}{2}$

Combining, $\frac{3}{2} < z < \frac{17}{3}$.

24. The lengths of the arms are the same in both positions, but the included ∠ measure is greater in the 2nd position. Therefore, by the Hinge Theorem, the distance from the cab to the bucket is greater in the 2nd position.

25.

Statements	Reasons
1. $\overline{JK} \cong \overline{NM}$, $\overline{KP} \cong \overline{MQ}$, $JQ > NP$	1. Given
2. $\overline{QP} \cong \overline{QP}$	2. Reflex. Prop. of ≅
3. $QP = QP$	3. Def. of ≅ segs.
4. $JQ + QP > NP + QP$	4. Add. Prop. of Inequal.
5. $JQ + QP = JP$, $NP + QP = NQ$	5. Segment Add. Post.
6. $JP > NQ$	6. Subst.
7. $m\angle K > m\angle M$	7. Con. of the Hinge Thm.

26. $6^2 + 8^2 = x^2$
$100 = x^2$
$x = 10$

27. $9^2 + x^2 = 13^2$
$81 + x^2 = 169$
$x^2 = 88$
$x = \sqrt{88} = 2\sqrt{22}$

28. $x^2 + 7^2 = (x + 1)^2$
$x^2 + 49 = x^2 + 2x + 1$
$48 = 2x$
$x = 24$

29. $(3x)^2 + (5x)^2 = 8^2$
$34x^2 = 64$
$x^2 = \frac{32}{17}$
$3x = 3\sqrt{\frac{32}{17}} \approx$ 4 ft 1 in.
$5x = 5\sqrt{\frac{32}{17}} \approx$ 6 ft 10 in.

30. $2.5^2 + b^2 = 6.5^2$
$6.25 + b^2 = 42.25$
$b^2 = 36$
$b = 6$
The side lengths cannot form a △ because 2.5 and 6.5 are not whole numbers.

31. $15^2 + 20^2 = c^2$
$625 = c^2$
$c = 25$
Yes; the three side lengths are nonzero whole numbers that satisfy $a^2 + b^2 = c^2$.

32. $2^2 + b^2 = 7^2$
$4 + b^2 = 49$
$b^2 = 45$
$b = \sqrt{45} = 3\sqrt{5}$
The side lengths cannot form a △ because $3\sqrt{5}$ is not a whole number.

33. $10 + 12 = 22 > 15$ ✓

$15^2 \stackrel{?}{=} 10^2 + 12^2$

$225 < 244$

The side lengths form an acute triangle.

34. $8 + 13 = 21 \not> 23$ X

The side lengths can not form a △.

35. $9 + 14 = 23 > 17$ ✓

$17^2 \stackrel{?}{=} 9^2 + 14^2$

$289 > 277$

The side lengths form an obtuse △.

36. $1\frac{1}{2} + 2 = 3\frac{1}{2} > 2\frac{1}{2}$ ✓

$\left(2\frac{1}{2}\right)^2 \stackrel{?}{=} \left(1\frac{1}{2}\right)^2 + 2^2$

$6\frac{1}{4} = 6\frac{1}{4}$

The side lengths form a rt. △.

37. $0.7 + 1.1 = 1.8 > 1.7$ ✓

$1.7^2 \stackrel{?}{=} 0.7^2 + 1.1^2$

$2.89 > 1.7$

The side lengths form an obtuse △.

38. $7 + 12 = 19 > 6\sqrt{5}$ ✓

$(6\sqrt{5})^2 \stackrel{?}{=} 7^2 + 12^2$

$180 < 193$

The side lengths form an acute △.

39. The △ is an isosc. rt. △, which is a 45°-45°-90° △.

$15 = x\sqrt{2}$

$\frac{15}{\sqrt{2}} = x$

$\frac{15\sqrt{2}}{2} = x$

40. The △ is an isosc. rt. △, which is a 45°-45°-90° △.

$x = (4\sqrt{2})\sqrt{2} = 8$

41. The △ is an isosc. rt. △, which is a 45°-45°-90° △.

$18\sqrt{2} = x\sqrt{2}$

$18 = x$

42. The tabletop is a 45°-45°-90° △.

$48 = w\sqrt{2}$

$48\sqrt{2} = 2w$

$w = 24\sqrt{2} \approx 33.9$ in.

43. $x = 2(24) = 48$

$y = 24\sqrt{3}$

44. $2x = 10\sqrt{3}$

$x = 5\sqrt{3}$

$y = x\sqrt{3} = (5\sqrt{3})\sqrt{3} = 15$

45. $2 = x\sqrt{3}$

$2\sqrt{3} = 3x$

$\frac{2\sqrt{3}}{3} = x$

$y = 2x = \frac{4\sqrt{3}}{3}$

46. a. The ramp forms a 30°-60°-90° △. Let the length of the ramp be x.

$x = 2(4.5) = 9$ ft

b. Length of the dog walk $= x + 12 + x$

Properties and Attributes of Polygons

1. polygon, decagon
2. not a polygon
3. polygon, quadrilateral
4. not a polygon
5. regular, convex
6. irregular, concave
7. irregular, convex
8. Think: Use Polygon ∠ Sum Thm.

$$(5-2)180° = 540°$$
$$3z + 4z + 5z + 3z + 5z = 540 \quad m\angle A = m\angle D = 3(27) = 81°$$
$$20z = 540 \quad m\angle B = 4(27) = 108°$$
$$z = 27 \quad m\angle C = m\angle E = 5(27) = 135°$$

9. Think: Use Polygon ∠ Sum Thm.

$$(12-2)180° = 1800°$$
$$m\angle 1 + m\angle 2 + \dots + m\angle 12 = 1800°$$
$$12m\angle 1 = 1800°$$
$$m\angle 1 = 150°$$

10. Think: Use Polygon ∠ Sum Thm.

$(n-2)180°$
$(20-2)180°$
$3240°$

11. Think: Use Polygon Ext. ∠ Sum Thm.

$$4y + 2y + 6y + 4y = 360$$
$$16y = 360$$
$$y = 22.5$$

12. Think: Use Polygon Ext. ∠ Sum Thm.

$$m\angle 1 + m\angle 2 + \dots + m\angle 5 = 360°$$
$$5m\angle 1 = 360°$$
$$m\angle 1 = 72°$$

13. pentagon
14. By Polygon ∠ Sum Thm., sum of ∠ measures is $(5-2)180 = 540°$. Think: $m\angle Q = m\angle S$ by def. of ≅ ∠s.

$$m\angle P + m\angle Q + m\angle R + m\angle S + m\angle T = 540$$
$$90 + m\angle S + 90 + m\angle S + 90 = 540$$
$$2m\angle S = 270$$
$$m\angle Q = m\angle S = 135°$$

Polygons and Quadrilaterals

Polygons and Parallelograms

Properties of Parallelograms

1. E is mdpt. of $\overline{BD}$
$BD = 2DE$
$= 2(18) = 36$

2. $\overline{CD} \cong \overline{AB}$
$CD = AB = 17.5$

3. E is mdpt. of $\overline{BD}$
$BE = DE = 18$

4. $\angle ABC$ and $\angle BCD$ are supp.
$m\angle ABC = 180 - m\angle BCD$
$= 180 - 110 = 70°$

5. $\angle ADC \cong \angle ABC$
$m\angle ADC = m\angle ABC$
$= 70°$

6. $\angle DAB \cong BCD$
$m\angle DAB = m\angle BCD$
$= 110°$

7. $\overline{JK} \cong \overline{LM}$
$JK = LM$
$7x = 3x + 14$
$4x = 14$
$x = 3.5$
$JK = 7(3.5) = 24.5$

8. $LM = 3(3.5) + 14$
$= 24.5$

9. $\angle L$ and $\angle M$ are supp.
$m\angle L + m\angle M = 180$
$2z - 3 + 5z - 6 = 180$
$7z = 189$
$z = 27$
$m\angle L = 2(27) - 3 = 51°$

10. $m\angle M = 5(27) - 6$
$= 129°$

11. **Step 1** Graph given pts.
Step 2 Find slope of $\overline{FG}$ by counting units from F to G.
Rise from 5 to 0 is -5.
Run from -1 to 2 is 3.
Step 3 Start at D and count same # of pts.
Rise of -5 from 4 is -1.
Run of 3 from -9 is -6.
Step 4 Use slope formula to verify that $\overline{DF} \parallel \overline{GH}$.

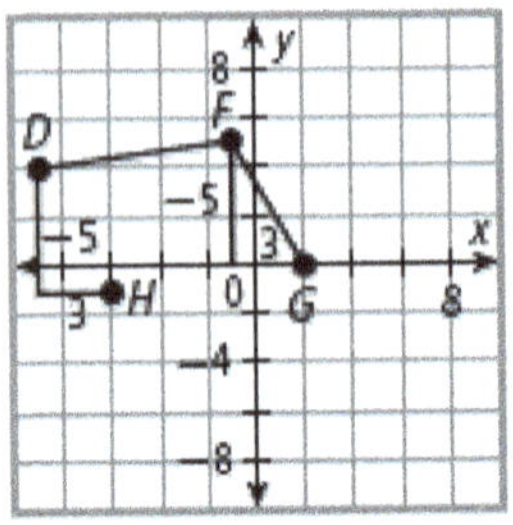

slope of $\overline{DF} = \frac{5-4}{-1+9} = \frac{1}{8}$

slope of $\overline{GH} = \frac{-1-0}{-6-2} = \frac{1}{8}$

Coords. of vertex H are $(-6, -1)$.

12.

Statements	Reasons
1. $PSTV$ is a ▱; $\overline{PQ} \cong \overline{RQ}$.	1. Given
2. $\angle STV \cong \angle P$	2. ▱ → opp. ∡ ≅
3. $\angle P \cong \angle R$	3. Isoc. △ Thm.
4. $\angle STV \cong \angle R$	4. Trans. Prop. of ≅

Polygons and Quadrilaterals

Polygons and Parallelograms

Conditions for Parallelograms

1. **Step 1** Find $\overline{EJ}$ and $\overline{JG}$.
   $EJ = t + 12$ $JG = 3t$
   $EJ = 6 + 12 = 18$ $JG = 3(6) = 18$
   Step 2 Find $\overline{FJ}$ and $\overline{JH}$.
   $FJ = 2s$ $JG = s + 5$
   $FJ = 2(5) = 10$ $JG = 5 + 5 = 10$
   Since $EJ = JG$ and $FJ = JH$, $EFGH$ is a ▱ by quad. with diagonals that bisect each other. → ▱ thm.
2. $\angle L = 5m + 36$
   $\angle L = 5(14) + 36 = 106°$
   $\angle P = 6n - 1$
   $\angle P = 6(12.5) - 1 = 74°$
   $\angle Q = 4m + 50$
   $\angle Q = 4(14) + 50 = 106°$
   Since $106° + 74° = 180°$, $\angle P$ is supp. to both $\angle L$ and $\angle Q$. $KLPQ$ is a ▱ by quad. with ∠ supp. to cons ∡ → ▱ thm. th its cons ∡. is supp. to both its cons ∡.
3. Yes; both pairs of opp. ∡ of the quad. are ≅. By quad. with opp. ∡ ≅. → ▱ thm., the quad. is a ▱.
4. No; 1 pair of opp. sides of quad. are ≅. 1 diag. is bisected by other diag. None of the conditions for a ▱ are met.
5. Yes; possible answer: a pair of alt. int. ∡ are ≅, so 1 pair of opp. sides are ∥. The same pair of opp. sides are ≅. By quad. with pair of opp. sides ∥ and ≅. → ▱ thm., the quad. is a ▱.
6. Possible answer: Find slopes of both pairs of opp. sides.
   slope of $\overline{WX} = \dfrac{3+2}{-3+5} = \dfrac{5}{2}$
   slope of $\overline{YZ} = \dfrac{0-5}{1-3} = \dfrac{5}{2}$
   slope of $\overline{XY} = \dfrac{5-3}{3+3} = \dfrac{1}{3}$
   slope $\overline{WZ} = \dfrac{0+2}{1+5} = \dfrac{1}{3}$
   Since both pairs of opp. sides are ∥, $WXYZ$ is a ▱ by definition.
7. Possible answer: Find slopes of both pairs of opp. sides.
   slope of $\overline{RS} = \dfrac{-1+5}{-2+1} = -4$
   slope of $\overline{TU} = \dfrac{-5+1}{5-4} = -4$
   slope of $\overline{ST} = \dfrac{-1+1}{4+2} = 0$
   slope of $\overline{RU} = \dfrac{-5+5}{5+1} = 0$
   Since both pairs of opp. sides are ∥, $RSTU$ is a ▱ by def.
8. Since $\overline{AD} \parallel \overline{BC}$ and $\overline{AD} \cong \overline{BC}$, $ABCD$ is a ▱ by quad. with pair of opp. sides ∥ and ≅. → ▱ thm. so $\overline{AB} \parallel \overline{CD}$ by def. of ▱.

Polygons and Quadrilaterals

Review 6.1 / Polygons and Parallelograms

1. polygon, hexagon

2. not a polygon

3. polygon, quadrilateral

4. irregular, concave

5. regular convex

6. irregular, convex

7. $2n + 6n + 2n + 5n = (4 - 2)180$
$15n = 360$
$n = 24$
$m\angle R = m\angle T = 2(24) = 48°$
$m\angle S = 6(24) = 144°$
$m\angle V = 5(24) = 120°$

8. $18m\angle = (18 - 2)180$
$18m\angle = 2880$
$m\angle = 160°$

9. $(7 - 2)180 = 900°$

10. $9m(\text{ext. }\angle) = 360$
$m(\text{ext. }\angle) = 40°$

11. $5a + 4a + 10a + 3a + 8a = 360$
$30a = 360$
$a = 12$

12. $6m\angle JKM = (6 - 2)180$
$6m\angle JKM = 720$
$m\angle JKM = 120°$

13. $6m\angle MKL = 360$
$m\angle MKL = 60°$

14. $JN = \frac{1}{2}JL$
$= \frac{1}{2}(165.8) = 82.9$

15. $LM = JK = 110$

16. $LN = JN = 82.9$

17. $m\angle JKL = m\angle JML$
$= 50°$

18. $m\angle KLM = 180 - m\angle JML$
$= 180 - 50 = 130°$

19. $m\angle MJK = m\angle KLM$
$= 130°$

20. $WV = VY$
$b + 8 = 5b$
$8 = 4b$
$b = 2$
$WV = (2) + 8 = 10$

21. $YW = 2WV$
$= 2(10) = 20$

22. $XV = ZV$
$3a - 7 = 2a$
$a = 7$
$XZ = 2ZV$
$= 2(2(7)) = 28$

23. $ZV = 2(7) = 14$

24. slope from P to R: rise $= 4$, run $= -6$
slope from V to T: rise of 4 from -1 is 3, run of -6 from 5 is -1
Coords. of T are $(-1, 3)$.

25.

Statements	Reasons
1. *ABCD* and *AFGH* are ▱s.	1. Given
2. $\angle C \cong \angle A$, $\angle A \cong \angle G$	2. ▱ → opp. ∠s ≅
3. $\angle C \cong \angle G$	3. Trans. Prop. of ≅

26. $BC = 3(3.2) + 7 = 16.6$, $GH = 8(3.2) - 9 = 16.6$
$BH = 3(7) + 7 = 28$, $CG = 6(7) - 14 = 28$
$BCGH$ is a ▱ by Thm. 6-3-2.

27. $UV = 10(19.5) - 6 = 189$, $TW = 8(19.5) + 33 = 189$
$m\angle V = 2(22) + 41 = 85°$, $m\angle W = 7(22) - 59 = 95°$
$\overline{UV} \cong \overline{TW}$; $\angle V$ and $\angle W$ are supp., so by Conv. of Same-Side Int. ∠s Thm., $\overline{UV} \parallel \overline{TW}$. $TUVW$ is a ▱ by quad. with pair of opp. sides ∥ and ≅ → ▱ thm.

28. Yes; both pairs of opp. sides are ≅, since all sides are ≅, so quad is a ▱ by quad. with opp. sides ≅ → ▱ thm.

29. Yes; by ∠ Add. Post., 1 pair of opp. ∠s are ≅, and by 3rd ∠s Thm., 2nd pair of opp. ∠s are ≅. So quad is a ▱ by quad. with opp. ∠s ≅ → ▱ thm.

30. No; by looking at the angles, we can be sure that one pair of sides are ∥. This is not enough.

31. slope of $\overline{JK} = \frac{7}{-2} = -\frac{7}{2}$; slope of $\overline{LM} = \frac{-7}{2} = -\frac{7}{2}$
slope of $\overline{KL} = \frac{-1}{5} = -\frac{1}{5}$; slope of $\overline{JM} = \frac{-1}{5} = -\frac{1}{5}$
both pairs of opp. sides have the same slope, so $\overline{JK} \parallel \overline{LM}$ and $\overline{KL} \parallel \overline{MJ}$; $JKLM$ is a ▱ by definition.

32. slope of $\overline{PQ} = \frac{5}{3}$; slope of $\overline{RS} = \frac{-5}{-3} = \frac{5}{3}$
slope of $\overline{QR} = \frac{-6}{6} = -1$; slope of $\overline{PS} = \frac{-6}{6} = -1$.
So, $\overline{PQ} \parallel \overline{RS}$ and $\overline{PS} \parallel \overline{QR}$.
$PQRS$ is a ▱ by def.

33. Possible answer: The brackets are always the same length, so it is always true that $AB = CD$. The bolts are always the same dist. apart, so it is always true that $BC = DA$. By quad. with opp. sides ≅ → ▱ thm. $ABCD$ is always a ▱.

The side $\overline{AD}$ stays horiz. no matter how you move the tray. Since $\overline{BC} \parallel \overline{AD}$, $\overline{BC}$ stays horiz. Since $\overline{BC}$ holds the tray in position, the tray will stay horiz. no matter how it is moved.

Polygons and Quadrilaterals

Other Special Quadrilaterals

Properties of Special Parallelograms

1. rect. ▱ → diags. bisect each other
$TQ = \frac{1}{2}QS$
$= \frac{1}{2}(380) = 190$ ft

2. $\overline{PQ} \cong \overline{RS}$
$PQ = RS = 160$ ft

3. T is mdpt. of $\overline{QS}$.
$ST = TQ = 190$ ft

4. rect. → diags. are ≅
$\overline{PR} \cong \overline{QS}$
$PR = QS = 380$ ft

5. $BC = CD$
$4x + 15 = 7x + 2$
$13 = 3x$
$x = 4\frac{1}{3}$

Substitute x back into either expression to find the length of the sides.

$CD = 7x + 2$

$7\left(\frac{13}{3}\right) + 2 =$

$\frac{91}{3} + \frac{6}{3} = \frac{97}{3} = 32\frac{1}{3}$

Since CD is equal to AB, the measure of AB is $32\frac{1}{3}$.

6. $AB = BC$
$= 4\left(4\frac{1}{3}\right) + 15 = 32\frac{1}{3}$
$\overline{AC} \perp \overline{BD}$
$m\angle AFB = 90$
$12y = 90$
$y = 7.5$
$m\angle ABC + m\angle BCD = 180$
$m\angle ABC + 2m\angle FCD = 180$
$m\angle ABC + 2(4(7.5) - 1) = 180$
$m\angle ABC + 58 = 180$
$m\angle ABC = 122°$

7. **Step 1** Show that $\overline{JL}$ and $\overline{KM}$ are ≅.
$JL = \sqrt{5^2 + 7^2} = \sqrt{74}$
$KM = \sqrt{7^2 + 5^2} = \sqrt{74}$
Since $JL = KM$, $\overline{JL} \cong \overline{KM}$.
Step 2 Show that $\overline{JL}$ and $\overline{KM}$ are ⊥.
slope of $\overline{JL} = \frac{7}{5}$; slope of $\overline{KM} = \frac{-5}{7} = -\frac{5}{7}$
Since $\left(\frac{7}{5}\right)\left(-\frac{5}{7}\right) = -1$, $\overline{JL}$ and $\overline{KM}$ are ⊥.
Step 3 Show that $\overline{JL}$ and $\overline{KM}$ bisect each other.
mdpt. of $\overline{JL} = \left(\frac{-3+2}{2}, \frac{-5+2}{2}\right) = \left(-\frac{1}{2}, -\frac{3}{2}\right)$
mdpt. of $\overline{KM} = \left(\frac{-4+3}{2}, \frac{1-4}{2}\right) = \left(-\frac{1}{2}, -\frac{3}{2}\right)$
Since $\overline{JL}$ and $\overline{KM}$ have same mdpt., they bisect each other.
Diags. are ≅ ⊥ bisectors of each other.

8. Possible answer:

Statements	Reasons
1. $RECT$ is a rect.; $\overline{RX} \cong \overline{TY}$	1. Given
2. $\overline{XY} \cong \overline{XY}$	2. Reflex. Prop. of ≅
3. $RX = TY$, $XY = XY$	3. Def. of ≅ segs.
4. $RX + XY = TY + XY$	4. Add. Prop. of =
5. $RX + XY = RY$, $TY + XY = TX$	5. Seg. Add. Post.
6. $RY = TX$	6. Subst.
7. $\overline{RY} \cong \overline{TX}$	7. Def. of ≅ segs.
8. $\angle R$ and $\angle T$ are rt. ∠s.	8. Def. of rect.
9. $\angle R \cong \angle T$	9. Rt. ∠ ≅ Thm.
10. $RECT$ is a ▱.	10. Rect. → ▱
11. $\overline{RE} \cong \overline{CT}$	11. ▱ → opp. sides ≅
12. $\triangle REY \cong \triangle TCX$	12. SAS

Polygons and Quadrilaterals

Other Special Quadrilaterals

Conditions for Special Parallelograms **Page [1 of 1]**

1. Possible answer: If *WXYZ* is both a rhombus and a rect., then it is a square. All 4 sides of *WXYZ* are ≅. So *WXYZ* is a rhombus and therefore a ▱. If the diags. of a ▱ are ≅, then by the ▱ with diagonals → rect. thm., ▱ is a rect. So the club members can measure the diags., and if they are equal, *WXYZ* is both a rhombus and a rect., and therefore it is a square.
2. Not valid; by ▱ with diagonals ≅ → rect. thm. if the diags. of a ▱ are ≅, then the ▱ is a rect. To apply this thm., you need to know that *ABCD* is a ▱.
3. Valid ($\overline{AB} \parallel \overline{CD}$, $\overline{AB} \cong \overline{CD} \rightarrow$ ▱, $\overline{AB} \perp \overline{BC} \rightarrow \angle B$ a rt. $\angle \rightarrow$ rect.)
4. **Step 1** Graph ▱*PQRS*.

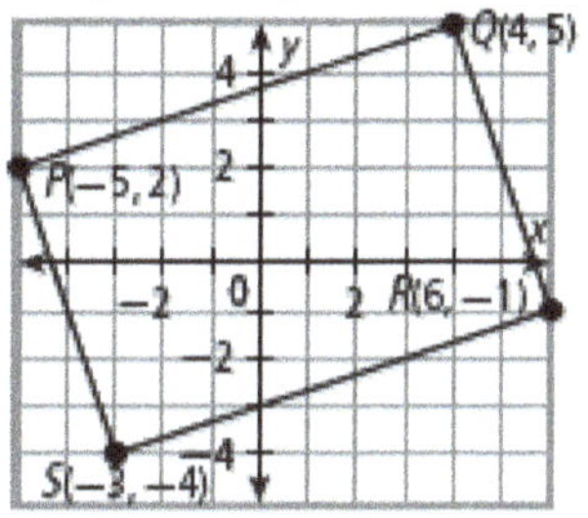

Step 2 Determine if *PQRS* is a rect.

$PR = \sqrt{11^2 + 3^2} = \sqrt{130}$

$QS = \sqrt{7^2 + 9^2} = \sqrt{130}$

Since $PR = QS$, diags. are ≅. *PQRS* is a rect.

Step 3 Determine if *PQRS* is a rhombus.

slope of $\overline{PR} = \frac{-3}{11} = -\frac{3}{11}$

slope of $\overline{QS} = \frac{-9}{-7} = \frac{9}{7}$

Since $\left(-\frac{3}{11}\right)\left(\frac{9}{7}\right) \neq -1$, *PQRS* is not a rhombus.

Thus *PQRS* is not a square.

5. **Step 1** Graph ▱*WXYZ*.

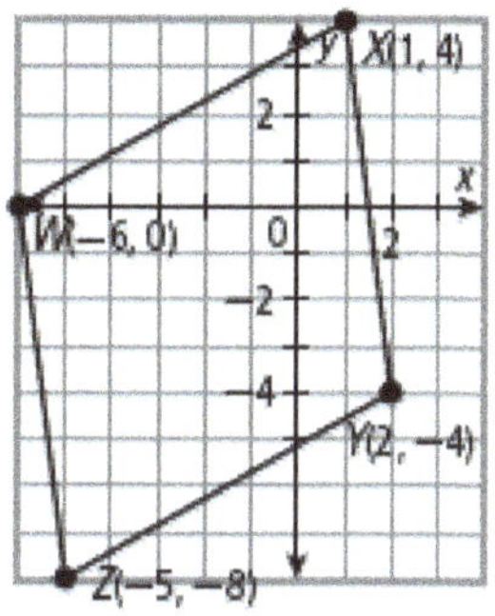

Step 2 Determine if *WXYZ* is a rect.

$WY = \sqrt{8^2 + 4^2}$
$= \sqrt{80} = 4\sqrt{5}$

$XZ = \sqrt{6^2 + 12^2}$
$= \sqrt{180} = 6\sqrt{5}$

Since $WY \neq XZ$, *WXYZ* is not a rect.

Thus *WXYZ* is not a square.

Step 3 Determine if *WXYZ* is a rhombus.

slope of $\overline{WY} = \frac{-4}{8} = -\frac{1}{2}$

slope of $\overline{XZ} = \frac{-12}{-6} = 2$

Since $\left(-\frac{1}{2}v\right)(2) = -1$, $\overline{WY} \perp \overline{XZ}$. *WXYZ* is a rhombus.

Polygons and Quadrilaterals

Other Special Quadrilaterals

Properties of Kites and Trapezoids

1. **1 Understand the Problem**

   Answer has 2 parts:
   - Total amount of lead needed
   - Number of suncatchers that can be sealed

   2 Make a Plan

   Diags. of a kite are $\perp$, so 4 $\triangle$s are rt. $\triangle$s. Use Pyth. Thm. and props. of kites to find unknown side lengths. Add these lengths to find perimeter of kite.

   3 Solve

   $JK = \sqrt{JH^2 + KH^2}$

   $= \sqrt{2.75^2 + 2.75^2} = \sqrt{15.125}$

   $KL = JK = \sqrt{15.125}$

   $JM = \sqrt{JH^2 + MH^2}$

   $= \sqrt{2.75^2 + 5.5^2} = \sqrt{37.8125}$

   $LM = JM = \sqrt{37.8125}$

   perimeter of $JKLM = 2\sqrt{15.125} + 2\sqrt{37.8125}$

   ≈ 20.1 in.

   20.1 in. of lead is needed to seal edges.

   One 3-ft length of lead contains 36 in.

   $\frac{2(36)}{20.1} \approx 3.6$

   3 sun catchers can be sealed.

   4 Look Back

   To estimate perimeter, change side lengths into decimals and round. $\sqrt{15.125} \approx 4$, and $\sqrt{37.8125} \approx 6$. Perimeter of sun catcher is approx. $2(4) + 2(6) = 20$. So 20.1 in. is a reasonable answer.

2. Think: Kite $\rightarrow$ diags. $\perp \rightarrow \angle VZY$ and $\angle VYZ$ are comp.

   $m\angle VZY + m\angle VYZ = 90$

   $m\angle VZY + 49 = 90$

   $m\angle VZY = 41°$

3. Think: $\triangle XYZ$ is isosc., so $\angle VXY \cong \angle VZY$.

   $m\angle VXW + m\angle VXY = m\angle WXY$

   $m\angle VXW + 41 = 104$

   $\angle VXW = 63°$

4. Think: Use Quad. $\angle$ Sum Thm., kite $\rightarrow$ 1 pair opp. $\angle$s $\cong$.

   $m\angle XWZ + m\angle WXY + m\angle XYZ + m\angle YZW = 360$

   $m\angle XWZ + 104 + 2m\angle VYZ + 104 = 360$

   $m\angle XWZ + 104 + 2(49) + 104 = 360$

   $m\angle XWZ = 54°$

Properties of Kites and Trapezoids

5. Think: Use Same-Side Int. ∡ Thm., isosc. trap. → base ∡ ≅.

$m\angle C + m\angle D = 180$
$74 + m\angle D = 180$
$m\angle D = 106°$
$\angle A \cong \angle D$
$m\angle A = m\angle D$
$= 106°$

6. Think: Isosc. trap. → diags. ≅.

$\overline{RT} \cong \overline{SV}$
$RT = SV$
$RW + TW = SV$
$17.7 + TW = 23.3$
$TW = 5.6$

7. Think: 1 pair base ∡ ≅ → trap. isosc.

$\angle E \cong \angle H$
$m\angle E = m\angle H$
$12z^2 = 7z^2 + 20$
$5z^2 = 20$
$z^2 = 4$
$z = 2 \text{ or } -2$

8. Think: Diags. ≅ → trap. isosc.

$\overline{MQ} \cong \overline{LP}$
$MQ = LP$
$7y - 6 = 4y + 11$
$3y = 17$
$y = \frac{17}{3} = 5\frac{2}{3}$

9. Think: Use Trap. Midseg. Thm.

$XY = \frac{1}{2}(PS + QR)$
$22 = \frac{1}{2}(30 + QR)$
$44 = 30 + QR$
$14 = QR$

10. Think: Use Trap. Midseg. Thm.

$AZ = \frac{1}{2}(DF + JK)$
$= \frac{1}{2}(11.9 + 7.1)$
$= \frac{1}{2}(19) = 9.5$

Polygons and Quadrilaterals

Review 6.2 / Other Special Quadrilaterals

1. $JL = 2JP$
$= 2(14.5) = 29$ in.

2. $KL = JM = 25$ in.

3. $KM = JL = 29$ in.

4. $MP = \frac{1}{2}KM$
$= \frac{1}{2}(29) = 14\frac{1}{2}$ in.

5. $WX = XY$
$9a - 18 = 3a + 15$
$6a = 33$
$a = 5.5$
$VW = WX$
$= 9(5.5) - 18$
$= 31.5$

6. $m\angle XZW = 90$
$10b - 5 = 90$
$10b = 95$
$b = 9.5$
$m\angle VWX + m\angle WVY = 180$
$m\angle VWX + 4(9.5) + 10 = 180$
$m\angle VWX = 132°$
$m\angle WYX = \frac{1}{2}m\angle VYX$
$= \frac{1}{2}m\angle VWX$
$= \frac{1}{2}(132) = 66°$

7. **Step 1** Show that $\overline{PR}$ and $\overline{QS}$ are $\cong$.
$PR = \sqrt{11^2 + 5^2} = \sqrt{146}$
$QS = \sqrt{5^2 + 11^2} = \sqrt{146}$
Since $PR = QS$, $\overline{PR} \cong \overline{QS}$.
Step 2 Show that $\overline{PR}$ and $\overline{QS}$ are $\perp$.
slope of $\overline{PR} = \frac{-5}{11} = -\frac{5}{11}$; slope of $\overline{QS} = \frac{-11}{-5} = \frac{11}{5}$
Since $\left(-\frac{5}{11}\right)\left(\frac{11}{5}\right) = -1$, $\overline{PR}$ and $\overline{QS}$ are $\perp$.
Step 3 Show that $\overline{PR}$ and $\overline{QS}$ bisect each other.
mdpt. of $\overline{PR} = \left(\frac{-4+7}{2}, \frac{0-5}{2}\right) = \left(\frac{3}{2}, -\frac{5}{2}\right)$
mdpt. of $\overline{QS} = \left(\frac{4-1}{2}, \frac{3-8}{2}\right) = \left(\frac{3}{2}, -\frac{5}{2}\right)$
Since $\overline{PR}$ and $\overline{QS}$ have same mdpt., they bisect each other.
Diags. are $\cong$ $\perp$ bisectors of each other.

8. Possible answer:

Statements	Reasons
1. *RHMB* is a rhombus. $\overline{HB}$ is a diag. of *RHMB*.	1. Given
2. $\overline{MH} \cong \overline{RH}$	2. Def. of rhombus
3. $\overline{HB}$ bisects $\angle RHM$.	3. Rhombus → each diag. bisects opp. ∠s
4. $\angle MHX \cong \angle RHX$	4. Def. of ∠ bisector
5. $\overline{HX} \cong \overline{HX}$	5. Reflex. Prop. of ≅
6. $\triangle MHX \cong \triangle RHX$	6. SAS
7. $\angle HMX \cong \angle HRX$	7. CPCTC

9. Both pairs of opp. sides of PQRS are ≅, so *PQRS* is a ▱. Since $PZ = QZ$ and $RZ = SZ$, it follows that $PR = QS$ by the Seg. Add. Post. Thus $\overline{PR} \cong \overline{QS}$. So the diags. of ▱*PQRS* are ≅. The frame is a rect. by ▱ with diagonals ≅ → rect. thm.

10. valid (by quad. with diagonals that bisect each other → ▱ thm. and ▱ with diagonals ⊥ → rhombus thm.)

11. Not valid; by ▱ with diagonals that bisect opp. ∠s → rhombus thm. if one diag. of a ▱ bisects a pair of opp. ∠s, then the ▱ is a rhombus. To apply this thm., you need to know that *EFGH* is a ▱.

12. **Step 1** Determine if *ABCD* is a rect.

$AC = \sqrt{14^2 + 2^2} = \sqrt{200} = 10\sqrt{2}$

$BD = \sqrt{2^2 + 14^2} = 10\sqrt{2}$

$AC = BD$, so diags. are ≅. *ABCD* is a rect.

Step 2 Determine if *ABCD* is a rhombus.

slope of $\overline{AC} = \frac{-2}{14} = -\frac{1}{7}$

slope of $\overline{BD} = \frac{-14}{-2} = 7$

Since $\left(-\frac{1}{7}\right)(7) = -1$, $AC \perp BD$. *ABCD* is a rhombus.

Since *ABCD* is a rect. and a rhombus, *ABCD* is a square.

13. Step 1 Determine if *JKLM* is a rect.

$JL = \sqrt{12^2 + 4^2} = 4\sqrt{10}$

$KM = \sqrt{2^2 + 6^2} = 2\sqrt{10}$

$JL \neq KM$, so *JKLM* is not a rect., and therefore not a square.

Step 2 Determine if *JKLM* is a rhombus.

slope of $\overline{JL} = \frac{4}{12} = \frac{1}{3}$

slope of $\overline{KM} = \frac{-6}{2} = -3$

Since $\left(\frac{1}{3}\right)(-3) = -1$, $JL \perp KM$. *JKLM* is a rhombus.

14. 1 Understand the Problem

Answer has 2 parts:

- Amount of iron needed to outline 1 kite
- Amount of iron needed for 1 complete section

2 Make a Plan

Diags. of a kite are ⊥., so 4 △ are rt. △. Use Pyth. Thm. and props. of kites to find unknown side lengths. Add these lengths to find perimeter of kite.

3 Solve

shorter side length $= \sqrt{7^2 + 7^2} = 7\sqrt{2}$

longer side length $= \sqrt{7^2 + 17^2} = 13\sqrt{2}$

perimeter $= 2(7\sqrt{2}) + 2(13\sqrt{2})$

$= 40\sqrt{2} \approx 56.6$ in.

56.6 in. of iron is needed for 1 kite.

For 1 complete section, need iron for 4 kites and 1 square of side length $2(7 + 17) = 48$ in.

Amount of iron $= 4(40\sqrt{2}) + 4(48) \approx 418.3$ in.

4 Look Back

To estimate perimeter, change side lengths into decimals and round. $7\sqrt{2} \approx 10$, and $13\sqrt{2} \approx 18$. Perimeter of one kite is approx. $2(10) + 2(18) = 56$. So 56.6 in. is a reasonable answer.

15. Think: Kite → diags. ⊥.

$m\angle XDA + m\angle DAX = 90$

$m\angle XDA + 32 = 90$

$m\angle XDA = 58°$

16. Think: Kite → 1 pair opp. ∠ ≅.

$m\angle ABC = m\angle ADC$

$= m\angle XDA + m\angle XDC$

$= 58 + 64 = 122°$

17. Think: Use Quad. ∠ Sum Thm.

$m\angle BCD + m\angle ADC + m\angle BAD + m\angle ABC = 360$

$m\angle BCD + 122 + 2m\angle DAX + 122 = 360$

$m\angle BCD + 122 + 2(32) + 122 = 360$

$m\angle BCD = 52°$

18. Think: Use Same-Side Int. ∠ Thm., isosc. trap. → base ∠ ≅ thm.

$m\angle L + m\angle M = 180$

$m\angle L + 118 = 180$

$m\angle L = 62°$

$\angle Q \cong \angle L$

$m\angle Q = m\angle L = 62°$

19. Think: Use isosc. trap. ↔ diagonals ≅ thm.

$\overline{RJ} \cong \overline{SK}$

$RJ = SK$

$= SZ + KZ$

$= 62.6 + 34 = 96.6$

20. Think: 1 pair base ∠ ≅ → trap. isosc.

$\angle X \cong \angle W$

$m\angle X = m\angle W$

$a^2 + 15 = 2a^2 - 65$

$80 = a^2$

$a = \pm\sqrt{80}$

$= \pm 4\sqrt{5}$

21. Think: Diags. ≅ → trap. isosc.

$\overline{GJ} \cong \overline{FH}$

$GJ = FH$

$4x - 1 = 9x - 15$

$14 = 5x$

$x = 2.8$

22. Think: Use Trap. Midseg. Thm.

$PQ = \frac{1}{2}(AV + BT)$

$= \frac{1}{2}(4.2 + 3) = 3.6$

23. Think: Use Trap. Midseg. Thm.

$MN = \frac{1}{2}(JT + KR)$

$52.5 = \frac{1}{2}(32.5 + KR)$

$105 = 32.5 + KR$

$72.5 = KR$

Similarity

Similarity Relationships

Ratio and Proportion

1. $\text{slope} = \frac{\text{rise}}{\text{run}} = \frac{y_2 - y_1}{x_2 - x_1}$
$= \frac{4-3}{1-(-1)} = \frac{1}{2}$

2. $\text{slope} = \frac{\text{rise}}{\text{run}} = \frac{y_2 - y_1}{x_2 - x_1}$
$= \frac{2-(-2)}{2-(-2)}$
$= \frac{4}{4} = \frac{1}{1}$

3. $\text{slope} = \frac{\text{rise}}{\text{run}} = \frac{y_2 - y_1}{x_2 - x_1}$
$= \frac{-1-1}{2-(-1)}$
$= \frac{-2}{3} = -\frac{2}{3}$

4. Let side lengths be $2x$, $4x$, $5x$, and $7x$. Then $2x + 4x + 5x + 7x = 36$. After like terms are combined, $18x = 36$. So $x = 2$. The shortest side measures $2x = 2(2) = 4$ m.

5. Let ∠ measures be $5x$, $12x$, and $19x$. Then $5x + 12x + 19x = 180$. After like terms are combined, $36x = 180$. So $x = 5$. The largest ∠ measures $19x = 19(5) = 95°$.

6. $\frac{x}{2} = \frac{40}{16}$
$x(16) = 2(40)$
$16x = 80$
$x = 5$

7. $\frac{7}{y} = \frac{21}{27}$
$7(27) = y(21)$
$189 = 21y$
$y = 9$

8. $\frac{6}{58} = \frac{t}{29}$
$6(29) = 58(t)$
$174 = 58t$
$t = 3$

9. $\frac{y}{3} = \frac{27}{y}$
$y(y) = 3(27)$
$y^2 = 81$
$y = \pm 9$

10. $\frac{16}{x-1} = \frac{x-1}{4}$
$16(4) = (x-1)(x-1)$
$64 = x^2 - 2x + 1$
$0 = x^2 - 2x - 63$
$0 = (x-9)(x+7)$
$x = 9 \text{ or } -7$

11. $\frac{x^2}{18} = \frac{x}{6}$
$x^2(6) = 18(x)$
$6x^2 - 18x = 0$
$6x(x-3) = 0$
$x = 0 \text{ or } 3$

12. $2a = 8b$
$\frac{a}{b} = \frac{8}{2}$
$\frac{a}{b} = \frac{4}{1} = 4{:}1$

13. $6x = 27y$
$\frac{6}{27} = \frac{y}{x}$
$\frac{y}{x} = \frac{2}{9} = 2{:}9$

14. 1 **Understand the Problem**
Answer will be height of Arkansas State Capitol.
2 Make a Plan
Let x be height of Arkansas State Capitol. Write a proportion that compares the ratios of height to width.

$$\frac{\text{height of U.S. Capitol}}{\text{width of U.S. Capitol}} = \frac{\text{height of Arkansas Capitol}}{\text{width of Arkansas Capitol}}$$

$$\frac{288}{752} = \frac{x}{564}$$

3 Solve

$$\frac{288}{752} = \frac{x}{564}$$

$$288(564) = 752(x)$$

$$162{,}432 = 752x$$

$$x = 216 \text{ ft}$$

4 Look Back
Check answer in original problem. Ratio of height to width for U.S. Capitol is 288:752, or 18:47. Ratio of height to width for Arkansas State Capitol is 216:564 In simplest form, this ratio is also 18:47. So ratios are equal, and answer is correct.

Similarity

Similarity Relationships

Ratios in Similar Polygons

1. $\angle M \cong \angle U$ and $\angle N \cong \angle V$. By 3rd ∠ Thm., $\angle P \cong \angle W$.
$\frac{MN}{UV} = \frac{4}{8} = \frac{1}{2}, \frac{MP}{UW} = \frac{3}{6} = \frac{1}{2}, \frac{NP}{VW} = \frac{2}{4} = \frac{1}{2}$

2. $\angle A \cong \angle H$ and $\angle C \cong \angle K$. By def. of ≅ ∠s, and taking vertices clockwise in both figures, $\angle B \cong \angle J$ and $\angle D \cong \angle L$.
$\frac{AB}{HJ} = \frac{8}{12} = \frac{2}{3}, \frac{BC}{JK} = \frac{4}{6} = \frac{2}{3}, \frac{CD}{KL} = \frac{4}{6} = \frac{2}{3},$
$\frac{DA}{LH} = \frac{8}{12} = \frac{2}{3}$

3. **Step 1** Identify pairs of ≅ ∠s. Think: All ∠s of a rect. are rt. ∠s and are ≅.
$\angle A \cong \angle E, \angle B \cong \angle F, \angle C \cong \angle G$, and $\angle D \cong \angle H$.
Step 2 Compare corr. sides.
$\frac{AB}{EF} = \frac{135}{90} = \frac{3}{2}, \frac{AD}{EH} = \frac{45}{30} = \frac{3}{2}$
Yes; since opp. sides of a rect. are ≅, corr. sides are proportional. Similarity ratio is $\frac{3}{2}$, and $ABCD \sim EFGH$.

4. **Step 1** Identify pairs of ≅ ∠s.
$\angle M \cong \angle W, \angle P \cong \angle U$ (Given)
$\angle R \cong \angle X$ (3rd ∠ Thm.)
Step 2 Compare corr. sides.
$\frac{RM}{XW} = \frac{8}{12} = \frac{2}{3}, \frac{MP}{WU} = \frac{10}{15} = \frac{2}{3}, \frac{RP}{XU} = \frac{4}{6} = \frac{2}{3}$
yes; similarity ratio is $\frac{2}{3}$, and $\triangle RMP \sim \triangle XWU$.

5. Let x be height of reproduction in feet. Reproduction is ~ to original, so corr. lengths are proportional.
$$\frac{\text{height of reproduction}}{\text{height of original}} = \frac{\text{width of reproduction}}{\text{width of original}}$$
$$\frac{x}{73} = \frac{24}{58}$$
$$58x = 73(24) = 1752$$
$$x = \frac{1752}{58} \approx 30 \text{ ft}$$

Similarity

Similarity Relationships

Triangle Similarity: AA, SSS, and SAS

1. By def. of $\angle \cong$, $\angle C \cong \angle H$. By $\triangle$ Sum Thm., $m\angle A = 47°$, so $\angle A \cong \angle F$. Therefore $\triangle ABC \sim \triangle FGH$ by AA $\sim$.

2. $\angle P \cong \angle T$ (given). $\angle QST$ is a rt. $\angle$ by the Lin. Pair Thm., so $\angle QST \cong \angle RSP$. Therefore $\triangle QST \sim \triangle RSP$ by AA $\sim$.

3. $\frac{DE}{JK} = \frac{8}{16} = \frac{1}{2}, \frac{DF}{JL} = \frac{6}{12} = \frac{1}{2}, \frac{EF}{KL} = \frac{10}{20} = \frac{1}{2}$
   Therefore $\triangle DEF \sim \triangle JKL$ by SSS $\sim$.

4. $\angle NMP \cong \angle RMQ$ (given)
   $\frac{MN}{MR} = \frac{4}{6} = \frac{2}{3}, \frac{MP}{MQ} = \frac{8}{4+8} = \frac{8}{12} = \frac{2}{3}$
   Therefore $\triangle MNP \sim \triangle MRQ$ by SAS $\sim$.

5. **Step 1** Prove $\triangle$s are $\sim$.
   It is given that $\angle C \cong \angle E$. $\angle A \cong \angle A$ by Reflex. Prop. of $\cong$. Therefore $\triangle AED \cong \triangle ACB$ by AA $\sim$.
   Step 2 Find AB.
   $\frac{AB}{AD} = \frac{BC}{DE}$
   $\frac{AB}{6} = \frac{15}{9}$
   $9AB = 15(6) = 90$
   $AB = 10$

6. **Step 1** Prove $\triangle$s are $\sim$.
   Since $\overline{UV} \parallel \overline{XY}$, by Alt. Int. $\angle$s Thm., $\angle U \cong \angle Y$ and $\angle V \cong \angle X$. Therefore $\triangle UVW \sim \triangle YXW$ by AA $\sim$.
   Step 2 Find WY.
   $\frac{WY}{WU} = \frac{WX}{WV}$
   $\frac{WY}{9} = \frac{8.75}{7}$
   $7WY = 9(8.75) = 78.75$
   $WY = 11.25$

7.

Statements	Reasons
1. $\overline{MN} \parallel \overline{KL}$	1. Given
2. $\angle JMN \cong \angle JKL$, $\angle JNM \cong \angle JLK$	2. Corr. $\angle$s Post.
3. $\triangle JMN \sim \triangle JKL$	3. AA $\sim$ *Step 2*

8.

Statements	Reasons
1. $SQ = 2QP$, $TR = 2RP$	1. Given
2. $SP = SQ + QP$, $TP = TR + RP$	2. Seg. Add. Post.
3. $SP = 2QP + QP$, $TP = 2RP + RP$	3. Subst.
4. $SP = 3QP$, $TP = 3RP$	4. Seg. Add. Post.
5. $\frac{SP}{QP} = 3$, $\frac{TP}{RP} = 3$	5. Div. Prop. of =
6. $\angle P \cong \angle P$	6. Reflex. Prop. of $\cong$
7. $\triangle PQR \sim \triangle PST$	7. SAS ~ *Steps 5, 6*

9. SAS or SSS ~ Thm.

10. **Step 1** Prove △s are ~.

$\angle S \cong \angle S$ by Reflex. Prop. of $\cong$

$\frac{SA}{SC} = \frac{733 + 586}{586} \approx 2.25$, $\frac{SB}{SD} = \frac{800 + 644}{644} \approx 2.24$

Therefore $\triangle SAB \sim \triangle SCD$ by SAS ~.

Step 2 Find AB.

$\frac{AB}{CD} = \frac{SA}{SC}$

$\frac{AB}{533} \approx 2.25$

$AB \approx 2.25(533)$

≈ 1200 m or 1.2 km

Similarity

Review 7.1 / Similarity Relationships

1. slope $= \frac{4-1}{1-0} = \frac{3}{1}$

2. slope $= \frac{-4+1}{3-0} = -\frac{1}{1}$

3. slope $= \frac{0+3}{3-1} = \frac{3}{2}$

4. Let side lengths be $4x$ and $4x$, and let base length be $7x$.
$4x + 4x + 7x = 52.5$
$15x = 52.5$
$x = 3.5$
length of base $= 7(3.5) = 24.5$ cm

5. Let ∠ measures be $2x$, $3x$, $2x$, and $3x$. By Quad. ∠ Sum Thm., sum of ∠ measures is 360°.
$2x + 3x + 2x + 3x = 360$
$10x = 360$
$x = 36$
∠ measures are $2(36) = 72°$, $3(36) = 108°$, 72°, and 108°.

6. $\frac{6}{8} = \frac{9}{y}$
$6y = 8(9) = 72$
$y = 12$

7. $\frac{x}{14} = \frac{50}{35}$
$35x = 14(50) = 700$
$x = 20$

8. $\frac{z}{12} = \frac{3}{8}$
$8z = 12(3) = 36$
$z = 4.5$

9. $\frac{2m+2}{3} = \frac{12}{2m+2}$
$(2m+2)^2 = 3(12)$
$4m^2 + 8m + 4 = 36$
$4m^2 + 8m - 32 = 0$
$m^2 + 2m - 8 = 0$
$(m-2)(m+4) = 0$
$m = 2$ or -4

10. $\frac{5y}{16} = \frac{125}{y}$
$5y^2 = 16(125)$
$5y^2 = 2000$
$y^2 = 400$
$y = \pm 20$

11. $\frac{x+2}{12} = \frac{5}{x-2}$
$(x+2)(x-2) = 12(5)$
$x^2 - 4 = 60$
$x^2 = 64$
$x = \pm 8$

12. $5y = 25x$
$\frac{5}{25} = \frac{x}{y}$
$\frac{x}{y} = \frac{1}{5}$

13. $35b = 21c$
$\frac{b}{c} = \frac{21}{35} = \frac{3}{5}$
Ratio is 3:5.

14. Let x represent height of actual windmill.
$\frac{\text{height of windmill}}{\text{width of windmill}} = \frac{\text{height of model}}{\text{width of model}}$
$\frac{x}{20} = \frac{1.2}{0.8}$
$0.8x = 20(1.2) = 24$
$x = 30$ m

15. $\angle K \cong \angle T$, $\angle L \cong \angle U$ (Given)
$\angle J \cong \angle S$, $\angle M \cong \angle V$ (Rt. $\angle \cong$ Thm.)
$\frac{JK}{ST} = \frac{20}{24} = \frac{5}{6}$, $\frac{KL}{TU} = \frac{14}{16.8} = \frac{5}{6}$, $\frac{LM}{UV} = \frac{30}{36} = \frac{5}{6}$,
$\frac{JM}{SV} = \frac{10}{12} = \frac{5}{6}$

16. $\angle A \cong \angle X$, $\angle C \cong \angle Z$ (Given)
$\angle B \cong \angle Y$ (3rd ∡ Thm.)
$\frac{AB}{XY} = \frac{8}{4} = 2$, $\frac{BC}{YZ} = \frac{6}{3} = 2$, $\frac{CA}{ZX} = \frac{12}{6} = 2$

17. **Step 1** Identify pairs of ≅ ∡.
$m\angle R = 90 - 53 = 37°$
$\angle R \cong \angle U$ (Def. of ≅ ∡)
$\angle S \cong \angle Z$ (Rt. $\angle \cong$ Thm.)
$\angle Q \cong \angle X$ (3rd ∡ Thm.)
Step 2 Compare corr. sides.
$\frac{QR}{XU} = \frac{35}{40} = \frac{7}{8}$, $\frac{QS}{XZ} = \frac{21}{24} = \frac{7}{8}$, $\frac{RS}{UZ} = \frac{28}{32} = \frac{7}{8}$
yes; similarity ratio $= \frac{7}{8}$; $\triangle RSQ \sim \triangle UZX$

18. **Step 1** Identify pairs of ≅ ∡.
$\angle A \cong \angle M$, $\angle B \cong \angle J$, $\angle C \cong \angle K$, $\angle D \cong \angle L$
(Rt. $\angle \cong$ Thm.)
Step 2 Compare corr. sides.
$\frac{AB}{MJ} = \frac{18}{24} = \frac{3}{4}$, $\frac{AD}{ML} = \frac{AD}{JK} = \frac{36}{54} = \frac{2}{3}$
no; the rectangles are not similar

19. $\frac{\text{model length}}{\text{car length}} = \frac{1}{56}$
$\frac{3}{\ell} = \frac{1}{56}$
$3(56) = \ell$
$\ell = 168 \text{ in.} = 14 \text{ ft}$

20. $\angle G \cong \angle G$ by Reflex. Prop. of ≅. $\angle GLH \cong \angle K$ by Rt. $\angle \cong$ Thm. Therefore $\triangle HLG \sim \triangle JKG$ by AA ~.

21. By Isosc. △ Thm., $\angle B \cong \angle C$ and $\angle E \cong \angle F$. By △ Sum Thm.,
$32 + 2m\angle B = 180$
$2m\angle B = 148°$
$m\angle B = 74°$
By def. of ≅ ∡, $\angle B \cong \angle E$ and $\angle C \cong \angle F$.
Therefore $\triangle ABC \sim \triangle DEF$ by AA ~.

22. $\angle K \cong \angle K$ by Reflex. Prop. of ≅
$\frac{KL}{KN} = \frac{6}{4} = \frac{3}{2}$, $\frac{KM}{KL} = \frac{5+4}{6} = \frac{3}{2}$
Therefore $\triangle KLM \sim \triangle KNL$ by SAS ~.

23. $\frac{UV}{XY} = \frac{VW}{YZ} = \frac{WU}{ZX} = \frac{4}{5.5} = \frac{8}{11}$
Therefore $\triangle UVW \sim \triangle XYZ$ by SSS ~.

24. **Step 1** Prove ∡ are ~.
It is given that $\angle ABD \cong \angle C$. $\angle A \cong \angle A$ by Reflex. Prop. of ≅. Therefore $\triangle ABD \cong \triangle ACB$ by AA ~.
Step 2 Find AB.
$\frac{AB}{AD} = \frac{AC}{AB}$
$\frac{AB}{4} = \frac{4+12}{AB}$
$AB^2 = 4(16) = 64$
$AB = +\sqrt{64} = 8$

Similarity

Review 7.1 / Similarity Relationships

25. **Step 1** Prove ⊿s are ~.
Since $\overline{ST} \parallel \overline{VW}$, $\angle PST \cong \angle V$ by Corr. ∠s Post. $\angle P \cong \angle P$ by Reflex. Prop. of ≅. Therefore $\triangle PST \sim \triangle PVW$ by AA ~.

Step 2 Find *PS*.

$$\frac{PS}{PV} = \frac{ST}{VW}$$
$$\frac{PS}{PS+6} = \frac{10}{17.5} = \frac{4}{7}$$
$$7PS = 4(PS+6)$$
$$7PS = 4PS + 24$$
$$3PS = 24$$
$$PS = 8$$

26.

Statements	Reasons
1. $CD = 3AC$, $CE = 3BC$	1. Given
2. $\frac{CD}{AC} = 3$, $\frac{CE}{BC} = 3$	2. Div. Prop. of ≅
3. $\angle ACB \cong \angle DCE$	3. Vert. ∠s Thm.
4. $\triangle ABC \sim \triangle DEC$	4. SAS ~ *Steps 2, 3*

27.

Statements	Reasons
1. $\frac{PR}{MR} = \frac{QR}{NR}$	1. Given
2. $\angle R \cong \angle R$	2. Reflex. Prop. of ≅
3. $\triangle PQR \sim \triangle MNR$	3. SAS ~ *Steps 1, 2*
4. $\angle 1 \cong \angle 2$	4. Def. of ~ ⊿s

28.

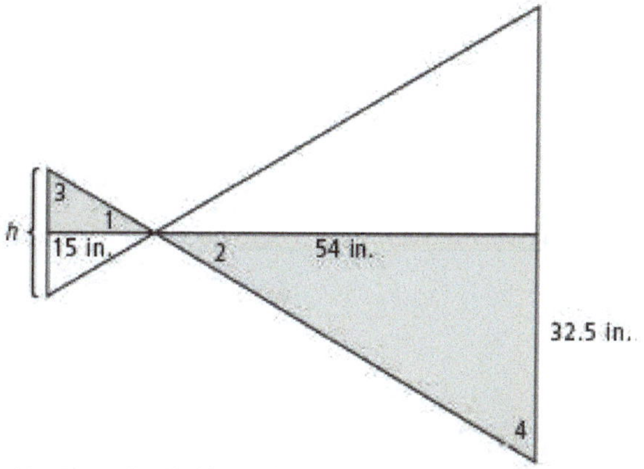

By Vert. ∠s Thm., $\angle 1 \cong \angle 2$. Since vert. sides are ∥, $\angle 3 \cong \angle 4$ by Corr. ∠s Post., so marked ⊿s are ~. Therefore,

$$\frac{h \div 2}{15} = \frac{32.5}{54}$$
$$54\left(\frac{h}{2}\right) = 15(32.5)$$
$$27h = 487.5$$
$$h \approx 18.1 \text{ in.}$$

Right Triangles and Trigonometry

Trigonometric Ratios

Similarity in Right Triangles

1. Sketch the 3 rt. △ with ∠ of △ in corr. positions.

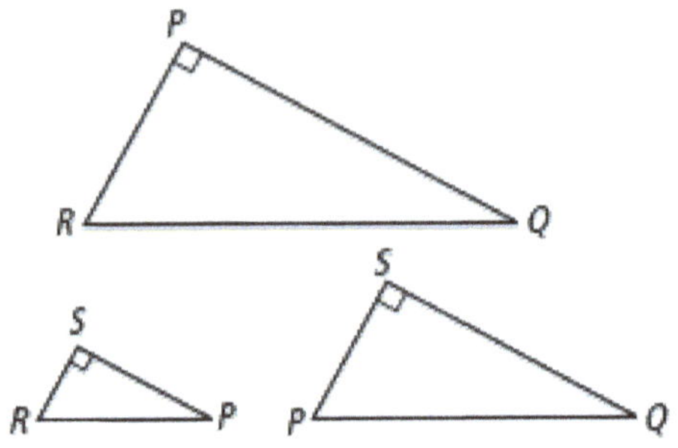

2. Sketch the 3 rt. △ with ∠ of △ in corr. positions.

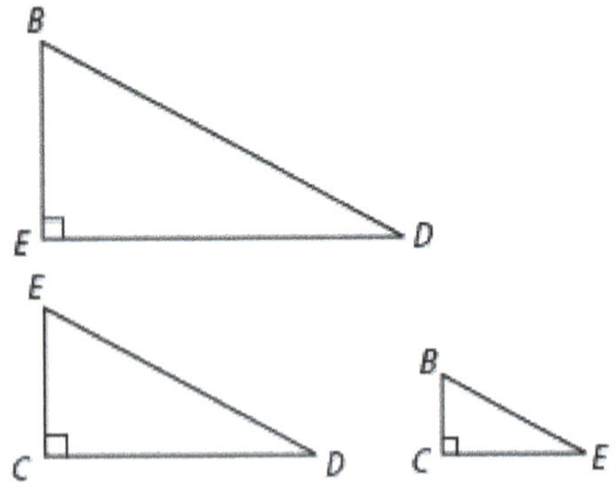

3. Sketch the 3 rt. △ with ∠ of △ in corr. positions.

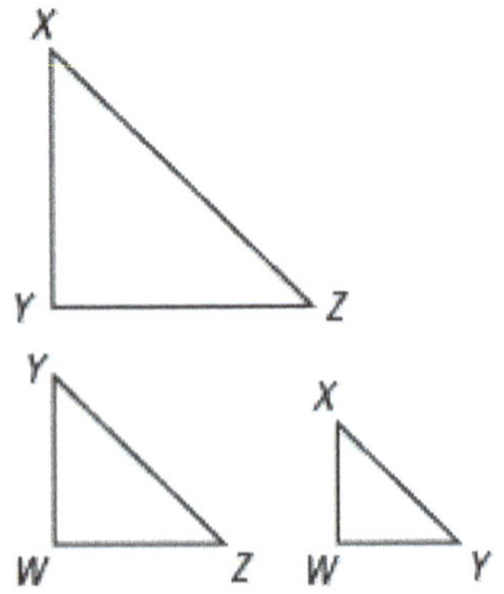

4. $x^2 = (2)(50) = 100$
   $x = 10$

5. $x^2 = (4)(16) = 64$
   $x = 8$

6. $x^2 = \left(\frac{1}{2}\right)(8) = 4$
   $x = 2$

7. $x^2 = (9)(12) = 108$
   $x = \sqrt{108} = 6\sqrt{3}$

8. $x^2 = (16)(25) = 400$
   $x = 20$

9. $x^2 = (7)(11) = 77$ $\quad y^2 = (6)(4) = 24$
   $x = \sqrt{77}$ $\quad y = \sqrt{24} = 2\sqrt{6}$

10. $x^2 = (10)(6) = 60$ $\quad z^2 = (10)(4) = 40$
    $x = \sqrt{60} = 2\sqrt{15}$ $\quad z = \sqrt{40} = 2\sqrt{10}$

11. $10^2 = 100 = 20x$ $\quad y^2 = (20)(20 + 5) = 500$
    $x = 5$ $\quad y = \sqrt{500} = 10\sqrt{5}$
    $z^2 = (5)(20 + 5) = 125$
    $z = \sqrt{125} = 5\sqrt{5}$

12. $(6\sqrt{13})^2 = (18)(18 + z)$ $\quad x^2 = (8)(18) = 144$
    $468 = 324 + 18z$ $\quad x = 12$
    $144 = 18z$
    $z = 8$
    $y^2 = (8)(18 + 8) = 208$
    $y = \sqrt{208} = 4\sqrt{13}$

13. $RS^2 = (64)(60) = 3840$
    $RS = \sqrt{3840} \approx 62.0$ m

Right Triangles and Trigonometry

Trigonometric Ratios

Trigonometric Ratios

1. $\sin C = \frac{4}{5} = 0.8$

2. $\tan A = \frac{3}{4} = 0.75$

3. $\cos A = \frac{4}{5} = 0.8$

4. $\cos C = \frac{3}{5} = 0.6$

5. $\tan C = \frac{4}{3} \approx 1.33$

6. $\sin A = \frac{3}{5} = 0.6$

7. 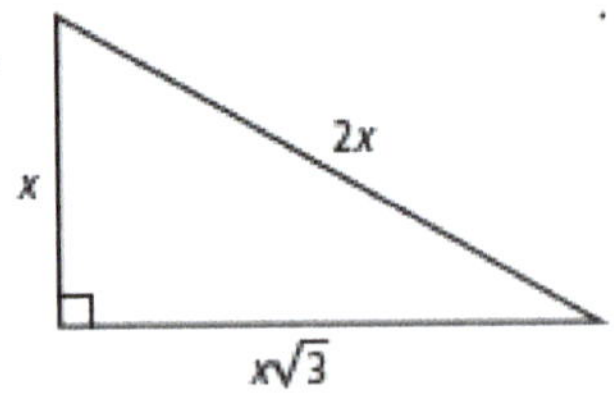

$\cos 60° = \frac{x}{2x} = \frac{1}{2}$

8. $\tan 30° = \frac{x}{x\sqrt{3}} = \frac{\sqrt{3}}{3}$

9. 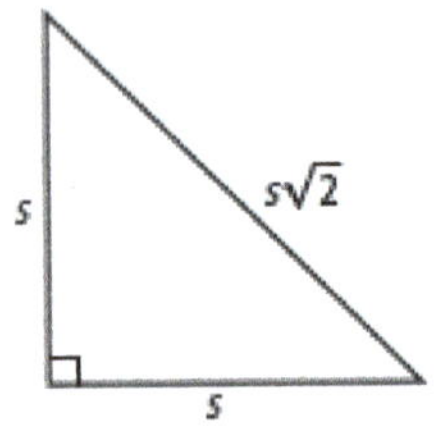

$\sin 45° = \frac{s}{s\sqrt{2}} = \frac{\sqrt{2}}{2}$

10. 

$\tan 67° \approx 2.36$

11. 

$\sin 23° \approx 0.39$

12. 

$\sin 49° \approx 0.75$

13. 

$\cos 88° \approx 0.03$

14. 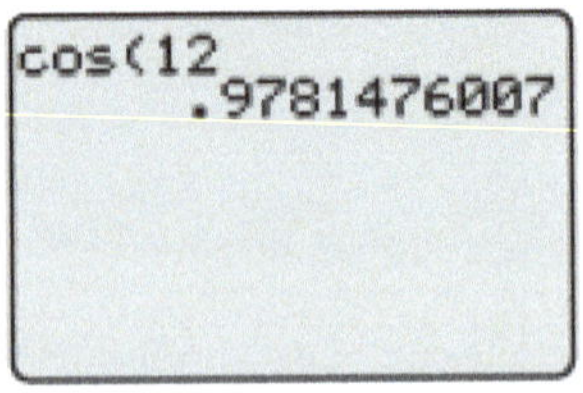

$\cos 12° \approx 0.98$

15.

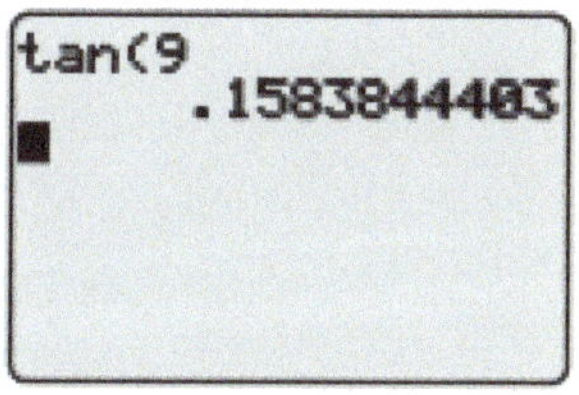

$\tan 9° \approx 0.16$

16. $\overline{BC}$ is opp. the given ∠. Given: AC, the hyp. Since opp. side and hyp. are involved, use a sine ratio.

$$\sin A = \frac{\text{opp. leg}}{\text{hyp.}} = \frac{BC}{AC}$$
$$\sin 23° = \frac{BC}{4}$$
$$4(\sin 23°) = BC$$
$$BC \approx 1.56 \text{ in.}$$

17. $\overline{QR}$ is opp. the given ∠. Given: PQ, adj. to given ∠. Since opp. and adj. sides are involved, use a tangent ratio.

$$\tan P = \frac{\text{opp. leg}}{\text{adj. leg}} = \frac{QR}{PQ}$$
$$\tan 50° = \frac{QR}{8.1}$$
$$8.1(\tan 50°) = QR$$
$$QR \approx 9.65 \text{ m}$$

18. $\overline{KL}$ is adj. to the given ∠. Given: JL, the hyp. Since adj. side and hyp. are involved, use a cosine ratio.

$$\cos L = \frac{\text{adj. leg}}{\text{hyp.}} = \frac{KL}{JL}$$
$$\cos 61° = \frac{KL}{2.5}$$
$$2.5(\cos 61°) = KL$$
$$KL \approx 1.21 \text{ cm}$$

19. 1 Understand the Problem

The **answer** is XY, opp. the given ∠.

2 Make a Plan

You are given WZ, which is twice WY, the leg adj. to $\angle W$. First, calculate WY. Then, since opp. and adj. legs are involved, write an equation using a tangent ratio.

3 Solve

$$WY = \frac{1}{2}WZ$$
$$= \frac{1}{2}(56) = 28 \text{ ft}$$
$$\tan W = \frac{XY}{WY}$$
$$\tan 15° = \frac{XY}{28}$$
$$XY = 28(\tan 15°) \approx 7.5028 \text{ ft}$$

4 Look Back

Problem asks for XY rounded to nearest inch. Height XY of pediment is 7 ft 6 in.

Right Triangles and Trigonometry

Trigonometric Ratios

Solving Right Triangles

1. $\frac{8}{10} = \frac{4}{5} = \sin A$
$\rightarrow \angle A$ is $\angle 1$

2. $\frac{8}{6} = 1\frac{1}{3} = \tan A$
$\rightarrow \angle A$ is $\angle 1$

3. $\frac{6}{10} = 0.6 = \cos A$
$\rightarrow \angle A$ is $\angle 1$

4. $\frac{8}{10} = 0.8 = \cos A$
$\rightarrow \angle A$ is $\angle 2$

5. $\frac{6}{8} = 0.75 = \tan A$
$\rightarrow \angle A$ is $\angle 2$

6. $\frac{6}{10} = 0.6 = \sin A$
$\rightarrow \angle A$ is $\angle 2$

7. 

$\tan^{-1}(2.1) \approx 65°$

8. 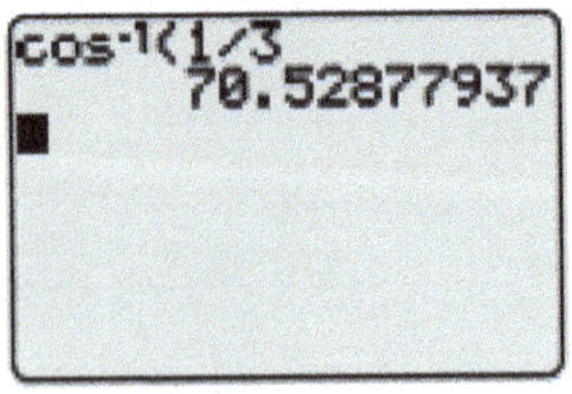

$\cos^{-1}\left(\frac{1}{3}\right) \approx 71°$

9. 

$\cos^{-1}\left(\frac{5}{6}\right) \approx 34°$

10. 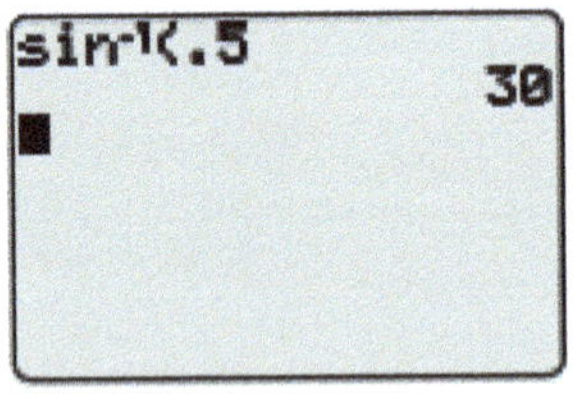

$\sin^{-1}(0.5) = 30°$

11. 

$\sin^{-1}(0.61) \approx 38°$

12. 

$\tan^{-1}(0.09) \approx 5°$

13. $\tan P = \frac{3.1}{8.9}$

$m\angle P = \tan^{-1}\left(\frac{3.1}{8.9}\right) \approx 19°$

Acute ∡ of a rt. △ are comp. So

$m\angle R \approx 90 - 19 \approx 71°$.

$RP = \frac{3.1}{\sin P}$

$= \frac{3.1}{\sin\left(\tan^{-1}\left(\frac{3.1}{8.9}\right)\right)} \approx 9.42$

Solving Right Triangles

14. $AB = 7.4\cos 32^\circ \approx 6.28$
$BC = 7.4\sin 32^\circ \approx 3.92$
Acute ∡ of a rt. △ are comp. So
$m\angle C = 90 - 32 = 58^\circ$.

15. By Pyth. Thm.,
$YZ = \sqrt{11^2 + 8.6^2}$
$= \sqrt{194.96} \approx 13.96$
$\tan Y = \frac{8.6}{11}$
$m\angle Y = \tan^{-1}\left(\frac{8.6}{11}\right) \approx 38^\circ$
$\tan Z = \frac{11}{8.6}$
$m\angle Z = \tan^{-1}\left(\frac{11}{8.6}\right) \approx 52^\circ$

16. Step 1 Find side lengths.
Plot pts. D, E, and F.
$DE = 3 \quad EF = 6$
$DF = \sqrt{(-2-4)^2 + (-2-1)^2}$
$= \sqrt{(-6)^2 + (-3)^2}$
$= \sqrt{45} \approx 6.71$

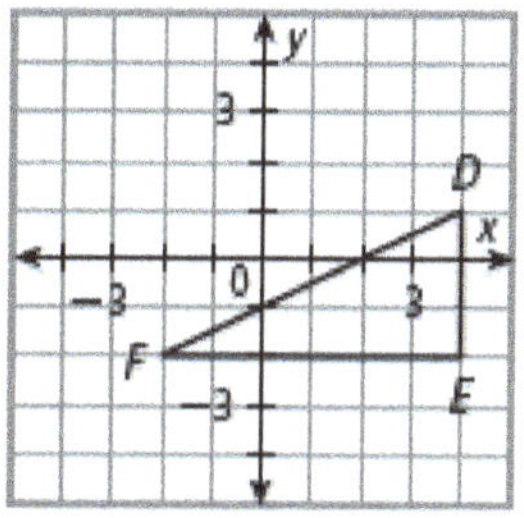

Step 2 Find ∠ measures.
$m\angle E = 90^\circ$
$m\angle D = \tan^{-1}\left(\frac{6}{3}\right) \approx 63^\circ$
$m\angle F \approx 90 - 63 \approx 27^\circ$

17. Step 1 Find side lengths.
Plot pts. R, S, and T.
$RS = 5 \quad ST = 6$
$RT = \sqrt{(-2-3)^2 + (-3-3)^2}$
$= \sqrt{(-5)^2 + (-6)^2}$
$= \sqrt{61} \approx 7.81$

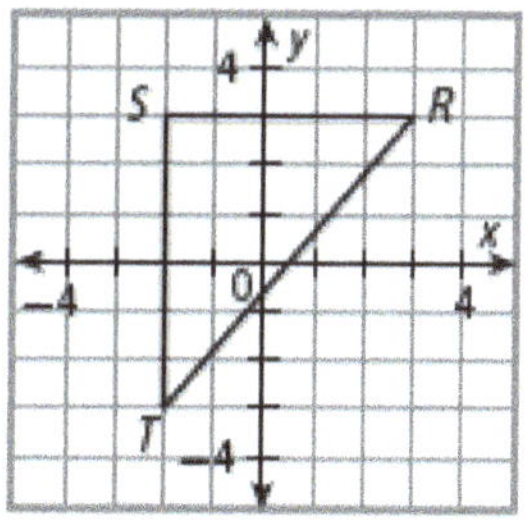

Step 2 Find ∠ measures.
$m\angle S = 90^\circ$
$m\angle R = \tan^{-1}\left(\frac{6}{5}\right) \approx 50^\circ$
$m\angle T \approx 90 - 50 \approx 40^\circ$

18. **Step 1** Find side lengths.
Plot pts. A, B, and C.
$AB = 2 \quad BC = 4$
$AC = \sqrt{(1+1)^2 + (5-1)^2}$
$= \sqrt{2^2 + 4^2}$
$= \sqrt{20} \approx 4.47$

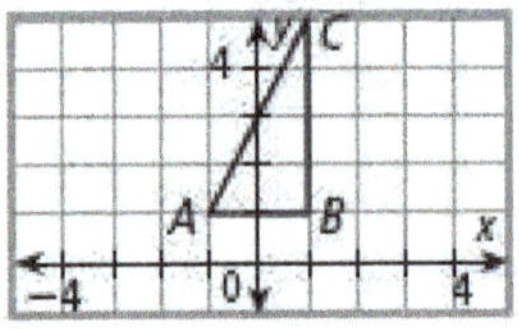

Step 2 Find ∠ measures.
$m\angle B = 90°$
$m\angle A = \tan^{-1}\left(\frac{4}{2}\right) \approx 63°$
$m\angle C = 90 - 63 \approx 27°$

19. $8\% = \frac{8}{100}$
An 8% grade means hill rises 8 m for every 100 m of horiz. dist.
$m\angle = \tan^{-1}\left(\frac{8}{100}\right) \approx 5°$

Right Triangles and Trigonometry

Review 8.1 / Trigonometric Ratios

1. By the thm. that states that the altitude to the hypotenuse of a right triangle forms two triangles that are similar to each other and to the original triangle, $\triangle MPN \sim \triangle PQN \sim \triangle MQP$.

2. By the thm. that states that the altitude to the hypotenuse of a right triangle forms two triangles that are similar to each other and to the original triangle, $\triangle CAB \sim \triangle ADB \sim \triangle CDA$.

3. By the thm. that states that the altitude to the hypotenuse of a right triangle forms two triangles that are similar to each other and to the original triangle, $\triangle RSU \sim \triangle RTS \sim \triangle STU$.

4. $x^2 = (5)(45) = 225$
   $x = 15$

5. $x^2 = (3)(15) = 45$
   $x = 3\sqrt{5}$

6. $x^2 = (5)(8) = 40$
   $x = 2\sqrt{10}$

7. $x^2 = \left(\frac{1}{4}\right)(80) = 20$
   $x = 2\sqrt{5}$

8. $x^2 = (1.5)(12) = 18$
   $x = 3\sqrt{2}$

9. $x^2 = \left(\frac{2}{3}\right)\left(\frac{27}{40}\right) = \frac{9}{20}$
   $x = \frac{3}{2\sqrt{5}} = \frac{3\sqrt{5}}{10}$

10. $12^2 = 4(4 + x)$
    $144 = 16 + 4x$
    $128 = 4x$
    $x = 32$
    $y^2 = 4(32) = 128$
    $y = 8\sqrt{2}$
    $z^2 = 32(4 + 32) = 1152$
    $z = 24\sqrt{2}$

11. $x^2 = (30)(40) = 1200$
    $x = 20\sqrt{3}$
    $y^2 = (30)(70) = 2100$
    $y = 10\sqrt{21}$
    $z^2 = (70)(40) = 2800$
    $z = 20\sqrt{7}$

12. $9.6^2 = (z)(12.8)$
    $92.16 = 12.8z$
    $z = 7.2$
    $y^2 = (12.8)(12.8 + 7.2) = 256$
    $y = 16$
    $x^2 = (7.2)(12.8 + 7.2) = 144$
    $x = 12$

13. Let h represent height of tower above eye level.
    91 ft 3 in. = 91.25 ft
    $(91.25)^2 = 5h$
    $h \approx 1665$ ft
    Tower is about $1665 + 5 = 1670$ ft high.

14. $\cos D = \frac{8}{17} \approx 0.47$

15. $\tan D = \frac{15}{8} \approx 1.88$

16. $\tan F = \frac{8}{15} \approx 0.53$

17. $\cos F = \frac{15}{17} \approx 0.88$

18. $\sin F = \frac{8}{17} \approx 0.47$

Right Triangles and Trigonometry

Review 8.1 / Trigonometric Ratios

19. $\sin D = \frac{15}{17} \approx 0.88$

20.

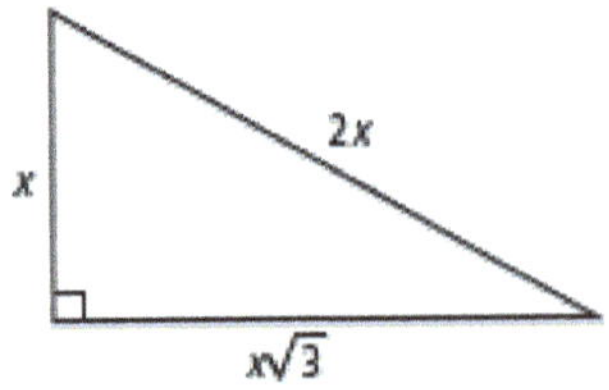

$\tan 60° = \frac{x\sqrt{3}}{x} = \sqrt{3}$

21. $\sin 30° = \frac{x}{2x} = \frac{1}{2}$

22.

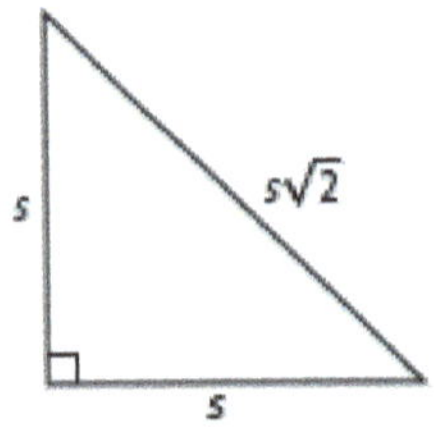

$\cos 45° = \frac{s}{s\sqrt{2}} = \frac{\sqrt{2}}{2}$

23. $\tan 51° \approx 1.23$

24. $\sin 80° \approx 0.98$

25. $\cos 77° \approx 0.22$

26. $\tan 14° \approx 0.25$

27. $\sin 55° \approx 0.82$

28. $\cos 48° \approx 0.67$

29. $PQ = 11 \sin 19°$
≈ 3.58 cm

30. $\cos 46° = \frac{19.2}{AC}$
$AC = \frac{19.2}{\cos 46°}$
≈ 27.64 in.

31. $\sin 34° = \frac{11}{GH}$
$GH = \frac{11}{\sin 34°}$
≈ 19.67 ft

32. $\cos 25° = \frac{33}{XZ}$
$XZ = \frac{33}{\cos 25°}$
≈ 36.41 in.

33. $\tan 61° = \frac{9.5}{KL}$
$KL = \frac{9.5}{\tan 61°}$
≈ 5.27 ft

34. $EF = 83.1 \tan 12°$
≈ 17.66 m

35. $\sin 15° = \frac{1.58}{\ell}$
$\ell = \frac{1.58}{\sin 15°} \approx 6.10$ m

36. $\frac{5}{12} = \frac{7.5}{18} = \tan\angle 2$
$\angle A = \angle 2$

37. $2.4 = \frac{18}{7.5} = \tan\angle 1$
$\angle A = \angle 1$

38. $\frac{12}{13} = \frac{18}{19.5} = \sin\angle 1$
$\angle A = \angle 1$

39. $\frac{5}{13} = \frac{7.5}{19.5} = \sin\angle 2$
$\angle A = \angle 2$

40. $\frac{12}{13} = \frac{18}{19.5} = \cos\angle 2$
$\angle A = \angle 2$

41. $\frac{5}{13} = \frac{7.5}{19.5} = \cos\angle 1$
$\angle A = \angle 1$

42. $\sin^{-1}(0.31) \approx 18°$

43. $\tan^{-1}(1) = 45°$

44. $\cos^{-1}(0.8) \approx 37°$

45. $\cos^{-1}(0.72) \approx 44°$

47. $\sin^{-1}\left(\frac{9}{17}\right) \approx 32°$

49. Step 1 Find unknown side length.

$DF = \sqrt{(\sqrt{5})^2 + (\sqrt{2})^2}$

$= \sqrt{5+2} = \sqrt{7} \approx 2.65$

Step 2 Find unknown ∠ measures.

$m\angle D = \tan^{-1}\left(\frac{\sqrt{2}}{\sqrt{5}}\right) \approx 32°$

$m\angle F = \tan^{-1}\left(\frac{\sqrt{5}}{\sqrt{2}}\right) \approx 58°$

51. Step 1 Find side lengths.

$\overline{AB}$ is vert., $AB = 5$; $\overline{BC}$ is horiz., $BC = 1$

By Pyth. Thm., $AC = \sqrt{5^2 + 1^2} = \sqrt{26} \approx 5.10$

Step 2 Find ∠ measures.

$m\angle B = 90°$

$m\angle A = \tan^{-1}\left(\frac{BC}{AB}\right) = \sin^{-1}\left(\frac{1}{5}\right) \approx 11°$

$m\angle C \approx 90 - 11 \approx 79°$

53. Range of ∠ measures is between $\tan^{-1}\left(\frac{1}{20}\right) \approx 3°$ and $\tan^{-1}\left(\frac{1}{16}\right) \approx 4°$.

46. $\tan^{-1}(1.55) \approx 57°$

48. Step 1 Find unknown side lengths.

$JK = 3.2\cos 26° \approx 2.88$

$LK = 3.2\sin 26° \approx 1.40$

Step 2 Find unknown ∠ measure.

$m\angle L = 90 - 26 = 64°$

50. Step 1 Find unknown ∠ measures.

$m\angle P = \cos^{-1}\left(\frac{6.7}{8.3}\right) \approx 36°$

$m\angle R = \sin^{-1}\left(\frac{6.7}{8.3}\right) \approx 54°$

Step 2 Find unknown side length.

$QR = 8.3\sin P$

$= 8.3\sin\left(\cos^{-1}\left(\frac{6.7}{8.3}\right)\right) \approx 4.90$

52. Step 1 Find side lengths.

$\overline{MN}$ is vert., $MN = 4$; $\overline{NP}$ is horiz., $NP = 4$

By Pyth. Thm., $MP = \sqrt{4^2 + 4^2} = \sqrt{32} \approx 5.66$

Step 2 Find ∠ measures.

$m\angle N = 90°$

$m\angle M = \tan^{-1}\left(\frac{NP}{MN}\right) = \tan^{-1}\left(\frac{4}{4}\right) = 45°$

$m\angle P = 90 - 45 = 45°$

Extending Perimeter, Circumference, and Area

Developing Geometric Formulas

Developing Formulas for Triangles and Quadrilaterals

1. $A = bh$
   $= (12)(10) = 120 \text{ cm}^2$

2. $A = bh$
   $10x^2 = (2x)h$
   $h = 5x \text{ ft}$

3. $A = s^2$
   $169 = s^2$
   $s = 13$
   $P = 4s$
   $= 4(13) = 52 \text{ cm}$

4. $A = \frac{1}{2}(b_1 + b_2)h$
   $= \frac{1}{2}(9 + 15)(20)$
   $= 240 \text{ m}^2$

5. $A = \frac{1}{2}bh$
   $58.5 = \frac{1}{2}b(9) = 4.5b$
   $b = 13 \text{ in.}$

6. $A = \frac{1}{2}(b_1 + b_2)h$
   $48x + 68 = \frac{1}{2}(b_1 + 9x + 12)(8)$
   $48x + 68 = 4b_1 + 36x + 48$
   $12x + 20 = 4b_1$
   $b_1 = (3x + 5) \text{ in.}$

7. $\left(\frac{d_1}{2}\right)^2 + \left(\frac{d_2}{2}\right)^2 = c^2$
   ${d_1}^2 + {d_2}^2 = 4c^2$
   ${d_1}^2 + 14^2 = 4(25)^2$
   ${d_1}^2 = 2500 - 196 = 2304$
   $d_1 = 48 \text{ in.}$
   $A = \frac{1}{2}d_1d_2$
   $= \frac{1}{2}(48)(14) = 336 \text{ in.}^2$

8. $A = \frac{1}{2}d_1d_2$
$187.5 = \frac{1}{2}(15)d_2$
$= 7.5d_2$
$d_2 = 25$ m

9. $A = \frac{1}{2}d_1d_2$
$12x^2y^3 = \frac{1}{2}(3xy)d_2$
$24x^2y^3 = (3xy)d_2$
$d_2 = 8xy^2$ cm

10. Each diagonal of a square has length $\sqrt{2}$.
The height of the parallelogram is 1.5 diagonals, or $1.5\sqrt{2}$.

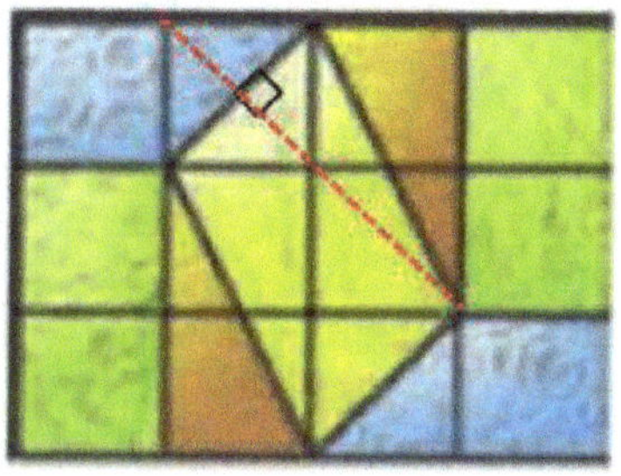

Rectangles: $A = bh = 1(2) = 2 \text{ ft}^2$
Triangles: $A = \frac{1}{2}bh = \frac{1}{2}(1)(2) = 1 \text{ ft}^2$
Trapezoids: $A = \frac{1}{2}(b_1 + b_2)h = \frac{1}{2}(1 + 2)(1) = 1.5 \text{ ft}^2$
Parallelogram: $A = bh = \sqrt{2}(1.5\sqrt{2}) \text{ ft}^2 = 3 \text{ ft}^2$

OR

Parallelogram: $A = A_{\text{Total}} - A_{\text{Rectangles}} - A_{\text{Triangles}} - A_{\text{Trapezoids}}$
$= 12 - 2(2) - 2(1) - 2(1.5) = 3 \text{ ft}^2$

Developing Formulas for Circles and Regular Polygons

1. $C = \pi d = \pi\left(\frac{10}{\pi}\right) = 10$ cm

2. $A = \pi r^2 = \pi(3x)^2 = 9x^2\pi$ in.^{2}

3. **Step 1** Use given area to solve for r.
$A = \pi r^2$
$36\pi = \pi r^2$
$36 = r^2$
$r = 6$ in.
Step 2 Use value of r to find circumference.
$C = 2\pi r$
$C = 2\pi(6) = 12\pi$ in.

4. $A = \pi\left(\frac{8}{2}\right)^2 \approx 50.3$ in.^{2}; $A = \pi\left(\frac{10}{2}\right)^2 \approx 78.5$ in.^{2};
$A = \pi\left(\frac{12}{2}\right)^2 \approx 113.1$ in.^{2}

5. **Step 1** Find perimeter.
$P = 6(10) = 60$ in.
Step 2 Use properties of 30°-60°-90° △ to find apothem.
$a = 5\sqrt{3}$ in.
Step 3 Use apothem and perimeter to find area.
$A = \frac{1}{2}aP$
$A = \frac{1}{2}(5\sqrt{3})(60) \approx 259.8$ in.^{2}

6. **Step 1** Find perimeter.
$P = 7(3) = 21$ cm
Step 2 Use tangent ratio to find apothem.
$\tan\left(\frac{360}{14}\right)^\circ = \frac{1.5}{a}$
$a = \frac{1.5}{\tan\left(\frac{360}{14}\right)^\circ}$
Step 3 Use apothem and perimeter to find area.
$A = \frac{1}{2}aP$
$A = \frac{1}{2}\left(\frac{1.5}{\tan\left(\frac{360}{14}\right)^\circ}\right)(21) \approx 32.7\text{ cm}^2$

7. **Step 1** Find side length.
$s = 2a\sqrt{3} = 4\sqrt{3}$ ft
Step 2 Find perimeter.
$P = 3s = 34\sqrt{3} = 12\sqrt{3}$ ft
Step 3 Find area.
$A = \frac{1}{2}aP$
$A = \frac{1}{2}(2)(12\sqrt{3}) \approx 20.8\text{ ft}^2$

8. **Step 1** Find perimeter.
$P = 12(5) = 60$ m
Step 2 Use tangent ratio to find apothem.
$\tan 15^\circ = \frac{2.5}{a}$
$a = \frac{2.5}{\tan 15^\circ}$
Step 3 Use apothem and perimeter to find area.
$A = \frac{1}{2}aP$
$A = \frac{1}{2}\left(\frac{2.5}{\tan 15^\circ}\right)(60) \approx 279.9\text{ m}^2$

Extending Perimeter, Circumference, and Area

Review 9.1 / Developing Geometric Formulas

1. $A = bh$
$7.5 = 6h$
$h = 1.25$ m

2. $P = 2\ell + 2w$
$= 2(x+2) + 2(x-1)$
$= 2x + 4 + 2x - 2$
$= (4x + 2)$ in.

3. $A = bh$
$= (3x+5)(7x-1)$
$= (21x^2 + 32x - 5)$ ft^2

4. Let $b = x + y$
$x^2 + 15^2 = 17^2$
$x^2 = 64$
$x = 8$ in.
$y^2 + 15^2 = 25^2$
$y^2 = 400$
$y = 20$ in.
$A = \frac{1}{2}bh$
$= \frac{1}{2}(8 + 20)(15)$
$= 210$ in.^{2}

5. $A = \frac{1}{2}(b_1 + b_2)h$
$280 = \frac{1}{2}(8 + 20)h = 14h$
$h = 20$ cm

6. $A = \frac{1}{2}bh$
$= \frac{1}{2}(x+1)(8x)$
$= (4x^2 + 4x)$ ft^2

7. Right triangles are 30°-60°-90° triangles; so $d_1 = 7 + 21 = 28$ in.
$A = \frac{1}{2}d_1d_2$
$= \frac{1}{2}(28)\left(2\left(7\sqrt{3}\right)\right)$
$= 196\sqrt{3}$ in.^{2}

8. $A = \frac{1}{2}d_1d_2$
$3x^2 + 6x = \frac{1}{2}(x+2)d_2$
$3x(x+2) = \frac{1}{2}(x+2)d_2$
$d_2 = 6x$ m

9. $A = \frac{1}{2}d_1d_2$
$= \frac{1}{2}(6x+5)(4x+8)$
$= (12x^2 + 34x + 20)$ ft

10. $A = bh = 6(3) = 18$ in.^{2}

11. $A = \frac{1}{2}bh = \frac{1}{2}(3)(3) = 4.5$ in.^{2}

12. $A = \frac{1}{2}(b_1 + b_2)h = \frac{1}{2}(3+6)(3) = 13.5$ in.^{2}

13. $A = \pi(7)^2 = 49\pi$ yd^2

14. $C = \pi(5) = 5\pi$ m

15. $C = \pi d$
$10 = \pi d$
$d = \frac{10}{\pi}$ ft

16. $A = \pi\left(\frac{35}{2}\right)^2 \approx 962.1$ ft^2;
$A = \pi\left(\frac{50}{2}\right)^2 \approx 1963.5$ ft^2;
$A = \pi\left(\frac{66}{2}\right)^2 \approx 3421.2$ ft^2

17. $A = (2(12))^2 = 576$ cm^2

Extending Perimeter, Circumference, and Area

Review 9.1 / Developing Geometric Formulas

18. Step 1 Find side length and perimeter.

$$\tan 22.5° = \frac{\frac{s}{2}}{a} = \frac{s}{4}$$
$$s = 4\tan 22.5°$$
$$P = 8s = 32\tan 22.5°$$

Step 2 Find area.

$$A = \frac{1}{2}aP$$
$$= \frac{1}{2}(2)(32\tan 22.5°) \approx 13.3 \text{ ft}^2$$

19. Step 1 Find side length and apothem.

$$P = 9s$$
$$144 = 9s$$
$$s = 16$$
$$\tan 20° = \frac{8}{a}$$
$$a = \frac{8}{\tan 20°}$$

Step 2 Find area.

$$A = \frac{1}{2}aP$$
$$= \frac{1}{2}\left(\frac{8}{\tan 20°}\right)(144) \approx 1582.5 \text{ in.}^2$$

20. Step 1 Find side length and perimeter.

$$\tan 36° = \frac{\frac{s}{2}}{a} = \frac{s}{4}$$
$$s = 4\tan 36°$$
$$P = 5s = 20\tan 36°$$

Step 2 Find area.

$$A = \frac{1}{2}aP$$
$$= \frac{1}{2}(2)(20\tan 36°) \approx 14.5 \text{ ft}^2$$

Spatial Reasoning

Surface Area and Volume

Surface Area of Prisms and Cylinders

1. $L = Ph$ $\quad$ $S = L + 2B$

   $= 24(3) = 72 \text{ ft}^2$ $\quad$ $= 72 + 2(5)(7) = 142 \text{ ft}^2$

   The base is a rt. △, because 3, 4, 5 is a Pythag. triple.

2. $L = Ph$ $\quad$ $S = L + 2B$

   $= 12(2) = 24 \text{ cm}^2$ $\quad$ $= 24 + 2\left(\frac{1}{2}(3)(4)\right)$

   $= 36 \text{ cm}^2$

3. $L = Ph$ $\quad$ $S = L + 2B$

   $= (4s)(s)$ $\quad$ $= L + 2s^2$

   $= 4(9)(9) = 324 \text{ in}^2$ $\quad$ $= 324 + 2(9)^2 = 486 \text{ in}^2$

4. $L = 2\pi rh = 2\pi(3)(4) = 24\pi \text{ ft}^2$

   $S = 2\pi rh + 2\pi r^2 = 24\pi + 2\pi(3)^2 = 42\pi \text{ ft}^2$

5. $L = 2\pi rh = 2\pi(7.5)(12) = 180\pi \text{ yd}^2$

   $S = 2\pi rh + 2\pi r^2 = 180\pi + 2\pi(7.5)^2 = 292.5\pi \text{ yd}^2$

6. **Step 1** Use the base area to find the radius.

   $A = \pi r^2$

   $64\pi = \pi r^2$

   $64 = r^2$

   $r = 8 \text{ m}$

   Step 2 Use the radius to find the lateral area and base area. The height is 3 less than the radius, or 5 m.

   $L = 2\pi rh = 2\pi(8)(5) = 80\pi \text{ m}^2$

   $S = 2\pi rh + 2\pi r^2 = 80\pi + 2\pi(8)^2 = 208\pi \text{ m}^2$

7. surface area of the right rectangular prism is

   $S = Ph + 2B$

   $= 44(12) + 2(14)(8) = 752 \text{ ft}^2.$

   A right cylinder is added to right rectangular prism. The lateral area of the cylinder is

   $L = 2\pi rh$

   $= 2\pi(4)(8) = 64\pi \text{ ft}^2.$

   The base area of the cylinder is not added, just raised through 8 ft.

   The surface area of the composite figure is the sum of the areas of all the surfaces on the exterior of the figure.

   $S = \text{(prism surface area)} + \text{(cylinder lateral area)}$

   $= 752 + 64\pi \approx 953.1 \text{ ft}^2$

8. surface area of cylinder is

$S = 2\pi rh + 2\pi r^2$

$= 2\pi(14)(14) + 2\pi(14)^2 = 784\pi \text{ ft}^2$

A right rectangular prism is removed from the cylinder.

The lateral area is $L = Ph = 40(14) = 560 \text{ ft}^2$.

The base area is $B = 14(6) = 84 \text{ ft}^2$.

The surface area of the composite figure is the sum of the areas of all the surfaces on the exterior of the figure.

$S =$ (cylinder surface area) + (prism rectangle area) − (prism base area)

$= 784\pi + 560 - 2(84)$

$= 784\pi + 392 \approx 2855.0 \text{ ft}^2$

9. original dimensions:

$S = 2\pi rh + 2\pi r^2$

$= 2\pi(8)(4) + 2\pi(8)^2$

$= 192\pi \text{ yd}^2$

dimensions multiplied by $\frac{1}{2}$:

$S = 2\pi rh + 2\pi r^2$

$= 2\pi(4)(2) + 2\pi(4)^2$

$= 48\pi \text{ yd}^2$

Notice $48\pi = \frac{1}{4}(192\pi)$. If the dimensions are multiplied by $\frac{1}{2}$, the surface area is multiplied by $\frac{1}{4}$.

10. original dimensions:

$S = Ph + 2B$

$= 32(6) + 2(8)(8) = 320 \text{ yd}^2$

dimensions multiplied by $\frac{2}{3}$:

$S = Ph + 2B$

$= \frac{64}{3}(4) + 2\left(\frac{16}{3}\right)\left(\frac{16}{3}\right) = \frac{1280}{9} \text{ yd}^2$

Notice $\frac{1280}{9} = \frac{4}{9}(320)$. If the dimensions are multiplied by $\frac{2}{3}$, the surface area is multiplied by $\frac{4}{9}$.

11. 16-in. bulb: $L = 2\pi rh = 2\pi(1)(16) = 32\pi \text{ in}^2$

23-in. bulb: $L = 2\pi rh = 2\pi 3/4(23) = 34.5\pi \text{ in}^2$

23-in. bulb will produce more light

Surface Area of Pyramids and Cones

1. **Step 1** Find the base perimeter and the apothem.
The base perimeter is 6(8) = 48 cm.
The apothem is $4\sqrt{3}$ cm. So, base area is $\frac{1}{2}aP = \frac{1}{2}(4\sqrt{3})(48) = 96\sqrt{3}$ cm^2.
Step 2 Find the lateral area.
$L = \frac{1}{2}P\ell$
$= \frac{1}{2}(48)(12) = 288$ cm^2
Step 3 Find the surface area.
$S = \frac{1}{2}P\ell + B$
$= 288 + 96\sqrt{3} \approx 454.3$ cm^2

2. **Step 1** Find the base perimeter and the area.
The base perimeter is 4(16) = 64 ft.
The base area is $(16)^2 = 256$ ft^2.
Step 2 Find the lateral area.
The slant height is $\sqrt{8^2 + 15^2} = 17$ ft.
$L = \frac{1}{2}P\ell$
$= \frac{1}{2}(64)(17) = 544$ ft^2
Step 3 Find the surface area.
$S = \frac{1}{2}P\ell + B$
$= 544 + 256 = 800$ ft^2

3. **Step 1** Find the base perimeter and the altitude.
The base perimeter is 3(15) = 45 in.
The altitude of base is $7.5\sqrt{3}$ in. So, base area is $\frac{1}{2}bh = \frac{1}{2}(15)(7.5\sqrt{3}) = 56.25\sqrt{3}$ in^2.
Step 2 Find the lateral area.
$L = \frac{1}{2}P\ell$
$= \frac{1}{2}(45)(20) = 450$ in^2
Step 3 Find the surface area.
$S = \frac{1}{2}P\ell + B$
$= 450 + 56.25\sqrt{3} \approx 547.4$ in^2

4. **Step 1** Use the Pythogorean Theorem to find the base radius r.
$r = \sqrt{25^2 - 24^2} = 7$ in.
Step 2 Find the lateral area and the surface area.
$L = \pi r\ell$
$= \pi(7)(25) = 175\pi$ in^2
$S = \pi r\ell + \pi r^2$
$= 175\pi + \pi(7)^2 = 224\pi$ in^2

5. $L = \pi r\ell$
$= \pi(14)(22) = 308\pi$ m^2
$S = \pi r\ell + \pi r^2$
$= 308\pi + \pi(14)^2 = 504\pi$ m^2

6. **Step 1** Use the base area to find the base radius r.
$A = 36\pi = \pi r^2$
$36 = r^2$
$r = 6$ ft
Step 2 Find the lateral area and the surface area.
$L = \pi r\ell$
$= \pi(6)(8) = 48\pi$ ft^2
$S = \pi r\ell + \pi r^2$
$= 48\pi + \pi(6)^2 = 84\pi$ ft^2

7. original:
$S = \frac{1}{2}P\ell + B$
$= \frac{1}{2}(24)(10) + (6)^2 = 156$ in^2
dimensions halved:
$S = \frac{1}{2}P\ell + B$
$= \frac{1}{2}(12)(5) + (3)^2 = 39$ in^2
Notice that $39 = \frac{1}{4}(156)$. If the dimensions are cut in half, the surface area is multiplied by $\frac{1}{4}$.

8. original:
$S = \pi r\ell + \pi r^2$
$= \pi(9)(15) + \pi(9)^2 = 216$ cm^2
dimensions tripled:
$S = \pi r\ell + \pi r^2$
$= \pi(27)(45) + \pi(27)^2 = 1944$ cm^2
Notice that 1944 = 9(216). If the dimensions are tripled, the surface area is multiplied by 9.

9. The lateral area of the upper pyramid is
$L = \frac{1}{2}Ph = \frac{1}{2}(32)(15) = 240 \text{ ft}^2$.
The lateral area of the upper pyramid is
$L = \frac{1}{2}Ph = \frac{1}{2}(32)(18) = 288 \text{ ft}^2$.
$S =$ (lateral area of upper pyramid)
+ (lateral area of lower pyramid)
$= 240 + 288 = 528 \text{ ft}^2$

10. The lateral area of upper cone is
$L = \pi r\ell = \pi(12)(26) = 312\pi \text{ m}^2$.
The lateral area of cylinder is
$L = 2\pi rh = 2\pi(12)(15) = 360\pi \text{ m}^2$.
The lateral area of upper cone is
$L = \pi r\ell = \pi(12)(32) = 384\pi \text{ m}^2$.
$S =$ (lateral area of upper cone)
+ (lateral area of cylinder)
+ (lateral area of lower cone)
$= 312\pi + 360\pi + 384\pi = 1056\pi \text{ m}^2$

11. The radius of the large circle used to create the hat is the slant height of the cone.
The area of the hat is the lateral area of the cone.
The area of the hat is also $\frac{3}{4}$ of the area of the large circle. So, $\pi r\ell = \frac{3}{4}\pi\ell^2$.
$\pi r(6) = \frac{3}{4}\pi(6)^2$
$r = 4.5$
$d = 2(4.5) = 9$ in. So, the hat will be too large.

Spatial Reasoning

Surface Area and Volume

Volume of Prisms and Cylinders

1. $V = \ell wh$
$= (9)(4)(6) = 216 \text{ cm}^3$

2. $V = Bh$
$= \frac{1}{2}aP \cdot h$
$= \frac{1}{2}(3\sqrt{3})(36)(8)$
$= 432\sqrt{3} \approx 748.2 \text{ m}^3$

3. $V = s^3$
$= (8)^3 = 512 \text{ ft}^3$

4. **Step 1** Find the volume of the ice cream cake in cubic feet.
$V = \ell wh = (19)(9)(2) = 342 \text{ ft}^3$
Step 2 Use the conversion factor $\frac{1 \text{ gal}}{0.134 \text{ ft}^3}$ to estimate the volume in gallons.
$342 \text{ ft}^3 \cdot \frac{1 \text{ gal}}{0.134 \text{ ft}^3} \approx 2552 \text{ gal}$
Step 3 Use the conversion factor $\frac{8.33 \text{ lb}}{1 \text{ gal}}$ to estimate the weight of water.
$2552 \text{ gal} \cdot \frac{4.73 \text{ lb}}{1 \text{ gal}} \approx 12{,}071 \text{ lb}$

5. $V = \pi r^2 h$
$= \pi(6)^2(10) = 360\pi \text{ ft}^3$
$\approx 1131.0 \text{ ft}^3$

6. $V = \pi r^2 h$
$= \pi(3)^2(5) = 45\pi \text{ m}^3$
$\approx 141.4 \text{ m}^3$

7. $B = \pi r^2$
$25\pi = \pi r^2$
$25 = r^2$
$r = 5 \text{ cm}$
$h = 5 + 3 = 8 \text{ cm}$
$V = \pi r^2 h$
$= \pi(5)^2(8) = 200\pi \text{ cm}^3$
$\approx 628.3 \text{ cm}^3$

8. original dimensions:
$V = \ell wh = (12)(4)(8) = 384 \text{ ft}^3$
dimensions multiplied by $\frac{1}{4}$:
$V = \ell wh = (3)(1)(2) = 6 \text{ ft}^3$
Notice $6 = \frac{1}{64}(384)$. If the dimensions are multiplied by $\frac{1}{4}$, the volume is multiplied by $\frac{1}{64}$.

9. original dimensions:
$V = \pi r^2 h = \pi(2)^2(7) = 28\pi \text{ in}^3$
dimensions tripled:
$V = \pi r^2 h = \pi(6)^2(21) = 756\pi \text{ in}^3$
Notice $756\pi = 27(28\pi)$. If the dimensions are tripled, the volume is multiplied by 27.

10. The volume of rect. prism is
$V = \ell wh = (12)(14)(6) = 1008 \text{ ft}^3$.
The volume of cylinder is
$V = \pi r^2 h = \pi(4)^2(4) = 624\pi \text{ ft}^3$.
The total volume of the figure is the sum of its volumes.
$V = 1008 + 64\pi \approx 1209.1 \text{ ft}^3$

11. The volume of outside cylinder is
$V = \pi r^2 h = \pi(10)^2(15) = 1500\pi \text{ in}^3$.
The volume of inside cylinder is
$V = \pi r^2 h = \pi(5)^2(15) = 375\pi \text{ in}^3$.
The net volume of the figure is the difference of the volumes.
$V = 1500\pi - 375\pi = 1125\pi \approx 3534.3 \text{ in}^3$

Spatial Reasoning

Review 10.1 / Surface Area and Volume

1. $L = Ph$
$= 20(10) = 200 \text{ cm}^2$

$S = L + 2B$
$= 200 + 2(5)(5)$
$= 250 \text{ cm}^2$

2. $L = Ph$
$= 6(12)(15)$
$= 1080 \text{ m}^2$

apothem of base is $6\sqrt{3}$ m
$S = L + 2B$
$= 1080 + 2\left(\frac{1}{2}(6\sqrt{3})(72)\right)$
$= 1080 + 432\sqrt{3}$
$\approx 1828.2 \text{ m}^2$

3. $L = Ph$
$= 3(8)(14)$
$= 336 \text{ ft}^2$

altitude of base is $4\sqrt{3}$ ft
$S = L + 2B$
$= 336 + 2\left(\frac{1}{2}(8)(4\sqrt{3})\right)$
$= 336 + 32\sqrt{3}$
$\approx 391.4 \text{ ft}^2$

4. $L = 2\pi rh = 2\pi(5.5)(7) = 77\pi \text{ in}^2$
$S = 2\pi rh + 2\pi r^2 = 77\pi + 2\pi(5.5)^2 = 137.5\pi \text{ in}^2$

5. $L = 2\pi rh = 2\pi(4)(23) = 184\pi \text{ cm}^2$
$S = 2\pi rh + 2\pi r^2 = 184\pi + 2\pi(4)^2 = 216\pi \text{ cm}^2$

6. **Step 1** Use the base circumference to find the radius.
$C = 2\pi r$
$16\pi = 2\pi r$
$r = 8$ yd
Step 2 Use the radius to find the lateral area and the base area. The height is 3 times the radius, or 24 yd.
$L = 2\pi rh = 2\pi(8)(24) = 384\pi \text{ yd}^2$
$S = 2\pi rh + 2\pi r^2 = 384\pi + 2\pi(8)^2 = 512\pi \text{ yd}^2$

7. The base of the right triangular prism is a rt. △, because 6, 8, 10 is a Pythagorean triple.
The surface area of the right triangular prism is
$S = Ph + 2B$
$= 24(9) + 2\left(\frac{1}{2}(6)(8)\right) = 264 \text{ cm}^2$.
A right cylinder is removed from the triangular prism.
The lateral area of cylinder is
$L = 2\pi rh$
$= 2\pi(2)(9) = 36\pi \text{ cm}^2$.
The base area of cylinder is
$B = \pi r^2 = \pi(2)^2 = 4\pi \text{ cm}^2$.
The surface area of the composite figure is the sum of areas of all surfaces on the exterior of the figure.
$S =$ (prism surface area) + (cylinder lateral area) − (cylinder base area)
$= 264 + 36\pi - 2(4\pi)$
$= 264 + 28\pi \approx 352.0 \text{ cm}^2$

8. The surface area of right rectangular prism is
$S = Ph + 2B$
$= 8(0.5) + 2(2)(2) = 12 \text{ ft}^2$.
A right cylinder is added to the rectangular prism.
The lateral area of cylinder is
$L = 2\pi rh$
$= 2\pi(0.5)(2) = 2\pi \text{ ft}^2$.
The base area of cylinder is not added, just lowered through 2 ft.
The surface area of the composite figure is the sum of the areas of all surfaces on the exterior of figure.
$S =$ (prism surface area) + (cylinder lateral area)
$= 12 + 2\pi \approx 18.3 \text{ ft}^2$

9. original:
$S = 2\pi rh + 2\pi r^2$
$= 2\pi(4.5)(11) + 2\pi(4.5)^2$
$= 139.5\pi \text{ ft}^2$
dimensions tripled:
$S = 2\pi rh + 2\pi r^2$
$= 2\pi(13.5)(33) + 2\pi(13.5)^2$
$= 1255.5\pi \text{ ft}^2$
$1255.5\pi = 9(139.5\pi)$. So, surface area is multiplied by 9.

10. original:
$S = Ph + 2B$
$= 42(3) + 2(12)(9) = 342 \text{ ft}^2$
dimensions doubled:
$S = Ph + 2B$
$= 84(6) + 2(24)(18) = 1368 \text{ ft}^2$
$1368 = 4(342)$. So, surface area is multiplied by 4.

11. left cell:
$S = Ph + 2B$
$= 90(7) + 2(35)(10) = 1330\ \mu\text{m}^2$
right cell:
$S = Ph + 2B$
$= 52(15) + 2(15)(11) = 1110\ \mu\text{m}^2$
The cell that measures 35 μm by 7 μm by 10 μm should absorb at a greater rate.

12. **Step 1** Find the base perimeter and the area.
The base perimeter is $4(6) = 24$ ft.
The base area is $(6)^2 = 36 \text{ ft}^2$.
Step 2 Find the lateral area.
The slant height is $\sqrt{3^2 + 4^2} = 5$ ft.
$L = \frac{1}{2}P\ell$
$= \frac{1}{2}(24)(5) = 60 \text{ ft}^2$
Step 3 Find the surface area.
$S = \frac{1}{2}P\ell + B$
$= 60 + 36 = 96 \text{ ft}^2$

13. **Step 1** Find the base perimeter and the altitude.
The base perimeter is $3(40) = 120$ cm.
The altitude of base is $20\sqrt{3}$ ft. So, the base area is $\frac{1}{2}bh = \frac{1}{2}(40)(20\sqrt{3}) = 400\sqrt{3} \text{ cm}^2$.
Step 2 Find the lateral area.
The slant height is $\sqrt{25^2 - 20^2} = 15$ cm.
$L = \frac{1}{2}P\ell$
$= \frac{1}{2}(120)(15) = 900 \text{ cm}^2$
Step 3 Find the surface area.
$S = \frac{1}{2}P\ell + B$
$= 900 + 400\sqrt{3} \approx 1592.8 \text{ cm}^2$

14. **Step 1** Find the base perimeter and the apothem.
The base perimeter is $6(7) = 42$ ft.
The apothem is $3.5\sqrt{3}$ ft. So, the base area is $\frac{1}{2}aP = \frac{1}{2}(3.5\sqrt{3})(42) = 73.5\sqrt{3} \text{ ft}^2$.
Step 2 Find the lateral area.
$L = \frac{1}{2}P\ell$
$= \frac{1}{2}(42)(15) = 315 \text{ ft}^2$
Step 3 Find the surface area.
$S = \frac{1}{2}P\ell + B$
$= 315 + 73.5\sqrt{3} \approx 442.3 \text{ ft}^2$

15. $L = \pi r\ell = \pi(11.5)(23) = 264.5\pi \text{ cm}^2$
$S = \pi r\ell + \pi r^2 = 264.5\pi + \pi(11.5)^2 = 396.75\pi \text{ cm}^2$

16. $\ell = \sqrt{12^2 + 35^2} = 37$ in.
$L = \pi r\ell = \pi(12)(37) = 444\pi \text{ in}^2$
$S = \pi r\ell + \pi r^2 = 444\pi + \pi(12)^2 = 588\pi \text{ in}^2$

17. $h = 2(8) - 1 = 15$ m; $\ell = \sqrt{8^2 + 15^2} = 17$ m
$L = \pi r \ell = \pi(8)(17) = 136\pi$ m^2
$S = \pi r \ell + \pi r^2 = 136\pi + \pi(8)^2 = 200\pi$ m^2

18. original:
$S = \frac{1}{2}P\ell + \frac{1}{2}aP$
$= \frac{1}{2}(24)(12) + \frac{1}{2}(2\sqrt{3})(24)$
$= (144 + 24\sqrt{3})$ ft^2
dimensions divided by 3:
$S = \frac{1}{2}P\ell + \frac{1}{2}aP$
$= \frac{1}{2}(8)(4) + \frac{1}{2}\left(\frac{2\sqrt{3}}{3}\right)(8)$
$= \left(16 + \frac{8\sqrt{3}}{3}\right)$ ft^2
$\left(16 + \frac{8\sqrt{3}}{3}\right) = (144 + 24\sqrt{3}) \div 9$. So, the surface area is divided by 9.

19. original:
$S = \pi r \ell + \pi r^2 = \pi(2)(5) + \pi(2)^2 = 14\pi$ m^2
dimensions doubled:
$S = \pi r \ell + \pi r^2 = \pi(4)(10) + \pi(4)^2 = 56\pi$ m^2
$56\pi = 4(14\pi)$. So, the surface area is multiplied by 4.

20. lateral area of left cone $= \pi(7)(24) = 168\pi$ in^2
lateral area of right cone $= \pi(7)(17) = 119\pi$ in^2
$S =$ (lateral area of left cone)
+ (lateral area of right cone)
$= 168\pi + 119\pi = 287\pi$ in^2

21. lateral area of left pyramid
$= \frac{1}{2}P\ell = \frac{1}{2}(36)(15) = 270$ cm^2
lateral area of cube $= Ph = (36)(9) = 324$ cm^2
lateral area of right pyramid
$= \frac{1}{2}P\ell = \frac{1}{2}(36)(19) = 342$ cm^2
$S =$ (lateral area of left pyramid)
+ (lateral area of cube)
+ (lateral area of right pyramid)
$= 270 + 324 + 342 = 936$ cm^2

22. $S = \pi r \ell = \frac{1}{2}\pi \ell^2$
$\pi r(6) = \frac{1}{2}\pi(6)^2$
$r = 3$
$d = 2(3) = 6$ in.

23. $V = Bh = \left(\frac{1}{2}(9)(15)\right)(12) = 810$ yd^3

24. $V = Bh$
$= \frac{1}{2}aP \cdot h$
$= \frac{1}{2}\left(\frac{5}{\tan 36°}\right)(50)(15) \approx 2580.7$ m^3

25. $B = s^2$
$49 = s^2$
$s = 7$ ft
$h = 7 - 2 = 5$ ft
$V = Bh = (49)(5) = 245$ ft^3

26. **Step 1** Find the volume in cubic feet.
$V = \ell wh = (9)(16)\left(\frac{1}{3}\right) = 48$ ft^3
Step 2 Use the conversion factor $\frac{1 \text{ yd}^3}{27 \text{ ft}^3}$ to find the volume in cubic yards. Then round up to nearest cubic yard.
$V = 48 \cdot \frac{1 \text{ yd}^3}{27 \text{ ft}^3} \approx 1.78$ yd^3
Colin must buy 2 yd^3 of dirt.
Step 3 Use the cost per yard to find the cost of dirt.
Cost $= 2(\$25) = \$50$

27. $V = \pi r^2 h$
$= \pi(14)^2(9) = 1764\pi \text{ cm}^3$
$\approx 5541.8 \text{ cm}^3$

28. $V = \pi r^2 h$
$= \pi(6)^2(3) = 108\pi \text{ in}^3$
$\approx 339.3 \text{ in}^3$

29. $V = Bh$
$= 24\pi(16) = 384\pi \text{ cm}^3$
$\approx 1206.4 \text{ cm}^3$

30. original dimensions:
$V = \pi r^2 h = \pi(2)^2(3) = 12\pi \text{ yd}^3$
dimensions multiplied by 5:
$V = \pi r^2 h = \pi(10)^2(15) = 1500\pi \text{ yd}^3$
$1500\pi = 125(12\pi)$. So, the volume is multiplied by 125.

31. original dimensions:
$V = Bh = (5)^2(10) = 250\pi \text{ m}^3$
dimensions multiplied by $\frac{3}{5}$:
$V = \pi r^2 h = \pi(3)^2(6) = 54\pi \text{ m}^3$
$54\pi = \frac{27}{125}(250\pi)$. So, the volume is multiplied by $\frac{27}{125}$.

32. volume of lower cube: $V = s^3 = (8)^3 = 512 \text{ cm}^3$
volume of middle cube: $V = s^3 = (6)^3 = 216 \text{ cm}^3$
volume of upper cube: $V = s^3 = (4)^3 = 64 \text{ cm}^3$
total volume: $V = 512 + 216 + 64 = 792 \text{ cm}^3$

33. volume of square-based prism:
$V = Bh = (4)^2(12) = 192 \text{ ft}^3$
volume of each half-cylinder:
$V = \frac{1}{2}\pi r^2 h = \frac{1}{2}\pi(2)^2(4) = 8\pi \text{ ft}^3$
total volume:
$V = 8\pi + 192 + 8\pi = 192 + 16\pi \approx 242.3 \text{ ft}^3$

Circles

Lines and Arcs in Circles

Lines That Intersect Circles

1. chord: $\overline{EF}$; tangent: m; radii: $\overline{DE}$, $\overline{DF}$; secant: ℓ; diameter: $\overline{EF}$
2. chord: $\overline{QS}$; tangent: $\overleftrightarrow{ST}$; radii: $\overline{PQ}$, $\overline{PR}$, $\overline{PS}$; secant: $\overleftrightarrow{QS}$; diameter: $\overline{QS}$
3. radius of circle A: $4 - 1 = 3$
   radius of circle B: $4 - 2 = 2$
   point of tangency: $(-1, 4)$
   equation of tangent line: $y = 4$
4. radius of circle R: $4 - 2 = 2$
   radius of circle S: $4 - 2 = 2$
   point of tangency: $(1, 2)$
   equation of tangent line: $x = 1$
5. **1 Understand the Problem**
   The answer will be the length of an imaginary segment from the ISS to the Earth's horizon.
   2 Make a Plan
   Let C be the center of the Earth, let E be ISS, and let H be the point on the horizon. Find the length of $\overline{EH}$, which is tangent to circle C at H. By Thm. 11-1-1, $\overline{EH} \perp \overline{CH}$. So CHE is a right $\triangle$.

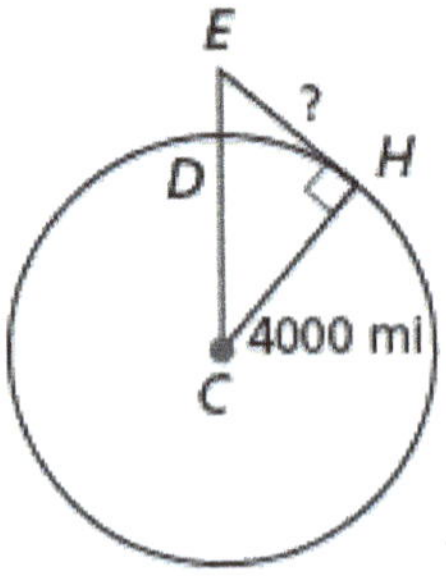

   3 Solve
   $EC = CD + ED$
   $\approx 4000 + 240 = 4240$ mi
   $EC^2 \approx EH^2 + CH^2$
   $4240^2 \approx EH^2 + 4000^2$
   $1{,}977{,}60 \approx EH^2$
   $1406 \text{ mi} \approx EH$
   4 Look Back
   The problem asks for the distance to the nearest mile. Check that answer is reasonable by using the Pythagorean Thm. Is $1406^2 + 4000^2 \approx 4240^2$? Yes, $17{,}976{,}836 \approx 17{,}977{,}600$.
6. By Thm. 11-1-3,
   $JK = JL$
   $4x - 1 = 2x + 9$
   $2x - 1 = 9$
   $2x = 10$
   $x = 5$
   $JK = 4(5) - 1 = 19$
7. By Thm. 11-1-3,
   $ST = SU$
   $y - 4 = \frac{3}{4}y$
   $4y - 16 = 3y$
   $y - 16 = 0$
   $y = 16$
   $ST = (16) - 4 = 12$

Circles

Lines and Arcs in Circles

Arcs and Chords

1. $m\angle PAQ = 0.45(360) = 162°$
2. $m\angle VAU = 0.07(360) = 25.2°$
3. $m\angle SAQ = (0.06 + 0.11)360 = 61.2°$
4. $m\widehat{UT} = m\angle UAT = 0.1(360) = 36°$
5. $m\widehat{RQ} = m\angle RAQ = 0.11(360) = 39.6°$
6. $m\widehat{UPT} = (1 - 0.1)360 = 324°$
7. $m\widehat{DE} = m\angle DAE = 90°$
   $m\widehat{EF} = m\angle EAF = m\angle BAC = 90 - 51 = 39°$
   $m\widehat{DF} = m\widehat{DE} + m\widehat{EF} = 90 + 39 = 129°$
8. $m\widehat{DEB} = m\angle DAE + m\angle EAB = 90 + 180 = 270°$
9. $m\angle HGJ + m\angle JGL = m\angle HGL$
   $72 + m\angle JGL = 180$
   $m\angle JGL = 108°$
   $m\widehat{JL} = 108°$
10. $m\widehat{HLK} = m\angle HGL + m\angle LGK = 180 + 30 = 210°$
11. $QR = RS$ (Thm. 11-2-2(1))
    $8y - 8 = 6y$
    $2y = 8$
    $y = 4$
    $QR = 8(4) - 8 = 24$
12. $m\angle CAD = m\angle EBF$ (Thm. 11-2-2(3))
    $45 - 6x = -9x$
    $3x = -45$
    $x = -15$
    $m\angle EBF = -9(-15) = 135°$
13. **Step 1** Draw radius $\overline{PR}$.
    $PR = 5 + 8 = 13$
    Step 2 Use the Pythagorean Thm. and

    Let the intersection of $\overline{PQ}$ and $\overline{RS}$ be T.
    $PT^2 + RT^2 = PR^2$
    $5^2 + RT^2 = 13^2$
    $RT = 12$
    Step 3 Find RS.
    $RS = 2(12) = 24$
14. **Step 1** Draw radius $\overline{CE}$.
    $CE = 50 + 20 = 70$
    Step 2 Use the Pythagorean Thm. and

    Let the intersection of $\overline{CD}$ and $\overline{EF}$ be G.
    $CG^2 + EG^2 = CE^2$
    $50^2 + EG^2 = 70^2$
    $RG = \sqrt{2400} = 20\sqrt{6}$
    Step 3 Find EF.
    $EF = 2(20\sqrt{6}) = 40\sqrt{6} \approx 98.0$

Circles

Review 11.1 / Lines and Arcs in Circles

1. chords: $\overline{RS}$, $\overline{VW}$; tangent: ℓ; radii: $\overline{PV}$, $\overline{PW}$; secant: $\overleftrightarrow{VW}$; diameter: $\overline{VW}$

2. chords: $\overline{AC}$, $\overline{DE}$; tangent: $\overleftrightarrow{CF}$; radii: $\overline{BA}$, $\overline{BC}$; secant: $\overleftrightarrow{DE}$; diameter: $\overline{AC}$

3. radius of circle C: $2 - 0 = 2$
radius of circle D: $4 - 0 = 4$
point of tangency: $(-4, 0)$
equation of tangent line: $x = -4$

4. radius of circle M: $3 - 2 = 1$
radius of circle N: $5 - 2 = 3$
point of tangency: $(2, 1)$
equation of tangent line: $y = 1$

5. **1 Understand the Problem**
The answer will be the length of an imaginary segment from the summit of Olympus Mons to Mars' horizon.

2 Make a Plan
Let C be the center of Mars, let E be summit of Olympus Mons, and let H be a point on the horizon. Find the length of $\overline{EH}$, which is tangent to circle C at H. By line tangent to circle → line ⊥ to radius thm., $\overline{EH} \perp \overline{CH}$. So triangle CHE is a right triangle.

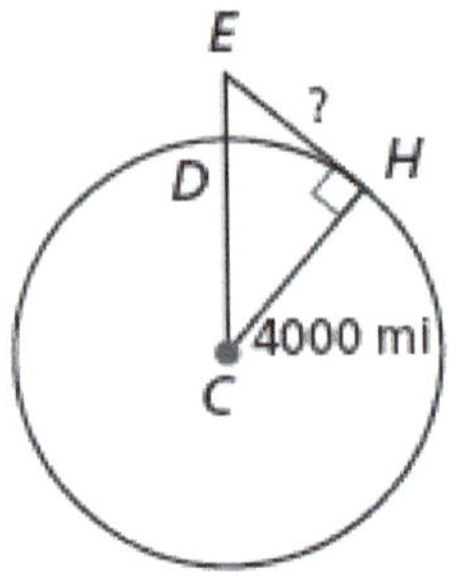

3 Solve

$EC = CD + ED$
$\approx 3397 + 25 = 3422$ km
$EC^2 \approx EH^2 + CH^2$
$3422^2 \approx EH^2 + 3397^2$
$170{,}475 \approx EH^2$
$413 \text{ km} \approx EH$

4 Look Back
The problem asks for the distance to the nearest km. Check that the answer is reasonable by using the Pythagorean Thm. Is $413^2 + 3397^2 \approx 3422^2$? Yes, $11{,}710{,}178 \approx 11{,}710{,}084$.

6. By 2 segments tangent to circle from same ext pt. → segments ≅ thm.
$AB = AC$
$2x^2 = 8x$
Since $x \neq 0$,
$2x = 8$
$x = 4$
$AB = 2(4)^2 = 32$

7. By 2 segments tangent to circle from same ext pt. → segments ≅ thm.
$RS = RT$
$y = \frac{y^2}{7}$
$7y = y^2$
Since $y \neq 0$,
$7 = y$
$RT = \frac{(7)^2}{7} = 7$

8. $m\angle ADB = \frac{35}{35 + 39 + 29}(360) = \frac{35}{103}(360) \approx 122.3°$

9. $m\angle ADC = \frac{29}{103}(360) = 101.4°$

10. $m\widehat{AB} = m\angle ADB \approx 122.3°$

11. $m\widehat{BC} = m\angle BDC = \frac{39}{103}(360) \approx 136.3°$

12. $m\widehat{ACB} = 360 - m\angle ADB \approx 360 - 122.3 = 237.7°$

13. $m\widehat{CAB} = 360 - m\angle BDC \approx 360 - 136.3 = 223.7°$

14. $m\widehat{MP} = m\angle MJP$
$= m\angle MJQ - m\angle PJQ$
$= 180 - 28 = 152°$

Circles

Review 11.1 / Lines and Arcs in Circles

15. $mQNL = mQNM + mML$
$= m\angle QJM + m\angle MJL$
$= 180 + 28 = 208°$

16. $m\widehat{WT} = m\widehat{WS} + m\widehat{ST}$
$= m\angle WXS + m\angle SXT$
$= 55 + 100 = 155°$

17. $m\widehat{WTV} = m\widehat{WS} + m\widehat{STV}$
$= m\angle WXS + m\angle SXV$
$= 55 + 180 = 235°$

18. $m\angle CAD = m\angle EBF$ (≅ arcs have ≅ central ∠s thm.)
$10x - 63 = 7x$
$3x = 63$
$x = 21$
$m\angle CAD = 10(21) - 63 = 147°$

19. $m\widehat{JK} = m\widehat{LM}$ (≅ chords have ≅ arcs thm.)
$4y + y = y + 68$
$y = 17$
$m\widehat{JK} = 4(17) + 17 = 85°$

20. $AC = AB = 2.4 + 1.7 = 4.1$
Let $\overline{AB}$ and $\overline{CD}$ meet at E.
$AE^2 + CE^2 = AC^2$
$2.4^2 + CE^2 = 4.1^2$
$CE = \sqrt{11.05}$
$CD = 2\sqrt{11.05} \approx 6.6$

21. $PR = PQ = 2(3) = 6$
Let $\overline{PQ}$ and $\overline{RS}$ meet at T.
$PT^2 + RT^2 = PR^2$
$3^2 + RT^2 = 6^2$
$RT = \sqrt{27} = 3\sqrt{3}$
$RS = 2(3\sqrt{3}) = 6\sqrt{3} \approx 10.4$

Circles

Angles and Segments in Circles

Inscribed Angles

1. $m\angle DEF = \frac{1}{2}m\widehat{DF}$

$= \frac{1}{2}(78) = 39°$

2. $m\angle EFG = \frac{1}{2}m\widehat{EG}$

$29 = \frac{1}{2}m\widehat{EG}$

$m\widehat{EG} = 58°$

3. $m\angle JNL = \frac{1}{2}m\widehat{JKL}$

$102 = \frac{1}{2}m\widehat{JKL}$

$m\widehat{JKL} = 204°$

4. $m\angle LKM = \frac{1}{2}m\widehat{LM}$

$= \frac{1}{2}(52) = 26°$

5. $m\angle QTR + m\angle Q + m\angle R = 180$

$m\angle QTR + \frac{1}{2}m\widehat{RS} + m\angle S = 180$

$m\angle QTR + \frac{1}{2}(90) + 25 = 180$

$m\angle QTR + 70 = 180$

$m\angle QTR = 110°$

6. $\angle DEF$ is a right $\angle$

$\frac{4x}{5} = 90$

$4x = 450$

$x = 112.5$

7. $\angle FHG$ is a right $\angle$. So $\triangle FGH$ is a 45°-45°-90° triangle.

$m\angle GFH = 45$

$3y + 6 = 45$

$3y = 39$

$y = 13$

8. $m\angle XYZ = m\angle XWZ$

$7y - 3 = 4 + 6y$

$y = 7$

$m\angle XYZ = 7(7) - 3 = 46°$

9. Step 1 Find the value of x.

$m\angle P + m\angle R = 180$

$5x + 20 + 7x - 8 = 180$

$12x = 168$

$x = 14$

Step 2 Find the measures.

$m\angle P = 5(14) + 20 = 90°$

$m\angle Q = 10(14) = 140°$

$m\angle R = 7(14) - 8 = 90°$

$m\angle S + m\angle Q = 180$

$m\angle S + 140 = 180$

$m\angle S = 40°$

10. Step 1 Find the value of z.

$m\angle A + m\angle C = 180$

$4z - 10 + 10 + 5z = 180$

$9z = 180$

$z = 20$

Step 2 Find the measures.

$m\angle A = 4(20) - 10 = 70°$

$m\angle B = 6(20) - 5 = 115°$

$m\angle C = 10 + 5(20) = 110°$

$m\angle B + m\angle D = 180$

$115 + m\angle D = 180$

$m\angle D = 65°$

Circles

Angles and Segments in Circles

Angle Relationships in Circles

1. $m\angle DAB = \frac{1}{2}m\overset{\frown}{AB} = \frac{1}{2}(140) = 70°$

2. $27 = \frac{1}{2}m\overset{\frown}{AC}$
$m\overset{\frown}{AC} = 54°$

3. $61 = \frac{1}{2}m\overset{\frown}{PN}$
$m\overset{\frown}{PN} = 122°$

4. $m\angle MNP = \frac{1}{2}(238) = 119°$

5. $m\angle STU = \frac{1}{2}(m\overset{\frown}{SU} + m\overset{\frown}{VW}) = \frac{1}{2}(104 + 30) = \frac{1}{2}(134) = 67°$

6. $m\angle HFG = \frac{1}{2}(m\overset{\frown}{EJ} + m\overset{\frown}{GH}) = \frac{1}{2}(59 + 23) = \frac{1}{2}(82) = 41°$

7. $m\angle NPL = \frac{1}{2}(m\overset{\frown}{KM} + m\overset{\frown}{NL}) = \frac{1}{2}(61 + 111) = \frac{1}{2}(172) = 86°$
$m\angle NPK + m\angle NPL = 180$
$m\angle NPK + 86 = 180$
$m\angle NPK = 94°$

8. $x = \frac{1}{2}(161 - 67) = \frac{1}{2}(94) = 47$

9. $x = \frac{1}{2}(238 - 122) = \frac{1}{2}(116) = 58$

10. $27 = \frac{1}{2}(x - 40)$
$54 = x - 40$
$x = 94°$

11. $m\angle S = \frac{1}{2}(m\overset{\frown}{ACB} - m\overset{\frown}{AB})$
$38 = \frac{1}{2}((360 - x) - x)$
$76 = 360 - 2x$
$2x = 360 - 76 = 284$
$x = 142°$

12. $m\angle E = \frac{1}{2}(m\overset{\frown}{BF} - m\overset{\frown}{DF})$
$50 = \frac{1}{2}(150 - m\overset{\frown}{DF})$
$100 = 150 - m\overset{\frown}{DF}$
$m\overset{\frown}{DF} = 50°$

13. $m\overset{\frown}{BC} + m\overset{\frown}{CD} + m\overset{\frown}{DF} + m\overset{\frown}{BF} = 360$
$64 + m\overset{\frown}{CD} + 50 + 150 = 360$
$m\overset{\frown}{CD} + 264 = 360$
$m\overset{\frown}{CD} = 96°$

14. $m\angle NPQ = \frac{1}{2}(m\overset{\frown}{JK} + m\overset{\frown}{PN})$
$79 = \frac{1}{2}(48 + m\overset{\frown}{PN})$
$158 = 48 + m\overset{\frown}{PN}$
$m\overset{\frown}{PN} = 110°$

15. $m\overset{\frown}{KN} + m\overset{\frown}{PN} + m\overset{\frown}{JP} + m\overset{\frown}{JL} = 360$
$m\overset{\frown}{KN} + 110 + 86 + 48 = 360$
$m\overset{\frown}{KN} + 244 = 360$
$m\overset{\frown}{KN} = 116°$

Circles

Angles and Segments in Circles

Segment Relationships in Circles

1. $HK \cdot KJ = LK \cdot KM$
   $8(3) = 4(y)$
   $24 = 4y$
   $y = 6$
   $LM = 4 + (6) = 10$
   $HJ = 8 + 3 = 11$

2. $AE \cdot EB = CE \cdot ED$
   $x(4) = 6(6)$
   $4x = 36$
   $x = 9$
   $AB = (9) + 4 = 13$
   $CD = 6 + 6 = 12$

3. $PS \cdot SQ = RS \cdot ST$
   $z(10) = 6(8)$
   $10z = 48$
   $z = 4.8$
   $PQ = 10 + (4.8) = 14.8$
   $RT = 6 + 8 = 14$

4. Let the diameter be d ft.
   $GF \cdot FH = EF \cdot (d - EF)$
   $25(25) = 20(d - 20)$
   $625 = 20d - 400$
   $225 = 20d$
   $d = 51\frac{1}{4}$ ft

5. $AC \cdot BC = EC \cdot DC$
   $(x + 7.2)7.2 = (9 + 7.2)7.2$
   $x + 7.2 = 9 + 7.2$
   $x = 9$
   $AC = (9) + 7.2 = 16.2$
   $EC = 9 + 7.2 = 16.2$

6. $PQ \cdot PR = PS \cdot PT$
   $5(5 + y) = 6(6 + 7)$
   $25 + 5y = 78$
   $5y = 53$
   $y = 10.6$
   $PR = 5 + (10.6) = 15.6$
   $PT = 6 + 7 = 13$

7. $GH \cdot GJ = GK \cdot GL$
   $10(10 + 10.7) = 11.5(11.5 + x)$
   $207 = 132.25 + 11.5x$
   $74.75 = 11.5x$
   $x = 6.5$
   $GJ = 10 + 10.7 = 20.7$
   $GL = 11.5 + (6.5) = 18$

8. $AB \cdot AC = AD^2$
   $2(2 + 6) = x^2$
   $16 = x^2$
   $x = 4$ (since $x > 0$)

9. $MP \cdot MQ = MN^2$
   $3(3 + y) = 4^2$
   $9 + 3y = 16$
   $3y = 7$
   $y = 2\frac{1}{3}$

10. $RT \cdot RU = RS^2$
    $3(3 + 8) = x^2$
    $33 = x^2$
    $x = \sqrt{33}$ (since $x > 0$)

Circles

Review 11.2 / Angles and Segments in Circles

1. $m\angle MNL = \frac{1}{2}m\overparen{ML}$

$43 = \frac{1}{2}m\overparen{ML}$

$m\overparen{ML} = 86°$

2. $m\angle KMN = \frac{1}{2}m\overparen{KN}$

$= \frac{1}{2}(95)$

$= 47.5°$

3. $m\angle EJH = \frac{1}{2}m\overparen{EGH}$

$139 = \frac{1}{2}m\overparen{EGH}$

$m\overparen{EGH} = 278°$

4. $m\angle GFH = \frac{1}{2}m\overparen{GH}$

$= \frac{1}{2}(95.2)$

$= 47.6°$

5. $m\angle ADC = m\angle ADB + m\angle BDC$

$= \frac{1}{2}mAB + m\angle BEC$

$= \frac{1}{2}(44) + 40 = 62°$

6. $m\angle SRT = 90$

$3y^2 - 18 = 90$

$3y^2 = 108$

$y^2 = 36$

$y = \pm 6$

7. $\triangle JKL$ is a 30°-60°-90°

$6z - 4 = 60$

$6z = 64$

$z = \frac{64}{6} = 10\frac{2}{3}$

8. $m\angle ADB = m\angle ACB$

$2x^2 = 10x$

$2x = 10 \quad (x \neq 0)$

$x = 5$

$m\angle ADB = \frac{1}{2}m\overparen{AB}$

$2(5)^2 = \frac{1}{2}m\overparen{AB}$

$50 = \frac{1}{2}m\overparen{AB}$

$m\overparen{AB} = 2(50) = 100°$

9. $\angle MPN$ and $\angle MNP$ are complementary.

$m\angle MPN + m\angle MNP = 90$

$\frac{11x}{3} + 3x - 10 = 90$

$20x - 30 = 270$

$20x = 300$

$x = 15$

$m\angle MPN = \frac{11(15)}{3} = 55°$

10. **Step 1** Find the value of x.

$m\angle B + m\angle D = 180$

$\frac{x}{2} + \frac{x}{4} + 30 = 180$

$\frac{3x}{4} = 150$

$3x = 600$

$x = 200$

Step 2 Find the measures.

$m\angle B = \frac{200}{2} = 100°$

$m\angle D = \frac{200}{4} + 30 = 80°$

$m\angle E = 200 - 59 = 141°$

$m\angle C + m\angle E = 180$

$m\angle C + 141 = 180$

$m\angle C = 39°$

Circles

Review 11.2 / Angles and Segments in Circles

11. Step 1 Find the value of x.

$$m\angle U + m\angle W = 180$$
$$14 + 4x + 6x - 14 = 180$$
$$10x = 180$$
$$x = 18$$

Step 2 Find the value of y.

$$m\angle T + m\angle V = 180$$
$$12y - 5 + 15y - 4 = 180$$
$$27y = 189$$
$$y = 7$$

Step 3 Find the measures.

$$m\angle T = 12(7) - 5 = 79°$$
$$m\angle U = 14 + 4(18) = 86°$$
$$m\angle V = 15(7) - 4 = 101°$$
$$m\angle W = 6(18) - 14 = 94°$$

12. $m\angle BCD = \frac{1}{2}m\overset{\frown}{BC} = \frac{1}{2}(112) = 56°$

13. $m\angle ABC = \frac{1}{2}(360 - 112) = \frac{1}{2}(248) = 124°$

14. $m\angle XZW = \frac{1}{2}m\overset{\frown}{XZ} = \frac{1}{2}(180) = 90°$

15. $m\overset{\frown}{XZV} = m\overset{\frown}{XZ} + m\overset{\frown}{ZV}$
$= 180 + 2m\angle VXZ$
$= 180 + 2(40) = 260°$

16. $m\angle QPR = \frac{1}{2}(m\overset{\frown}{QR} + m\overset{\frown}{ST}) = \frac{1}{2}(31 + 98) = 64.5°$

17. $m\angle ABC = 180 - m\angle ABD$
$= 180 - \frac{1}{2}(m\overset{\frown}{AD} + m\overset{\frown}{CE})$
$= 180 - \frac{1}{2}(100 + 45) = 107.5°$

18. $m\angle MKJ = 180 - m\angle MKL$
$= 180 - \frac{1}{2}(m\overset{\frown}{JN} + m\overset{\frown}{ML})$
$= 180 - \frac{1}{2}(38.5 + 51.5) = 135°$

19. $x = \frac{1}{2}(170 - (360 - (170 + 135)))$
$= \frac{1}{2}(170 - 55) = 57.5°$

20. $x = \frac{1}{2}(220 - 140) = 40°$

21. $x = \frac{1}{2}((180 - (20 + 104)) - 20)$
$= \frac{1}{2}(56 - 20) = 18°$

22. $m\angle ABV = \frac{1}{2}(180 - m\overset{\frown}{AB}) = \frac{1}{2}(180 - 48) = 66°$

23. $m\angle DJH = \frac{1}{2}(m\overset{\frown}{DH} + m\overset{\frown}{EG})$
$180 - 89 = \frac{1}{2}(137 + m\overset{\frown}{EG})$
$91 = \frac{1}{2}(137 + m\overset{\frown}{EG})$
$182 = 137 + m\overset{\frown}{EG}$
$m\overset{\frown}{EG} = 45°$

24. $m\angle DJE = \frac{1}{2}(m\overset{\frown}{DE} + m\overset{\frown}{GH})$
$89 = \frac{1}{2}(m\overset{\frown}{DE} + 61)$
$178 = m\overset{\frown}{DE} + 61$
$m\overset{\frown}{DE} = 117°$

25. Step 1 Find $m\angle P$.

$m\angle P = \frac{1}{2}m\overset{\frown}{LR} = \frac{1}{2}(170) = 85°$

Step 2 Find $m\angle QPR$.

$m\angle QPR + 50 + 85 = 180$
$m\angle QPR = 45°$

Step 3 Find $m\overset{\frown}{PR}$.

$m\angle QPR = \frac{1}{2}m\overset{\frown}{PR}$
$45 = \frac{1}{2}m\overset{\frown}{PR}$
$m\overset{\frown}{PR} = 90°$

26. $50 = \frac{1}{2}(m\overset{\frown}{PS} - m\overset{\frown}{ML})$
$= \frac{1}{2}(m\overset{\frown}{PM} - m\overset{\frown}{ML})$
$= \frac{1}{2}m\overset{\frown}{LP}$
$m\overset{\frown}{LP} = 100°$

Circles

Review 11.2 / Angles and Segments in Circles

27. $DH \cdot HE = FH \cdot HG$

$3(4) = 2(y)$

$12 = 2y$

$y = 6$

$DE = 3 + 4 = 7$

$FG = 2 + (6) = 8$

28. $JK \cdot KL = MN \cdot KN$

$10(x) = 6(7)$

$10x = 42$

$x = 4.2$

$JL = 10 + (4.2) = 14.2$

$MN = 6 + 7 = 13$

29. $UY \cdot YV = WY \cdot YZ$

$8(5) = x(11)$

$40 = 11x$

$x = 3\frac{7}{11}$

$UV = 8 + 5 = 13$

$WZ = 3\frac{7}{11} + 11 = 14\frac{7}{11}$

30. $590(590) = 225.4(d - 225.4)$

$348{,}100 = 225.4d - 50{,}805.16$

$398{,}905.16 = 225.4d$

$d \approx 1770$ ft

31. $AB \cdot AC = AD \cdot AE$

$5(5 + x) = 6(6 + 6)$

$25 + 5x = 72$

$5x = 47$

$x = 9.4$

$AC = 5 + (9.4) = 14.4$

$AE = 6 + 6 = 12$

32. $HL \cdot JL = NL \cdot ML$

$(y + 10)10 = (18 + 9)9$

$10y + 100 = 243$

$10y = 143$

$y = 14.3$

$HL = (14.3) + 10 = 24.3$

$NL = 18 + 9 = 27$

33. $PQ \cdot PR = PS \cdot PT$

$3(3 + 6) = 2(2 + x)$

$27 = 4 + 2x$

$23 = 2x$

$x = 11.5$

$PR = 3 + 6 = 9$

$PT = 2 + (11.5) = 13.5$

34. $UW \cdot UX = UV^2$

$6(6 + 8) = z^2$

$84 = z^2$

$z = \sqrt{84} = 2\sqrt{21}$ (since $z > 0$)

35. $BC \cdot BD = BA^2$

$2(2 + x) = 5^2$

$4 + 2x = 25$

$2x = 21$

$x = 10.5$

36. $GE \cdot GF = GH^2$

$8(8 + 12) = y^2$

$160 = y^2$

$y = 4\sqrt{10}$ (since $y > 0$)

Transformational Geometry

Congruence Transformations

Reflections

1. Yes; the image appears to be flipped across a line.
2. No; the image does not appear to be flipped.
3. Yes; the image appears to be flipped across a line.
4. No; the image does not appear to be flipped.
5. **Step 1** Through each vertex draw a line ⊥ to the line of reflection.

   Step 2 Measure the distance from each vertex to the line of reflection. Locate the image of each vertex on the opposite side of the line of reflection and the same distance from it.

   Step 3 Connect the images of the vertices.

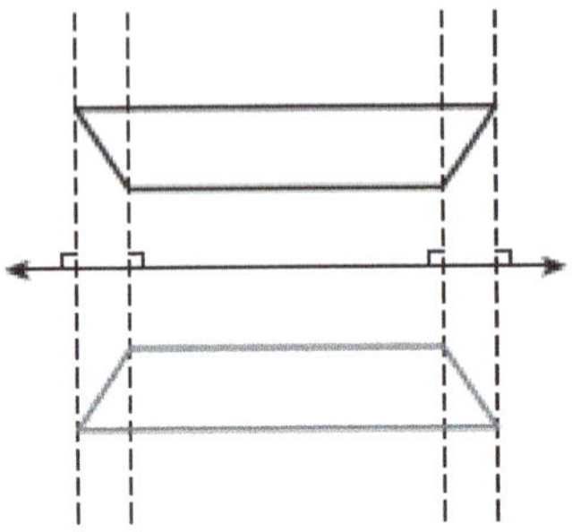

6. **Step 1** Through each vertex draw a line ⊥ to the line of reflection.

   Step 2 Measure the distance from each vertex to the line of reflection. Locate the image of each vertex on the opposite side of the line of reflection and the same distance from it.

   Step 3 Connect the images of the vertices.

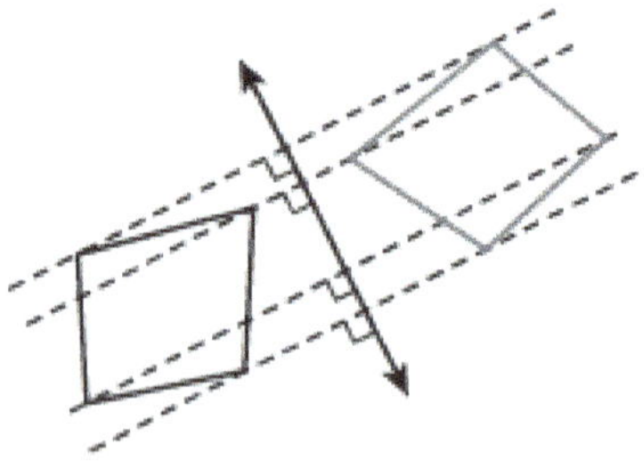

7. **1 Understand the Problem**
   The problem asks you to draw a diagram, locating point *P* on a highway so that $\overline{SP} + \overline{PT}$ has the least possible value.

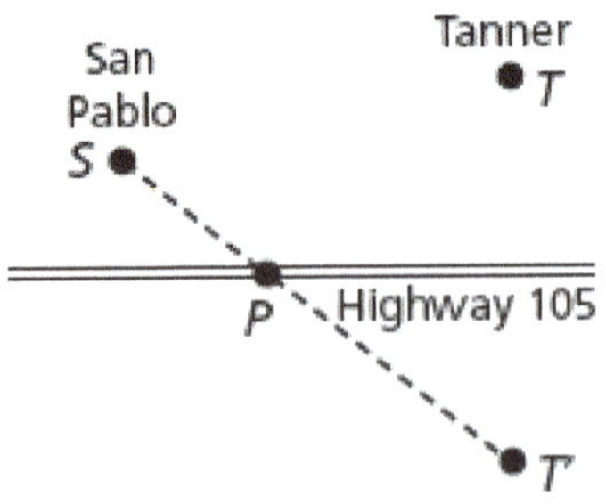

2 Make a Plan
Let *T'* be the reflection of *T* across the highway. For any point *P* on the highway, $\overline{PT'} \cong \overline{PT}$. So $\overline{SP} + \overline{PT} = \overline{SP} + \overline{PT'}$. $\overline{SP} + \overline{PT'}$ is the least when *S*, *P*, and *T'* are collinear.

3 Solve
Reflect *T* across the highway to locate *T'*. Draw $\overline{ST'}$ and locate *P* at the intersection of $\overline{ST'}$ and the highway.

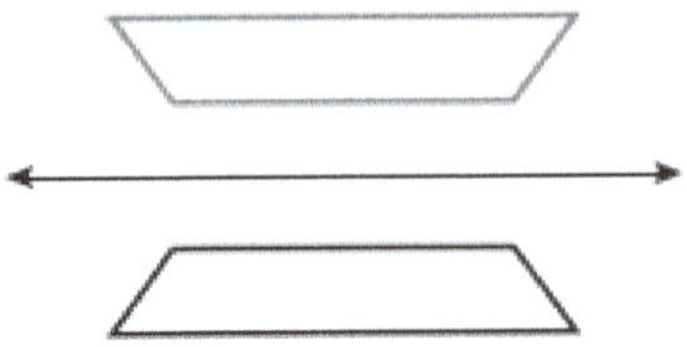

4 Look Back
To verify the answer, choose several possible locations for *P* and measure the total length of the access roads for each location.

Transformational Geometry

Congruence Transformations

Reflections

8. The reflection of (x, y) is $(x, -y)$.
$A(-2, 1) \rightarrow A'(-2, -1)$
$B(2, 3) \rightarrow B'(2, -3)$
$C(5, 2) \rightarrow C'(5, -2)$

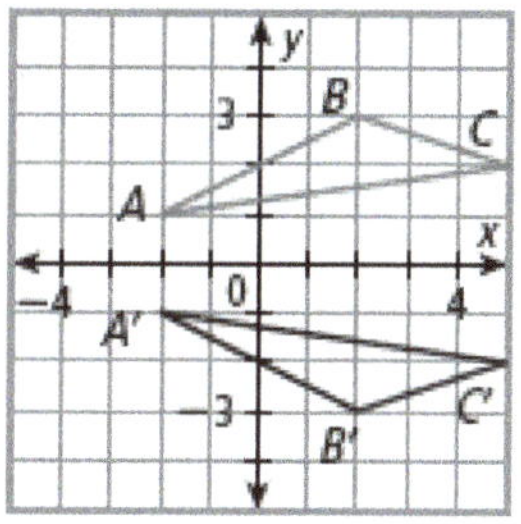

9. The reflection of (x, y) is $(-x, y)$.
$R(0, -1) \rightarrow R'(0, -1)$
$S(2, 2) \rightarrow S'(-2, 2)$
$T(3, 0) \rightarrow T'(-3, 0)$

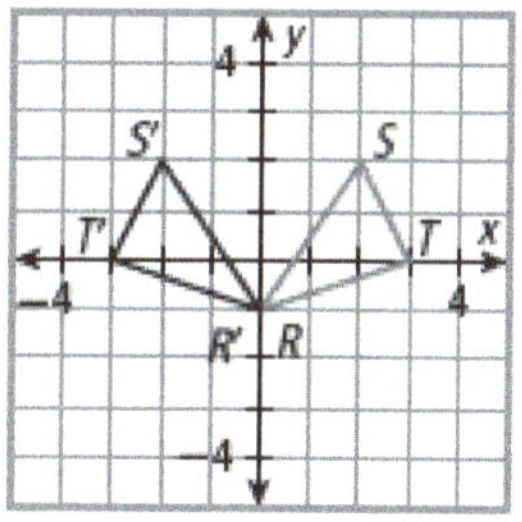

10. The reflection of (x, y) is (y, x).
$M(2, 1) \rightarrow M'(1, 2)$
$N(3, 1) \rightarrow N'(1, 3)$
$P(2, -1) \rightarrow P'(-1, 2)$
$Q(1, -1) \rightarrow Q'(-1, 1)$

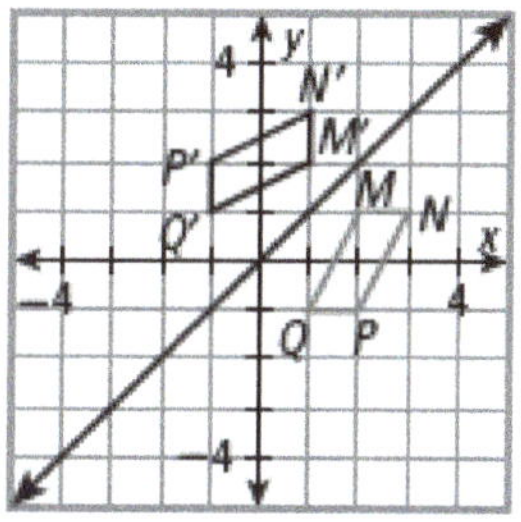

11. The reflection of (x, y) is (y, x).
$A(-2, 2) \rightarrow A'(2, -2)$
$B(-1, 3) \rightarrow B'(3, -1)$
$C(1, 2) \rightarrow C'(2, 1)$
$D(-2, -2) \rightarrow D'(-2, -2)$

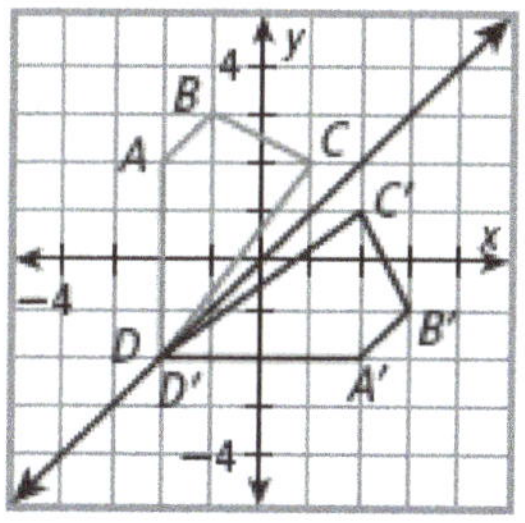

Transformational Geometry

Congruence Transformations

Translations

1. No; not all the points have moved the same distance.
2. No; not all the points have moved the same distance.
3. Yes; all the points have moved the same distance in the same direction.
4. No; not all the points have moved the same distance.
5. **Step 1** Draw a line || to the vector through each vertex of the square.

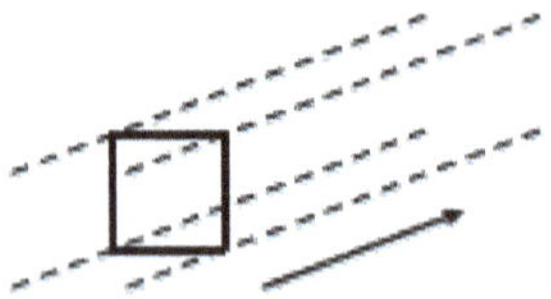

Step 2 Measure the length of the vector. Then, from each vertex mark off this distance in the same direction as the vector, on each of the || lines.

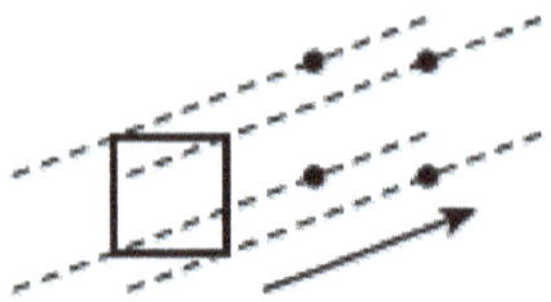

Step 3 Connect the images of the vertices.

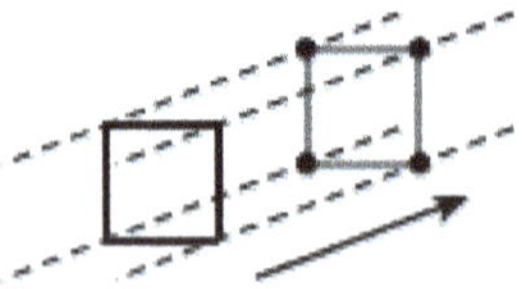

6. **Step 1** Draw a line || to the vector through each vertex of the triangle.

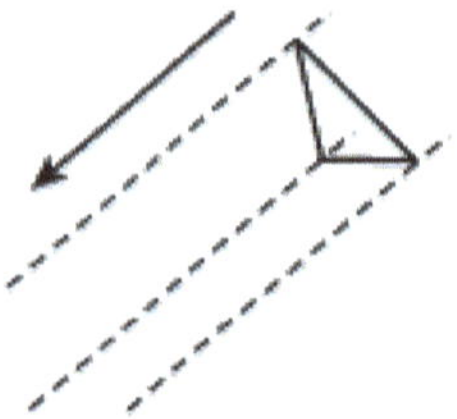

Step 2 Measure the length of vector. Then, from each vertex mark off this distance in the same direction as the vector, on each of the || lines.

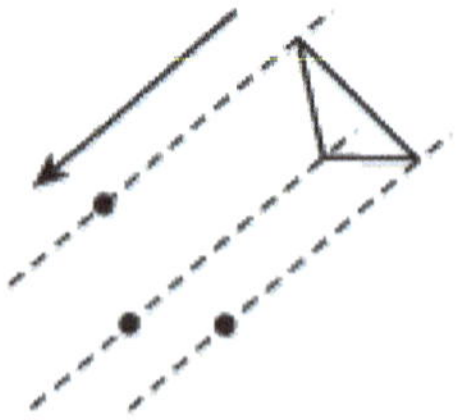

Step 3 Connect the images of the vertices.

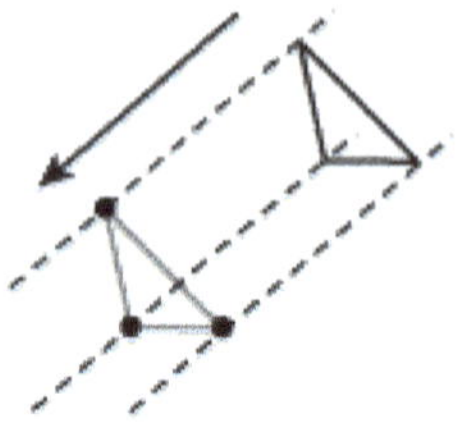

7. The image of (x, y) is $(x + 5, y)$.
$A(-4, -4) \rightarrow A'(-4 + 5, -4) = A'(1, -4)$
$B(-2, -3) \rightarrow B'(-2 + 5, -3) = S'(3, -3)$
$C(-1, 3) \rightarrow C'(-1 + 5, 3) = C'(4, 3)$
Graph image and preimage.

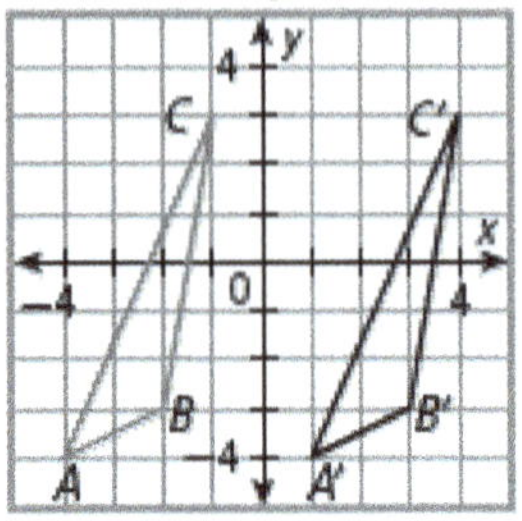

8. The image of (x, y) is $(x, y - 4)$.
$R(-3, 1) \rightarrow R'(-3, 1 - 4) = R'(-3, -3)$
$S(-2, 3) \rightarrow S'(-2, 3 - 4) = S'(-2, -1)$
$T(2, 3) \rightarrow T'(2, 3 - 4) = T'(2, -1)$
$U(3, 1) \rightarrow U'(3, 1 - 4) = U'(3, -3)$
Graph image and preimage.

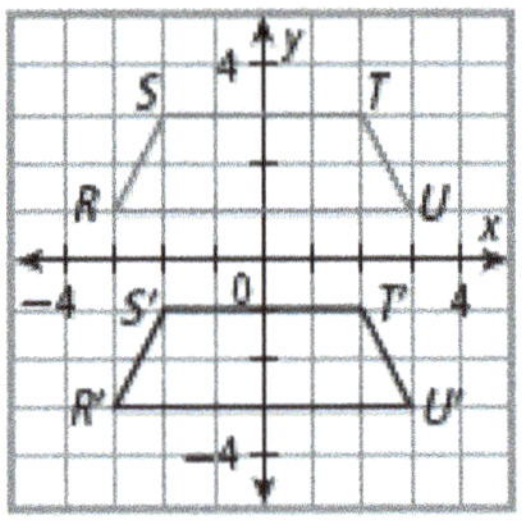

9. The image of (x, y) is $(x + 3, y + 2)$.
$J(-2, 2) \rightarrow J'(-2 + 3, 2 + 2) = J'(1, 4)$
$K(-1, 2) \rightarrow K'(-1 + 3, 2 + 2) = K'(2, 4)$
$L(-1, -2) \rightarrow L'(-1 + 3, -2 + 2) = L'(2, 0)$
$M(-3, -1) \rightarrow M'(-3 + 3, -1 + 2) = U'(0, 1)$
Graph image and preimage.

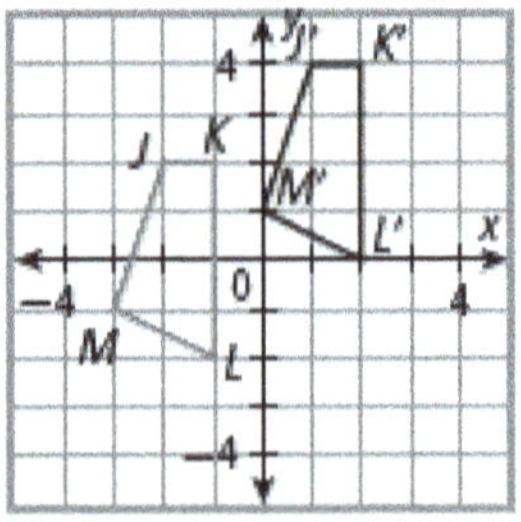

10. The second polygon has coordinates
$(1, 5 - 4) = (1, 1)$, $(2, 3 - 4) = (2, -1)$,
$(1, 1 - 4) = (1, -3)$, and $(0, 3 - 4) = (0, -1)$.
The third polygon has coordinates
$(1, 1 - 4) = (1, -3)$, $(2, -1 - 4) = (2, -5)$,
$(1, -3 - 4) = (1, -7)$, and $(0, -1 - 4) = (0, -5)$.
The fourth polygon has coordinates
$(1, -3 - 4) = (1, -7)$, $(2, -5 - 4) = (2, -9)$,
$(1, -7 - 4) = (1, -11)$, and $(0, -5 - 4) = (0, -9)$.

Transformational Geometry

Congruence Transformations

Rotations

1. Yes; the figure appears to be turned around a point.
2. No; the figure appears to be flipped, not turned.
3. No; the figure appears to be translated, not turned.
4. Yes; the figure appears to be turned around a point.
5. **Step 1** Draw a segment from each vertex to point P.

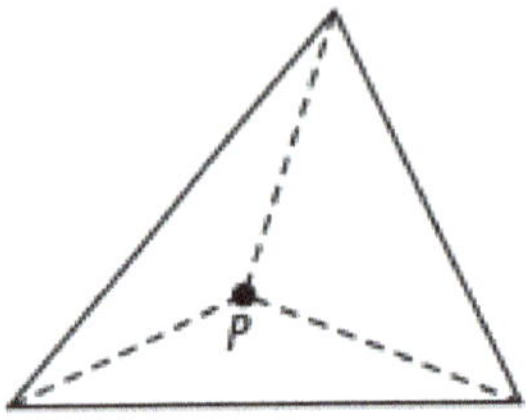

Step 2 Construct an $\angle \cong$ to $\angle A$ onto each segment. Measure distance from each vertex to point P and mark off this distance on the corresponding ray to locate image of each vertex.

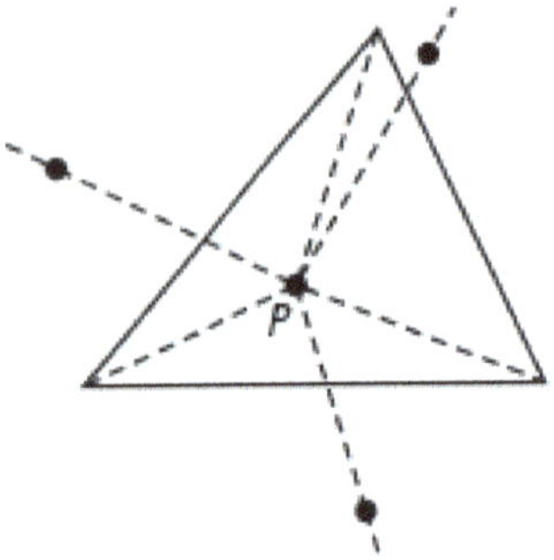

Step 3 Connect the images of the vertices.

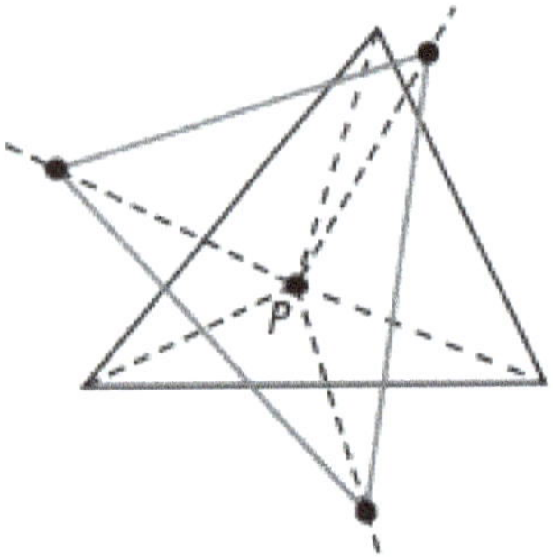

6. **Step 1** Draw a segment from each vertex to point P.

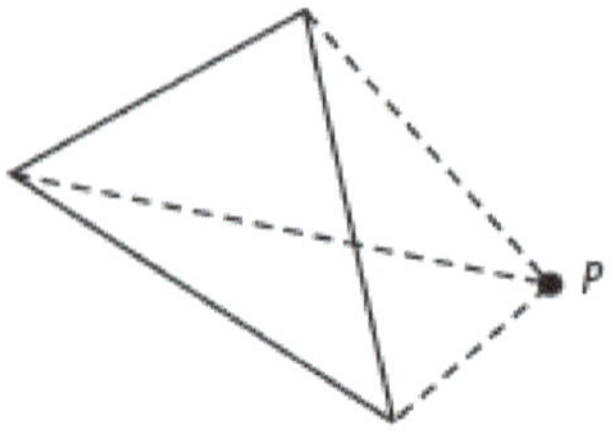

Step 2 Construct an $\angle \cong$ to $\angle A$ onto each segment. Measure the distance from each vertex to point P and mark off this distance on the corresponding ray to locate image of each vertex.

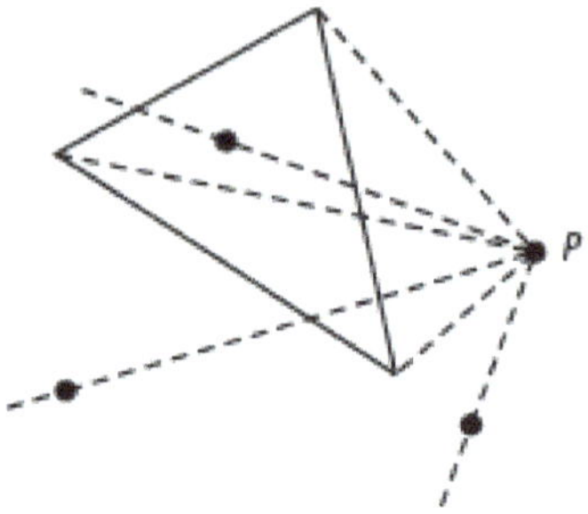

Step 3 Connect the images of the vertices.

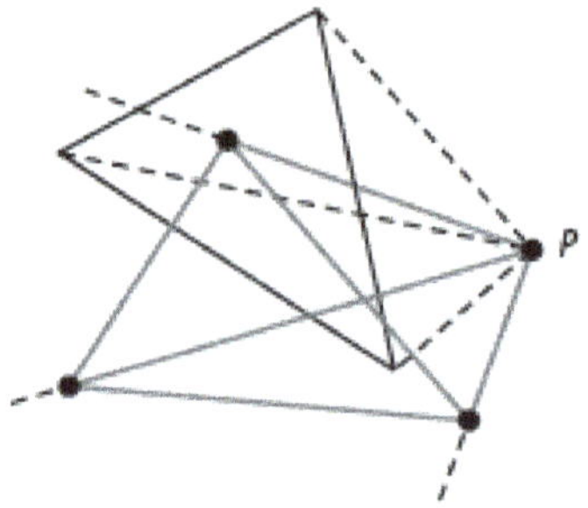

7. The rotation of (x, y) is $(-y, x)$.
$A(1, 0) \rightarrow A'(0, 1)$
$B(3, 2) \rightarrow B'(-2, 3)$
$C(5, 0) \rightarrow C'(0, 5)$

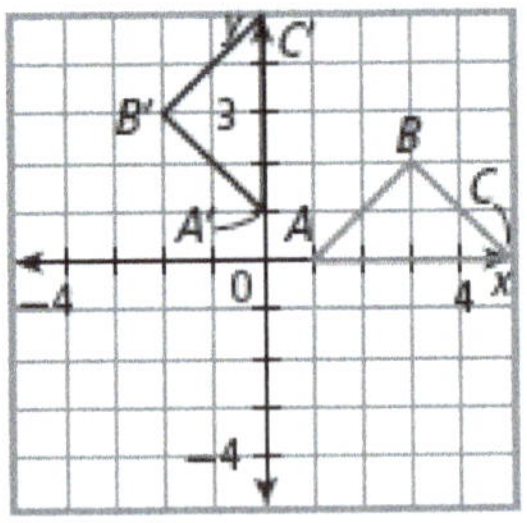

8. The rotation of (x, y) is $(-y, x)$.
$J(2, 1) \rightarrow J'(-1, 2)$
$K(4, 3) \rightarrow K'(-3, 4)$
$L(2, 4) \rightarrow L'(-4, 2)$
$M(-1, 2) \rightarrow M'(-2, -1)$

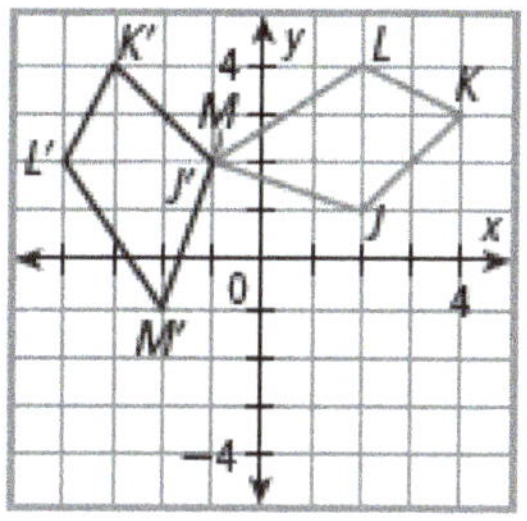

9. The rotation of (x, y) is $(-x, -y)$.
$D(2, 3) \rightarrow D'(-2, -3)$
$E(-1, 2) \rightarrow E'(1, -2)$
$F(2, 1) \rightarrow F'(-2, -1)$

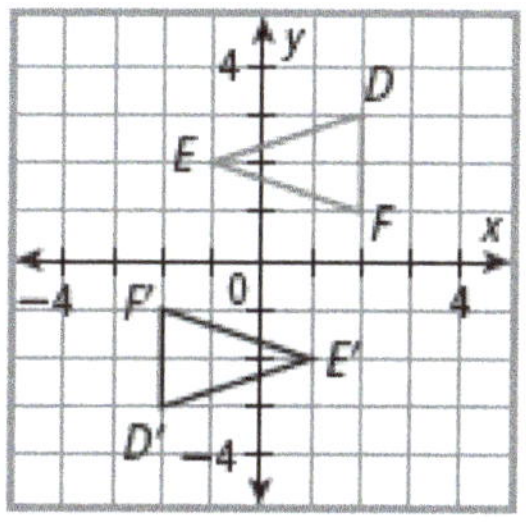

10. The rotation of (x, y) is $(-x, -y)$.
$P(-1, -1) \rightarrow P'(1, 1)$
$Q(-4, -2) \rightarrow Q'(4, 2)$
$R(0, -2) \rightarrow R'(0, 2)$

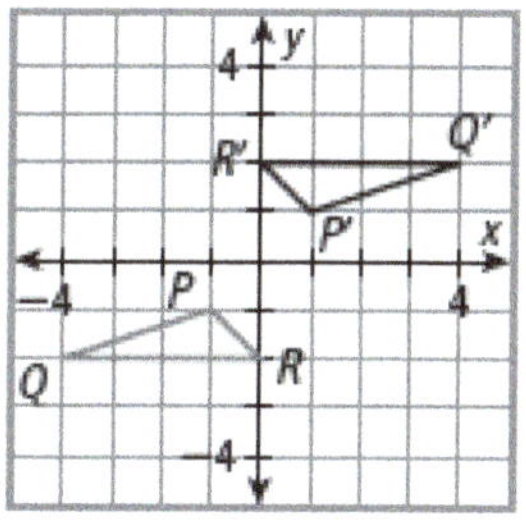

11. **Step 1** Draw a right △ to represent the car's location (x, y) after a rotation of 30° about the origin.

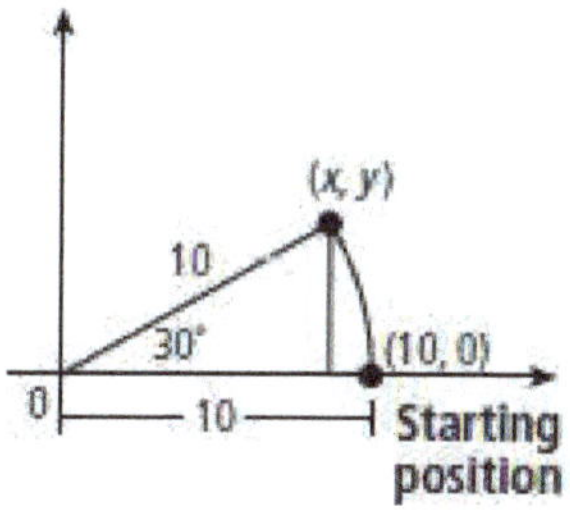

Step 2 Use the cosine ratio to find the x-coordinate

$$\cos 30^\circ = \frac{x}{10}$$

$$x = 10\cos 30^\circ \approx 8.7$$

Step 3 Use the sine ratio to find the y-coordinate

$$\sin 30^\circ = \frac{y}{10}$$

$$y = 10\sin 30^\circ = 5$$

The car's final position is approximately (8.7, 5).

1. No; the image does not appear to be flipped.

2. Yes; the image appears to be flipped across a line.

3. Yes; the image appears to be flipped across a line.

4. No; the image does not appear to be flipped.

5. **Step 1** Through each vertex draw a line ⊥ to the line of reflection.
Step 2 Measure the distance from each vertex to the line of reflection. Locate the image of each vertex on the opposite side of the line of reflection and the same distance from it.
Step 3 Connect the images of the vertices.

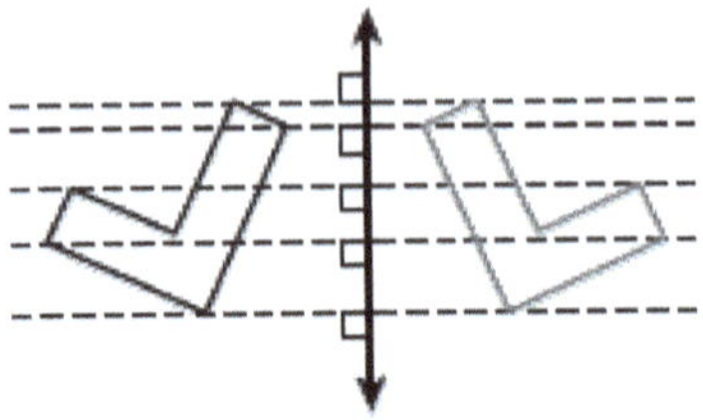

6. **Step 1** Through each vertex draw a line ⊥ to the line of reflection.
Step 2 Measure the distance from each vertex to the line of reflection. Locate image of each vertex on the opposite side of the line of reflection and the same distance from it.
Step 3 Connect the images of the vertices.

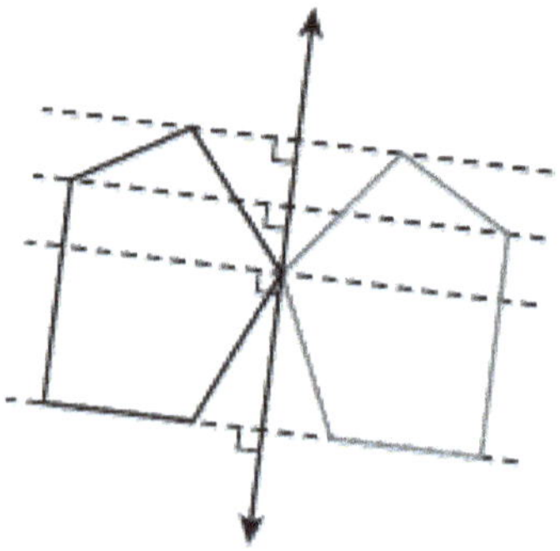

7. **1 Understand the Problem**
The problem asks you to draw a diagram, locating point X on the side rail so that $\overline{AX}$ and $\overline{XC}$ make the same ∠ with the side rail.
2 Make a Plan
Let C' be the reflection of C across the side rail. For any point X on side rail, the ∡ of $\overline{AX}$ and $\overline{XC'}$ with the side rail are ≅ when A, X, and C' are collinear.
3 Solve
Reflect C across the side rail to locate C'. Draw $\overline{AC'}$ and locate P at the intersection of $\overline{AC'}$ and the side rail.
4 Look Back
To verify the answer, choose several possible locations for X and measure the ∡ for each location.

Transformational Geometry

Review 12.1 / Congruence Transformations

8. The reflection of (x, y) is $(-x, y)$.
$A(-3, 2) \rightarrow A'(3, 2)$
$B(0, 2) \rightarrow B'(0, 2)$
$C(-2, 0) \rightarrow C'(2, 0)$

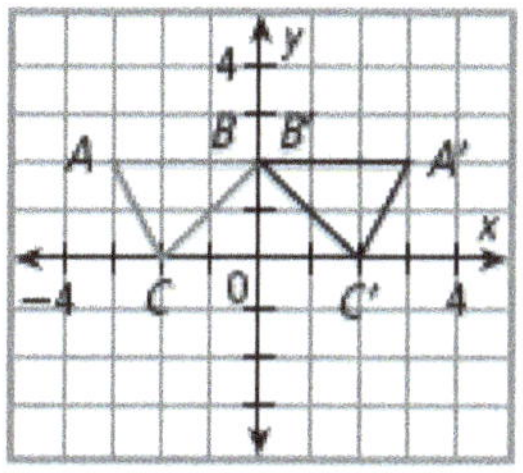

9. The reflection of (x, y) is (y, x).
$M(-4, -1) \rightarrow M'(-1, -4)$
$N(-1, -1) \rightarrow N(-1, -1)$
$P(-2, -2) \rightarrow P'(-2, -2)$

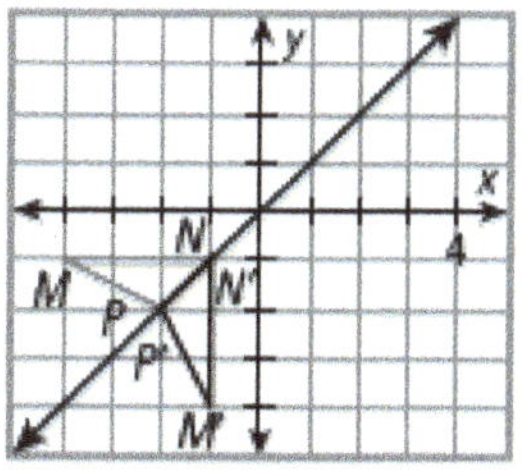

10. The reflection of (x, y) is $(x, -y)$.
$J(1, 2) \rightarrow J'(1, -2)$
$K(-2, -1) \rightarrow K'(-2, 1)$
$L(3, -1) \rightarrow L'(3, 1)$

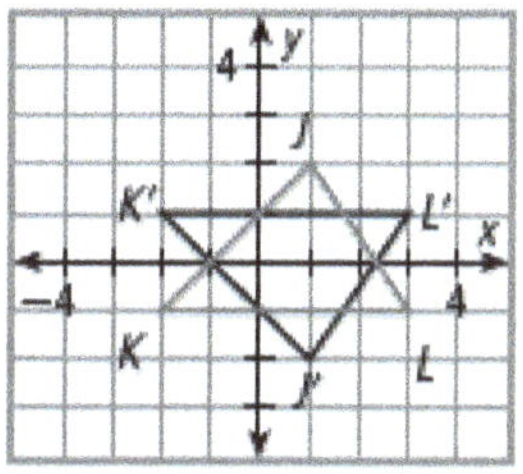

11. The reflection of (x, y) is (y, x).
$S(-1, 1) \rightarrow S'(1, -1)$
$T(1, 4) \rightarrow T'(4, 1)$
$U(3, 2) \rightarrow U'(2, 3)$
$V(1, -3) \rightarrow V'(-3, 1)$

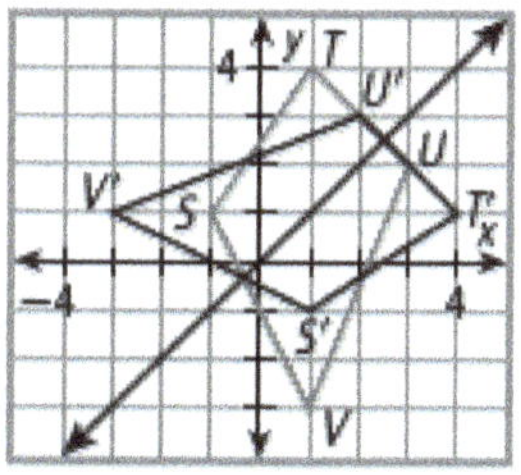

12. Yes; all the points have moved the same distance in the same direction.

13. No; not all points have moved the same distance.

14. No; not all the points have moved the same distance.

15. Yes; all the points have moved the same distance in the same direction.

16. **Step 1** Draw a line || to the vector through each vertex of the square.

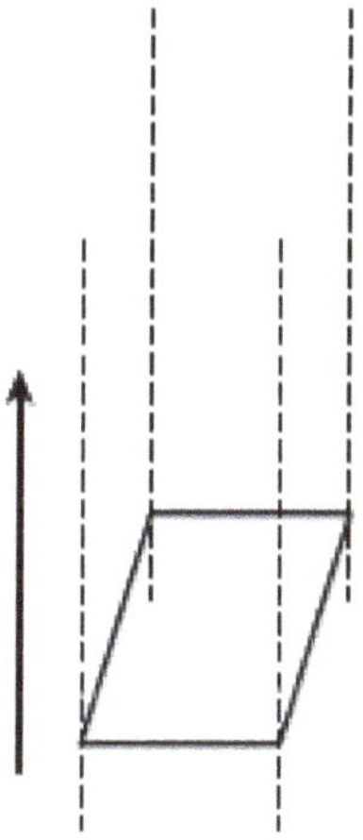

Step 2 Measure the length of the vector. Then, from each vertex mark off this distance in the same direction as the vector, on each of the || lines.

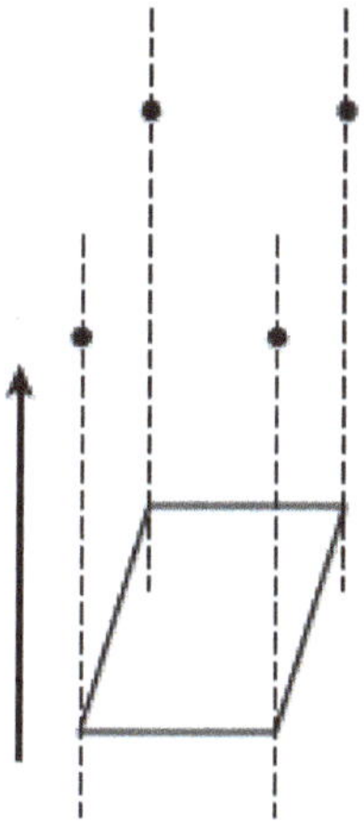

Step 3 Connect the images of the vertices.

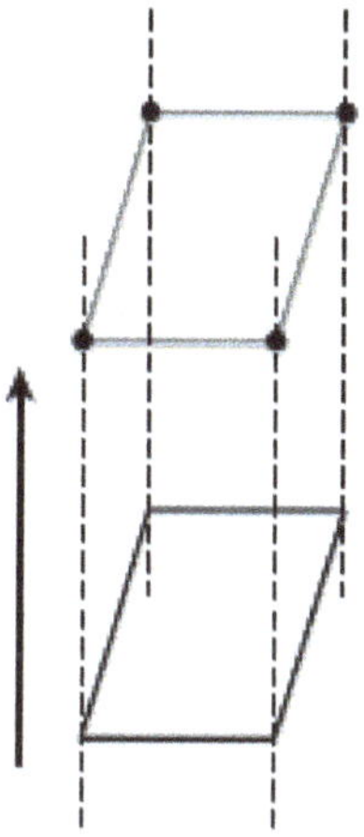

Transformational Geometry

Review 12.1 / Congruence Transformations

17. Step 1 Draw a line ∥ to the vector through each vertex of the square.

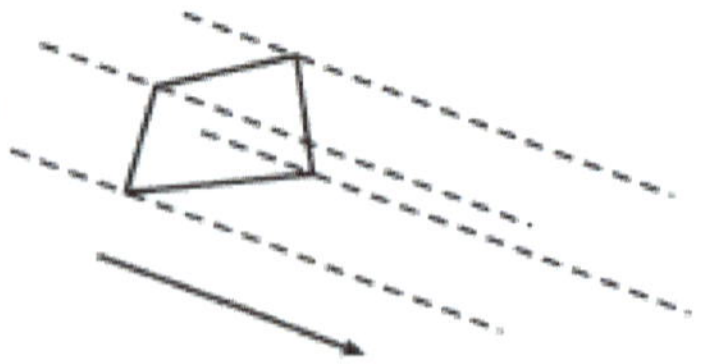

Step 2 Measure the length of the vector. Then, from each vertex mark off this distance in the same direction as the vector, on each of the ∥ lines.

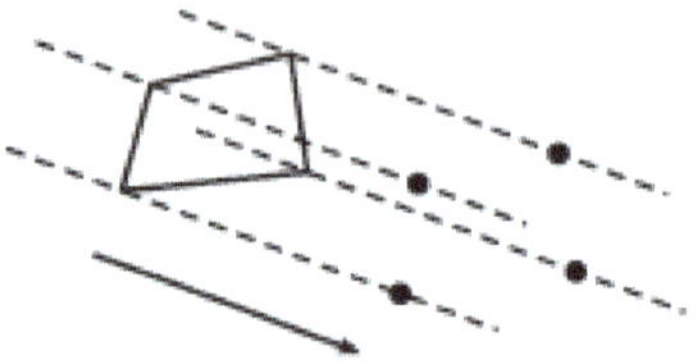

Step 3 Connect the images of the vertices.

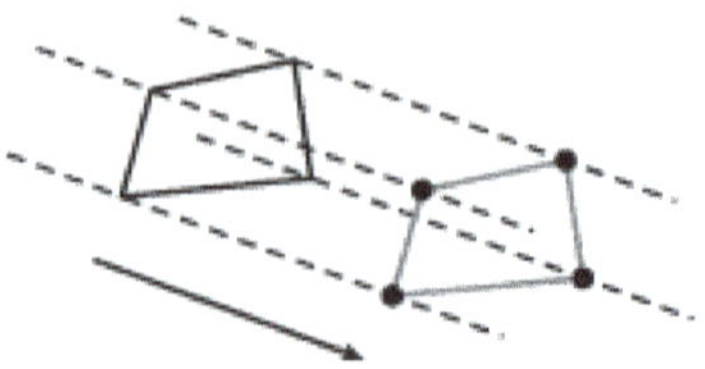

18. The image of (x, y) is $(x - 3, y)$.
$P(-1, 2) \rightarrow P'(-1 - 3, 2) = P'(-4, 2)$
$Q(1, -1) \rightarrow Q'(1 - 3, -1) = Q'(-2, -1)$
$R(3, 1) \rightarrow R'(3 - 3, 1) = R'(0, 1)$
$S(2, 3) \rightarrow S'(2 - 3, 3) = S'(-1, 3)$
Graph the image and the preimage.

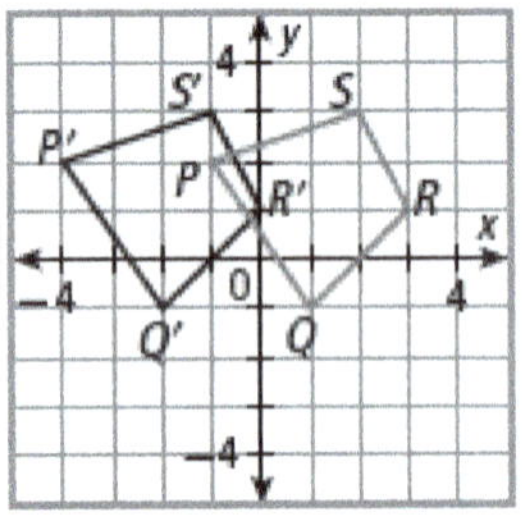

19. The image of (x, y) is $(x - 3, y - 3)$.
$A(1, 3) \rightarrow A'(1 - 3, 3 - 3) = A'(-2, 0)$
$B(-1, 2) \rightarrow B'(-1 - 3, 2 - 3) = B'(-4, -1)$
$C(2, 1) \rightarrow C'(2 - 3, 1 - 3) = C'(-1, -2)$
$D(4, 2) \rightarrow D'(4 - 3, 2 - 3) = D'(1, -1)$
Graph the image and the preimage.

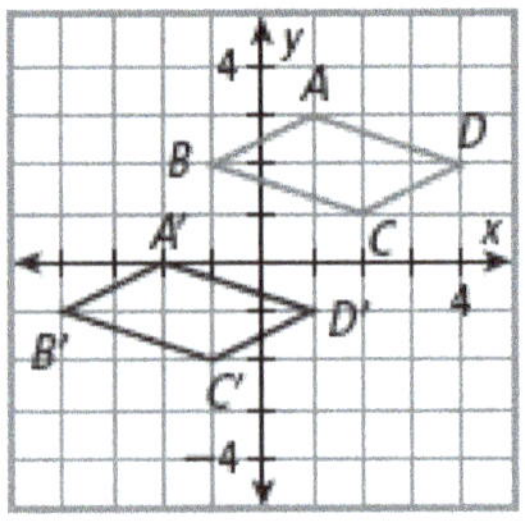

20. The image of (x, y) is $(x + 5, y - 20)$.
$D(0, 15) \rightarrow D'(0 + 5, 15 - 20) = D'(5, -5)$
$E(-10, 5) \rightarrow E'(-10 + 5, 5 - 20) = E'(-5, -15)$
$F(10, -5) \rightarrow F'(10 + 5, -5 - 20) = F'(15, -25)$
Graph the image and the preimage.

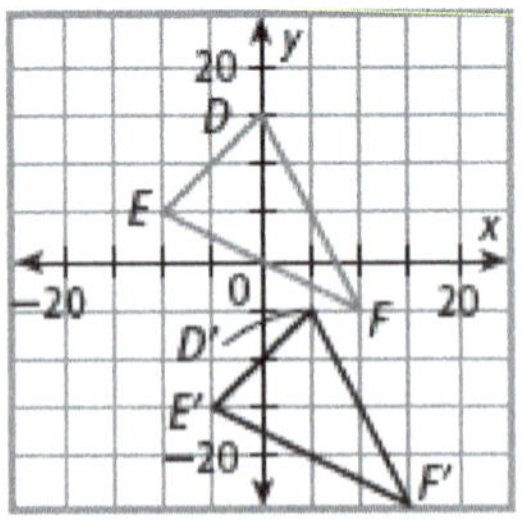

21. a. The ladybug starts at $(-2.5, -1.5)$.
The second position is at
$(-2.5 + 1, -1.5 + 1) = (-1.5, -0.5)$.
The third position is at $(-1.5 + 2, -0.5 + 2)$
$= (0.5, 1.5)$.
The final position is at $(0.5 + 3, 1.5 + 3)$
$= (3.5, 4.5)$.

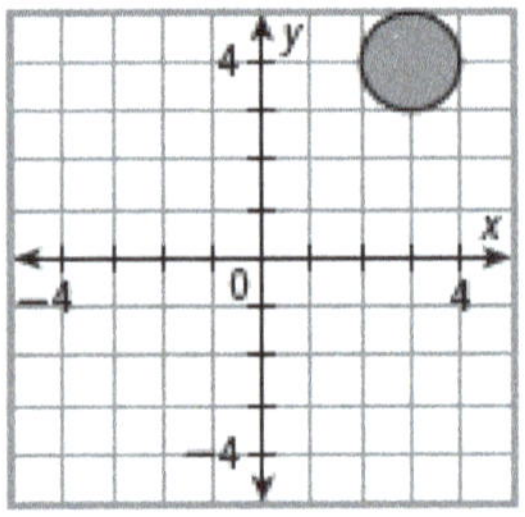

b. $\langle 1, 1\rangle + \langle 2, 2\rangle + \langle 3, 3\rangle = \langle 6, 6\rangle$

Transformational Geometry

Review 12.1 / Congruence Transformations

22. No; the figure appears to be flipped and translated, not turned.

23. Yes; the figure appears to be turned around a point.

24. Yes; the figure appears to be turned around a point.

25. No; the figure appears to be enlarged, not turned.

26. **Step 1** Draw a segment from each vertex to point *P*.

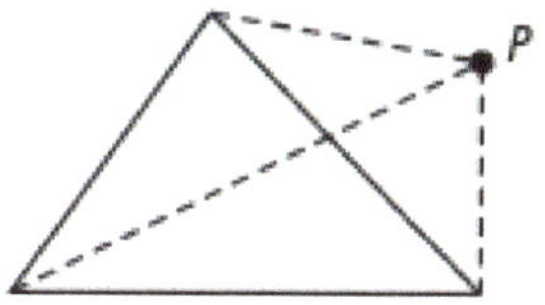

Step 2 Construct an ∠ ≅ to ∠*A* onto each segment. Measure the distance from each vertex to point *P* and mark off this distance on the corresponding ray to locate the image of each vertex.

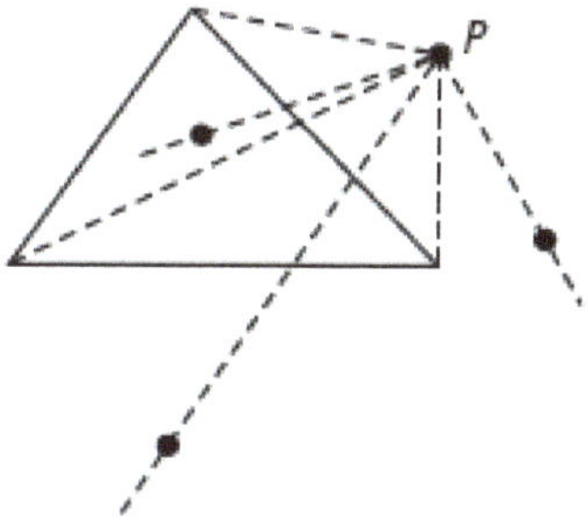

Step 3 Connect the images of the vertices.

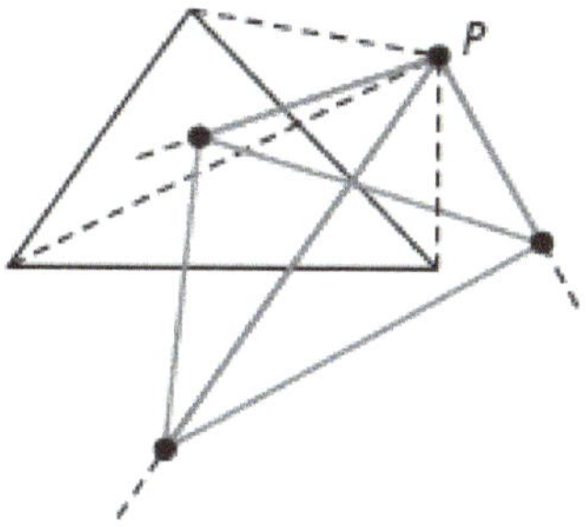

27. **Step 1** Draw a segment from each vertex to point *P*.

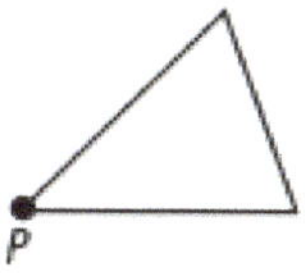

Step 2 Construct an ∠ ≅ to ∠*A* onto each segment. Measure the distance from each vertex to point *P* and mark off this distance on the corresponding ray to locate the image of each vertex.

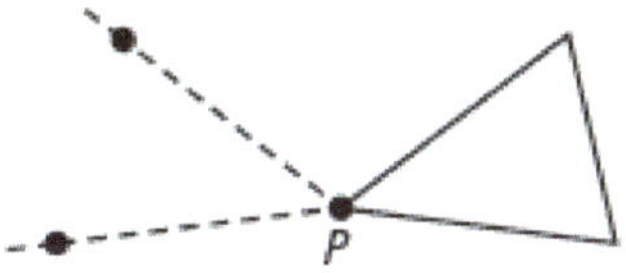

Step 3 Connect the images of the vertices.

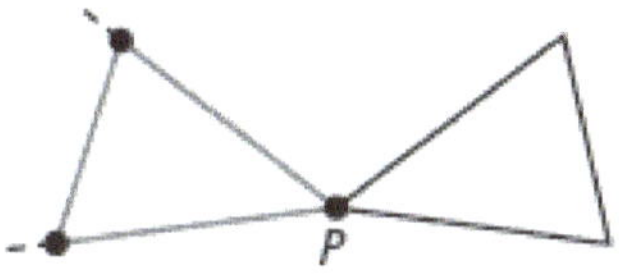

28. The rotation of (x, y) is $(-y, x)$.
$E(-1, 2) \rightarrow E'(-2, -1)$
$F(3, 1) \rightarrow F'(-1, 3)$
$G(2, 3) \rightarrow G'(-3, 2)$

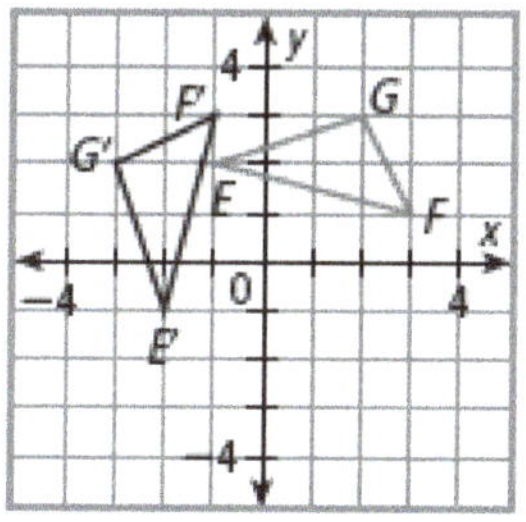

29. The rotation of (x, y) is $(-y, x)$.
$A(-1, 0) \rightarrow A'(0, -1)$
$B(-1, -3) \rightarrow B'(3, -1)$
$C(1, -3) \rightarrow C'(3, 1)$
$D(1, 0) \rightarrow D'(0, 1)$

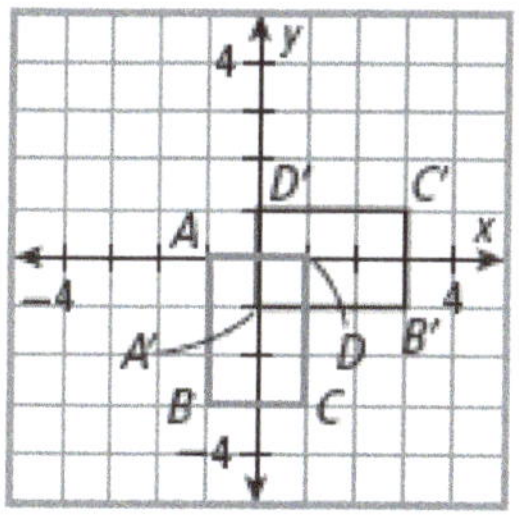

30. The rotation of (x, y) is $(-x, -y)$.
$P(0, -2) \rightarrow P'(0, 2)$
$Q(2, 0) \rightarrow Q'(-2, 0)$
$R(3, -3) \rightarrow R'(-3, 3)$

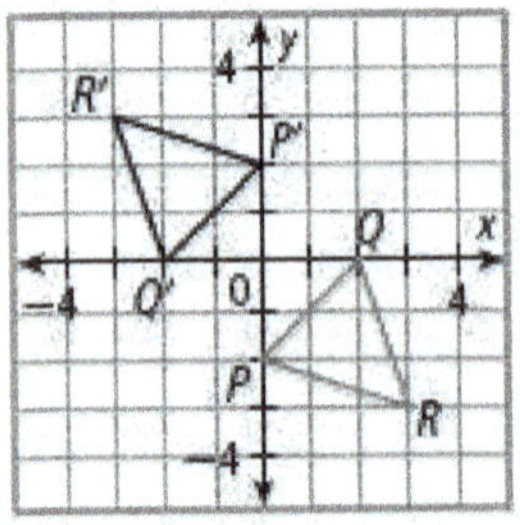

31. The rotation of (x, y) is $(-x, -y)$.
$L(2, 0) \rightarrow L'(-2, 0)$
$M(-1, -2) \rightarrow M'(1, 2)$
$N(2, -2) \rightarrow N'(-2, 2)$

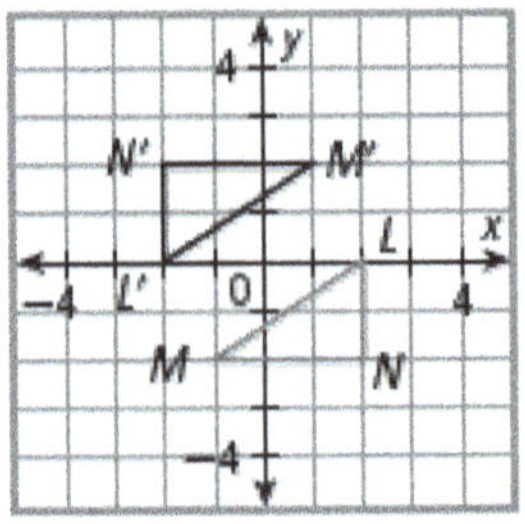

32. Step 1 Find the $\angle$ of rotation. 6 min is $\frac{6}{72} = \frac{1}{12}$ of a complete rotation, or $\frac{1}{12}(360) = 30°$.

Step 1 Draw a right $\triangle$ to represent the table's location (x, y) after a rotation of 30° about the origin.

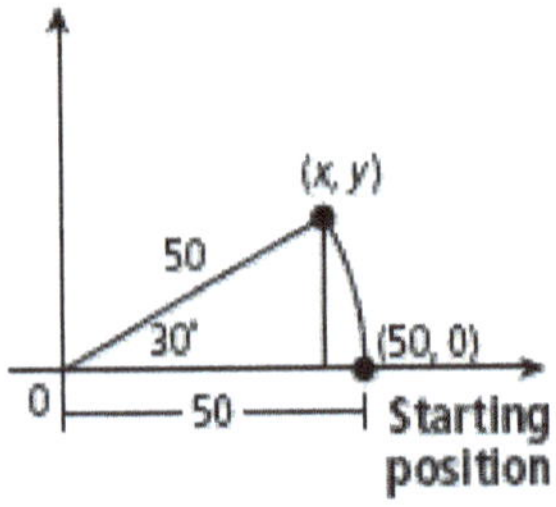

Step 2 Use the cosine ratio to find the x-coordinate

$\cos 30° = \frac{x}{50}$

$x = 50 \cos 30° \approx 43.3$

Step 3 Use the sine ratio to find the y-coordinate

$\sin 30° = \frac{y}{50}$

$y = 50 \sin 30° = 25$

The table's final position is approximately (43.3, 25).

Simplifying Expressions

1. $-12 + 67 + 12 + 23$
   $= (-12 + 12) +$
   $(67 + 23)$
   $= 0 + 90 = 90$
2. $16 + 2\frac{1}{2} + 4 + 1\frac{1}{2}$
   $= 16 + 4 + 2\frac{1}{2} + 1\frac{1}{2}$
   $= 20 + 4 = 24$
3. $27 + 98 + 73$
   $= 73 + 27 + 98$
   $= 100 + 98 = 198$
4. $\frac{1}{3} \cdot 8 \cdot 21$
   $= \frac{1}{3} \cdot 21 \cdot 8$
   $= 7 \cdot 8 = 56$
5. $2 \cdot 38 \cdot 50$
   $= 2 \cdot 50 \cdot 28$
   $= 100 \cdot 28 = 2800$
6. $50 \cdot 118 \cdot 20$
   $= 50 \cdot 20 \cdot 118$
   $= 1000 \cdot 118 = 118{,}000$
7. $14(1002)$
   $= 14(1000 + 2)$
   $= 14(1000) + 14(2)$
   $= 14{,}000 + 28 = 14028$
8. $16(19)$
   $= 16(20 - 1)$
   $= 16(20) - 16(1)$
   $= 320 - 16 = 304$
9. $9(38)$
   $= 9(40 - 2)$
   $= 9(40) - 9(2)$
   $= 360 - 18 = 342$
10. $8(57)$
   $= 8(60 - 3)$
   $= 8(60) - 8(3)$
   $= 480 - 24 = 456$
11. $12(112)$
   $= 12(100 + 12)$
   $= 12(100) + 12(12)$
   $= 1200 + 144 = 1344$

Appendix: Essential Algebra Review

Solving Equations and Inequalities

Simplifying Expressions

12. $7(109)$
$= 7(100 + 9)$
$= 7(100) + 7(9)$
$= 700 + 63 = 763$

13. $6x + 10x = 16x$

14. $35x - 15x = 20x$

15. $-3a + 9a = 6a$

16. $-8r - r$
$= -8r - 1r = -9r$

17. $17x^2 + x$

18. $3.2x + 4.7x = 7.9x$

19.

$5(x + 3) - 7x$	
$5(x) + 5(3) - 7x$	Distribute the 5
$5x - 7x + 15$	Multiply
$-2x + 15$	Add like terms

20.

$9(a - 3) - 4$	
$9(a) - 9(3) - 4$	Distribute the 9
$9a - 27 - 4$	Multiply
$9a - 31$	Add like terms

21.

$5x^2 - 2(x - 3x^2)$	
$5x^2 - 2(x) + 2(3x^2)$	Distribute the 2
$5x^2 - 2x + 6x^2$	Multiply
$11x^2 - 2x$	Add like terms

22.

$5x^2 - 2(x - 3x^2)$	
$5x^2 - 2(x) - 2(-3x^2)$	Distribute the -2
$5x^2 - 2x + 6x^2$	Multiply
$11x^2 - 2x$	Add like terms

23.

$12x + 8x + t - 7x$	
$12x + 8x - 7x + t$	Comm. Prop.
$13x + t$	Add like terms

24.

$4a - 2(a - 1)$	
$4a - 2(a) - 2(-1)$	Distribute the -2
$4a - 2a + 2$	Multiply
$2a + 2$	Add like terms

Appendix: Essential Algebra Review

Solving Equations and Inequalities

Writing and Solving Two-Step Equations

1. $20 + (-x) = 7$

2. $0 = 4 + \frac{x}{3}$

3. $4a + 3 = 11$
$-3 \quad -3$
$4a = 8$
$\frac{4a}{4} = \frac{8}{4}$
$a = 2$

4. $15y + 31 = 61$
$-31 \quad -31$
$15y = 30$
$\frac{15y}{15} = \frac{30}{15}$
$y = 2$

5. $8 = 3r - 1$
$+1 \quad +1$
$9 = 3r$
$\frac{9}{3} = \frac{3r}{3}$
$3 = r$

6. $42 = -2d + 6$
$-6 \quad -6$
$36 = -2d$
$\frac{36}{-2} = \frac{-2d}{-2}$
$-18 = d$

7. $9 - c = -13$
$-9 \quad -9$
$-c = -22$
$(-1)(-c) = (-1)(-22)$
$c = 22$

8. $\frac{x}{6} + 4 = 15$
$-4 \quad -4$
$\frac{x}{6} = 11$
$6\left(\frac{x}{6}\right) = 6(11)$
$x = 66$

9. $15 = \frac{a}{3} - 2$
$+2 \quad +2$
$17 = \frac{a}{3}$
$3(17) = 3\left(\frac{a}{3}\right)$
$51 = a$

10. $4 - \frac{m}{2} = 10$
$-4 \quad -4$
$-\frac{m}{2} = 6$
$(-2)\left(-\frac{m}{2}\right) = (-2)(6)$
$m = -12$

11. Let x represent the number of daily bus passes Paul bought.
$1.50x + 7 = 29.50$
$-7 \quad -7$
$1.50x = 22.50$
$\frac{1.50x}{1.50} = \frac{22.50}{1.50}$
$x = 15$

Paul bought 15 daily bus passes.

12. $12.3 = \frac{m}{0.4} - 0.2$
$+0.2 \quad +0.2$
$12.5 = \frac{m}{0.4}$
$0.4(12.5) = \left(\frac{m}{0.4}\right)0.4$
$m = 5$

13. $\frac{2.5 + d}{5} = 1.1$
$5\left(\frac{2.5 + d}{5}\right) = (1.1)5$
$2.5 + d = 5.5$
$-2.5 \quad -2.5$
$d = 3$

Solving Two-Step Inequalities

1. $5x + 3 < 18$

$\underline{\quad -3 \quad -3}$

$5x < 15$

$\frac{5x}{5} < \frac{15}{5}$

$x < 3$

2. $-19 \geq \frac{z}{7} + 23$

$\underline{-23 \qquad -23}$

$-42 \geq \frac{z}{7}$

$(7)-42 \geq (7)\frac{z}{7}$

$-294 \geq z$

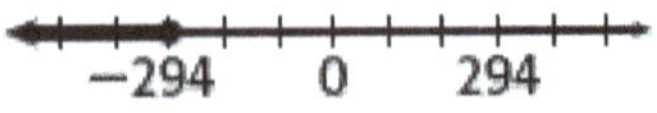

3. $3y - 4 \geq 14$

$\underline{\quad +4 \quad +4}$

$3y \geq 18$

$\frac{3y}{3} \geq \frac{18}{3}$

$y \geq 6$

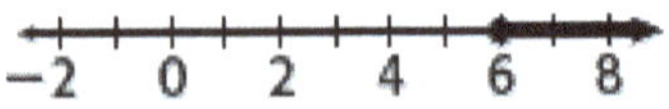

4. $\frac{m}{4} - 2 > -3$

$\underline{\quad +2 \quad +2}$

$\frac{m}{4} > -1$

$(4)\frac{m}{4} > (4) - 1$

$m > -4$

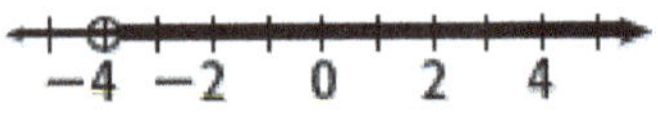

5. $-11p - 13 \geq 42$

$\underline{\quad +13 \quad +13}$

$-11p \geq 55$

$\frac{-11p}{-11} \leq \frac{55}{-11}$

$p \leq -5$

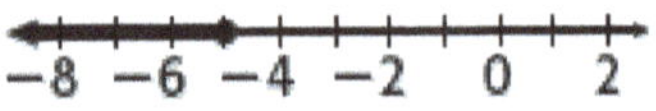

6. $\frac{n}{-3} - 4 > 4$

$\underline{\quad +4 \quad +4}$

$\frac{n}{-3} > 8$

$(-3)\frac{n}{-3} < (-3)8$

$n < -24$

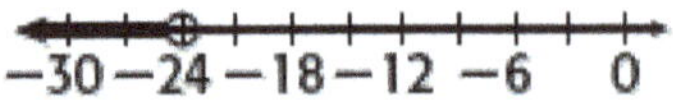

7. Let x represent the amount each student earned.

$3x + 15 > 93$

$\underline{\quad -15 \quad -15}$

$3x > 78$

$\frac{3x}{3} > \frac{78}{3}$

$x > 26$

They earned more than \$26 each.

Appendix: Essential Algebra Review

Solving Equations and Inequalities

Writing and Solving Multi-Step Equations

1. $14n + 2 - 7n = 37$

$$7n + 2 = 37$$
$$-2 \quad -2$$
$$7n = 35$$
$$\frac{7n}{7} = \frac{35}{7}$$
$$n = 5$$

2. $10x - 11 - 4x = 43$

$$6x - 11 = 43$$
$$+11 \quad +11$$
$$6x = 54$$
$$\frac{6x}{6} = \frac{54}{6}$$
$$x = 9$$

3. $1 = -3 + 4p - 2p$

$$1 = -3 + 2p$$
$$+3 \quad +3$$
$$4 = 2p$$
$$\frac{4}{2} = \frac{2p}{2}$$
$$2 = p$$

4. $12 - (x + 3) = 10$

$$12 - x - 3 = 10$$
$$9 - x = 10$$
$$-9 \quad -9$$
$$-x = 1$$
$$\frac{-x}{-1} = \frac{1}{-1}$$
$$x = -1$$

5. $15 = 2(q + 4) + 3$

$$15 = 2q + 8 + 3$$
$$15 = 2q + 11$$
$$-11 \quad -11$$
$$4 = 2q$$
$$\frac{4}{2} = \frac{2q}{2}$$
$$2 = q$$

6. $5(m - 2) + 36 = -4$

$$5m - 10 + 36 = -4$$
$$5m + 26 = -4$$
$$-26 \quad -26$$
$$5m = -30$$
$$\frac{5m}{5} = \frac{-30}{5}$$
$$m = -6$$

7. Let x represent the number of books Ben read. Then $2x$ represents the number Keisha read, and $\frac{2x - 4}{2}$ represents the number Sheldon read, which equals 10.

$$\frac{2x - 4}{2} = 10$$
$$(2)\frac{2x - 4}{2} = (2)10$$
$$2x - 4 = 20$$
$$+4 \quad +4$$
$$2x = 24$$
$$\frac{2x}{2} = \frac{24}{2}$$
$$x = 12$$

Ben read 12 books.

Appendix: Essential Algebra Review

Solving Equations and Inequalities

Solving Equations with Variables on Both Sides

1.
$$\begin{aligned} 2c - 5 &= c + 4 \\ -c \quad & \quad -c \\ c - 5 &= 4 \\ +5 \quad & \quad +5 \\ c &= 9 \end{aligned}$$

2.
$$\begin{aligned} 8r + 4 &= 10 + 2r \\ -2r \quad & \quad -2r \\ 6r + 4 &= 10 \\ -4 \quad & \quad -4 \\ 6r &= 6 \\ \frac{6r}{6} &= \frac{6}{6} \\ r &= 1 \end{aligned}$$

3.
$$\begin{aligned} 2x - 1 &= x + 11 \\ -x \quad & \quad -x \\ x - 1 &= 11 \\ +1 \quad & \quad +1 \\ x &= 12 \end{aligned}$$

4.
$$\begin{aligned} 28 - 0.3y &= 0.7y - 12 \\ +0.3y \quad & \quad +0.3y \\ 28 &= y - 12 \\ +12 \quad & \quad +12 \\ 40 &= y \end{aligned}$$

5.
$$\begin{aligned} -2(x + 3) &= 4x - 3 \\ -2(x) - 2(3) &= 4x - 3 \\ -2x - 6 &= 4x - 3 \\ +2x \quad & \quad +2x \\ -6 &= 6x - 3 \\ +3 \quad & \quad +3 \\ -3 &= 6x \\ \frac{-3}{6} &= \frac{6x}{6} \\ -\frac{1}{2} &= x \end{aligned}$$

6.
$$\begin{aligned} 3c - 4c + 1 &= 5c + 2 + 3 \\ -c + 1 &= 5c + 5 \\ +c \quad & \quad +c \\ 1 &= 6c + 5 \\ -5 \quad & \quad -5 \\ -4 &= 6c \\ \frac{-4}{6} &= \frac{6c}{6} \\ -\frac{2}{3} &= c \end{aligned}$$

7.
$$\begin{aligned} 5 + 3(q - 4) &= 2(q + 1) \\ 5 + 3(q) + 3(-4) &= 2(q) + 2(1) \\ 5 + 3q - 12 &= 2q + 2 \\ 3q - 7 &= 2q + 2 \\ -2q \quad & \quad -2q \\ q - 7 &= 2 \\ +7 \quad & \quad +7 \\ q &= 9 \end{aligned}$$

8.
$$\begin{aligned} 5 - (t + 3) &= -1 + 2(t - 3) \\ 5 - (t) - (3) &= -1 + 2(t) + 2(-3) \\ 5 - t - 3 &= -1 + 2t - 6 \\ 2 - t &= 2t - 7 \\ +t \quad & \quad +t \\ 2 &= 3t - 7 \\ +7 \quad & \quad +7 \\ 9 &= 3t \\ \frac{9}{3} &= \frac{3t}{3} \\ 3 &= t \end{aligned}$$

9.
$$\begin{aligned} 7x - 4 &= -2x + 1 + 9x - 5 \\ 7x - 4 &= 7x - 4 \\ -7x \quad & \quad -7x \\ -4 &= -4 \checkmark \end{aligned}$$

all real numbers

10.
$$\begin{aligned} 8x + 6 - 9x &= 2 - x - 15 \\ 6 - x &= -13 - x \\ +x \quad & \quad +x \\ 6 &= -13 \text{ ✗} \end{aligned}$$

no solution

Solving Equations with Variables on Both Sides

11.
$$
\begin{aligned}
6y &= 8 - 9 + 6y \\
6y &= -1 + 6y \\
\underline{-6y} &\quad \underline{\quad -6y} \\
0 &= -1 \text{ ✗}
\end{aligned}
$$
no solution

12.
$$
\begin{aligned}
6 - 2x - 1 &= 4x + 8 - 6x - 3 \\
5 - 2x &= 5 - 2x \\
\underline{+2x} &\quad \underline{+2x} \\
5 &= 5 \text{ ✓}
\end{aligned}
$$
all real numbers

13. a. Let x represent the number of hours.
$$
\begin{aligned}
376 + 12x &= 280 + 15x \\
\underline{\quad -12x} &\quad \underline{\quad -12x} \\
376 &= 280 + 3x \\
\underline{-280} &\quad \underline{-280} \\
96 &= 3x \\
\frac{96}{3} &= \frac{3x}{3} \\
32 &= x
\end{aligned}
$$
The two companies cost the same amount for a 32-hour job.

b. $376 + 12x$
$= 376 + 12(32)$
$= 376 + 384$
$= \$760$

Systems of Equations

1. $y = x + 1$
   $y = 2x - 1$
   The expressions $x + 1$ and $2x - 1$ both equal y, so they equal each other.
   Solve the equation $x + 1 = 2x - 1$ to find x.
   $$\begin{aligned} x + 1 &= 2x - 1 \\ -x\quad &\quad -x \\ 1 &= x - 1 \\ +1\quad &\quad +1 \\ 2 &= x \end{aligned}$$
   To find y, substitute 2 for x in one of the original equations.
   $y = x + 1 = 2 + 1 = 3$
   The solution is $(2, 3)$.

2. $y = -2x + 3$
   $y = 5x - 4$
   The expressions $-2x + 3$ and $5x - 4$ both equal y, so they equal each other.
   Solve the equation $-2x + 3 = 5x - 4$ to find x.
   $$\begin{aligned} -2x + 3 &= 5x - 4 \\ +2x\quad &\quad +2x \\ 3 &= 7x - 4 \\ +4\quad &\quad +4 \\ 7 &= 7x \\ \frac{7}{7} &= \frac{7x}{7} \\ x &= 1 \end{aligned}$$
   To find y, substitute 1 for x in one of the original equations.
   $y = -2x + 3 = -2(1) + 3 = -2 + 3 = 1$
   The solution is $(1, 1)$.

3. $y = 3x - 5$
   $y = 6x + 7$
   The expressions $3x - 5$ and $6x + 7$ both equal y, so they equal each other.
   Solve the equation $3x - 5 = 6x + 7$ to find x.
   $$\begin{aligned} 3x - 5 &= 6x + 7 \\ -3x\quad &\quad -3x \\ -5 &= 3x + 7 \\ -7\quad &\quad -7 \\ -12 &= 3x \\ \frac{-12}{3} &= \frac{3x}{3} \\ x &= -4 \end{aligned}$$
   To find y, substitute -4 for x in one of the original equations.
   $y = 3x - 5 = 3(-4) - 5 = -12 - 5 = -17$
   The solution is $(-4, -17)$.

4. $y = 6x - 12$
   $y = -9x + 3$
   The expressions $6x - 12$ and $-9x + 3$ both equal y, so they equal each other.
   Solve the equation $6x - 12 = -9x + 3$ to find x.
   $$\begin{aligned} 6x - 12 &= -9x + 3 \\ +9x\quad &\quad +9x \\ 15x - 12 &= 3 \\ +12\quad &\quad +12 \\ 15x &= 15 \\ \frac{15x}{15} &= \frac{15}{15} \\ x &= 1 \end{aligned}$$
   To find y, substitute 1 for x in one of the original equations.
   $y = 6x - 12 = 6(1) - 12 = 6 - 12 = -6$
   The solution is $(1, -6)$.

5. $y = 5x + 7$
   $y = -3x + 7$
   The expressions $5x + 7$ and $-3x + 7$ both equal y, so they equal each other.
   Solve the equation $5x + 7 = -3x + 7$ to find x.
   $$\begin{aligned} 5x + 7 &= -3x + 7 \\ +3x\quad &\quad +3x \\ 8x + 7 &= 7 \\ -7\quad &\quad -7 \\ 8x &= 0 \\ \frac{8x}{8} &= \frac{0}{8} \\ x &= 0 \end{aligned}$$
   To find y, substitute 0 for x in one of the original equations.
   $y = 5x + 7 = 5(0) + 7 = 0 + 7 = 7$
   The solution is $(0, 7)$.

Systems of Equations

6. $y = 3x + 5$

$y = 3x - 10$

The expressions $3x + 5$ and $3x - 10$ both equal y, so they equal each other.

Solve the equation $3x + 5 = 3x - 10$ to find x.

$$3x + 5 = 3x - 10$$
$$-3x \quad -3x$$
$$5 = -10$$

Because the variables were eliminated on both sides of the equation, there is no solution to this system of equations.

7. $2x + 2y = 16$ $\quad$ $2x + 6y = 28$

These two equations can be reduced to:

$$x + y = 8 \qquad x + 3y = 14$$
$$-y \quad -y \qquad -3y \quad -3y$$
$$x = 8 - y \qquad x = 14 - 3y$$
$$8 - y = 14 - 3y$$
$$+3y \quad +3y$$
$$8 + 2y = 14$$
$$-8 \quad -8$$
$$2y = 6$$
$$\frac{2y}{2} = \frac{6}{2}$$
$$y = 3$$
$$x = 8 - y$$
$$= 8 - 3 = 5$$

The solution is $(5, 3)$.

8. $x + y = 20$ $\quad$ $x = y - 4$

$$-y \quad -y$$
$$x = 20 - y$$
$$20 - y = y - 4$$
$$+y \quad +y$$
$$20 = 2y - 4$$
$$+4 \quad +4$$
$$24 = 2y$$
$$\frac{24}{2} = \frac{2y}{2}$$
$$y = 12$$
$$x = 20 - y$$
$$= 20 - 12 = 8$$

The solution is $(8, 12)$.

9. $x + 2y = 21$ $\quad$ $-x + 3y = 29$

$x = 3y - 29$

$$-2y \quad -2y$$
$$x = 21 - 2y$$
$$21 - 2y = 3y - 29$$
$$+2y \quad +2y$$
$$21 = 5y - 29$$
$$+29 \quad +29$$
$$50 = 5y$$
$$\frac{50}{5} = \frac{5y}{5}$$
$$y = 10$$
$$x = 21 - 2y$$
$$= 21 - 2(10)$$
$$= 21 - 20 = 1$$

The solution is $(1, 10)$.

10. $5x - 2y = 4$ $\quad$ $11x + 4y = -8$

$$+2y \quad +2y \qquad -4y \quad -4y$$
$$5x = 4 + 2y \qquad 11x = -8 - 4y$$
$$\frac{5x}{5} = \frac{4}{5} + \frac{2y}{5} \qquad \frac{11x}{11} = -\frac{8}{11} - \frac{4y}{11}$$
$$x = \frac{4}{5} + \frac{2y}{5} \qquad x = -\frac{8}{11} - \frac{4y}{11}$$
$$\frac{4}{5} + \frac{2y}{5} = -\frac{8}{11} - \frac{4y}{11}$$
$$55\left(\frac{4}{5} + \frac{2y}{5}\right) = 55\left(-\frac{8}{11} - \frac{4y}{11}\right)$$
$$55\left(\frac{4}{5}\right) + 55\left(\frac{2y}{5}\right) = 55\left(-\frac{8}{11}\right) - 55\left(\frac{4y}{11}\right)$$
$$44 + 22y = -40 - 20y$$
$$+20y \quad +20y$$
$$44 + 42y = -40$$
$$-44 \quad -44$$
$$42y = -84$$
$$\frac{42y}{42} = \frac{-84}{42}$$
$$y = -2$$
$$x = \frac{4}{5} + \frac{2y}{5}$$
$$= \frac{4}{5} + \frac{2(-2)}{5} = \frac{4}{5} + \left(-\frac{4}{5}\right) = 0$$

The solution is $(0, -2)$.

11. $x = -3y$

$$\begin{aligned} 7x - 2y &= -69 \\ +2y \quad & \quad +2y \\ 7x &= -69 + 2y \\ \frac{7x}{7} &= -\frac{69}{7} + \frac{2y}{7} \\ x &= -\frac{69}{7} + \frac{2y}{7} \end{aligned}$$

$$\begin{aligned} -3y &= -\frac{69}{7} + \frac{2y}{7} \\ 7(-3y) &= 7\left(-\frac{69}{7} + \frac{2y}{7}\right) \\ 7(-3y) &= 7\left(-\frac{69}{7}\right) + 7\left(\frac{2y}{7}\right) \\ -21y &= -69 + 2y \\ -2y \quad & \quad -2y \\ -23y &= -69 \\ \frac{-23y}{-23} &= \frac{-69}{-23} \\ y &= 3 \\ x &= -3y \\ &= -3(3) = -9 \end{aligned}$$

The solution is $(-9, 3)$.

12. $-4x - 5y = -7$

$$\begin{aligned} -4x - 5y &= -7 \\ +5y \quad & \quad +5y \\ -4x &= -7 + 5y \\ \frac{-4x}{-4} &= \frac{-7}{-4} + \frac{5y}{-4} \\ x &= \frac{-5y}{4} + \frac{7}{4} \end{aligned}$$

$$\begin{aligned} 11y &= 2x + 37 \\ -37 \quad & \quad -37 \\ 11y - 37 &= 2x \text{ or } 2x = 11y - 37 \\ \frac{2x}{2} &= \frac{11y}{2} - \frac{37}{2} \\ x &= \frac{11y}{2} - \frac{37}{2} \end{aligned}$$

$$\begin{aligned} \frac{-5y}{4} + \frac{7}{4} &= \frac{11y}{2} - \frac{37}{2} \\ 4\left(\frac{-5y}{4} + \frac{7}{4}\right) &= 4\left(\frac{11y}{2} - \frac{37}{2}\right) \\ 4\left(\frac{-5y}{4}\right) + 4\left(\frac{7}{4}\right) &= 4\left(-\frac{11y}{2}\right) - 4\left(\frac{37}{2}\right) \\ -5y + 7 &= 22y - 74 \\ +5y \quad & \quad +5y \\ 7 &= 27y - 74 \\ +74 \quad & \quad +74 \\ 81 &= 27y \\ \frac{81}{27} &= \frac{27y}{27} \\ y &= 3 \\ x &= \frac{-5y}{4} + \frac{7}{4} \\ &= \frac{-5(3)}{4} + \frac{7}{4} = \frac{-15}{4} + \frac{7}{4} = \frac{-8}{4} = -2 \end{aligned}$$

The solution is $(-2, 3)$.

Appendix: Essential Algebra Review

Review 13.1 / Solving Equations and Inequalities, Part 1

1. $53 + 28 + 17 + 12$
$= 53 + 17 + 28 + 12$
$= 70 + 40 = 110$

2. $5 \cdot 14 \cdot 20$
$= 5 \cdot 20 \cdot 14$
$= 100 \cdot 14 = 1400$

3. $6 \cdot 3 \cdot 5$
$= 6 \cdot 5 \cdot 3$
$= 30 \cdot 3 = 90$

4. $4.5 + 7.1 + 8.5 + 3.9$
$= 4.5 + 8.5 + 7.1 + 3.9$
$= 13 + 11 = 24$

5. $9(62)$
$= 9(60 + 2)$
$= 9(60) + 9(2)$
$= 540 + 18 = 558$

6. $8(29)$
$= 8(30 - 1)$
$= 8(30) - 8(1)$
$= 240 - 8 = 232$

7. $11(25)$
$= (10 + 1)(25)$
$= 10(25) + 1(25)$
$= 250 + 25 = 275$

8. $6(53)$
$= 6(50 + 3)$
$= 6(50) + 6(3)$
$= 300 + 18 = 318$

9. $3x + 9x = 12x$

10. $14x^2 - 5x^2 = 9x^2$

11. $-7x + 8x = x$

12. $3x^2 - 4$

13.

$4(y + 6) + 9$	
$4(y) + 4(6) + 9$	Distribute the 4
$6y + 24 + 9$	Multiply
$6y + 33$	Add like terms

14.

$-7(x + 2) + 4x$	
$-7(x) - 7(2) + 4x$	Distribute the -7
$-7x - 14 + 4x$	Multiply
$-3x - 14$	Add like terms

15.

$3x + 2 - 2x - 1$	
$3x - 2x + 2 - 1$	Comm. Prop.
$x + 1$	Add like terms

16.

$5x - 3x + 3x^2 + 9x$	
$5x - 3x + 9x + 3x^2$	Comm. Prop.
$11x + 3x^2$	Add like terms

17.

$8x + 2x - 3y - 9x$	
$8x + 2x - 9x - 3y$	Comm. Prop.
$x - 3y$	Add like terms

18.

$7y - 3 + 6y - 7$	
$7y + 6y - 3 - 7$	Comm. Prop.
$13y - 10$	Add like terms

19.
$$\begin{aligned} x - 8 &= 17 \\ +8 &\quad +8 \\ x &= 25 \end{aligned}$$

20.
$$10x = 30$$
$$\frac{10x}{10} = \frac{30}{10}$$
$$x = 3$$

21.
$$\begin{aligned} 5h - 20 &= 8 \\ +20 &\quad +20 \\ 5h &= 28 \end{aligned}$$
$$\frac{5h}{5} = \frac{28}{5}$$
$$h = \frac{28}{5}$$

22.
$$\begin{aligned} 6 &= -14 + 2c \\ +14 &\quad +14 \\ 20 &= 2c \end{aligned}$$
$$\frac{20}{2} = \frac{2c}{2}$$
$$10 = c$$

23.
$$\begin{aligned} 2x + 6 &= 10 \\ -6 &\quad -6 \\ 2x &= 4 \end{aligned}$$
$$\frac{2x}{2} = \frac{4}{2}$$
$$x = 2$$

24.
$$\begin{aligned} 17 &= 3p - 7 \\ +7 &\quad +7 \\ 24 &= 3p \end{aligned}$$
$$\frac{24}{3} = \frac{3p}{3}$$
$$8 = p$$

25.

$$2d + \frac{1}{5} = \frac{3}{5}$$
$$-\frac{1}{5} \quad -\frac{1}{5}$$
$$2d = \frac{2}{5}$$
$$\frac{1}{2}(2d) = \frac{1}{2}\left(\frac{2}{5}\right)$$
$$d = \frac{1}{5}$$

26.

$$1 = 2x + \frac{1}{2}$$
$$-\frac{1}{2} \quad -\frac{1}{2}$$
$$\frac{1}{2} = 2x$$
$$\frac{1}{2}\left(\frac{1}{2}\right) = \frac{1}{2}(2x)$$
$$\frac{1}{4} = x$$

27.

$$\frac{z}{2} + 1 = \frac{3}{2}$$
$$-1 \quad -1$$
$$\frac{z}{2} = \frac{1}{2}$$
$$2\left(\frac{z}{2}\right) = 2\left(\frac{1}{2}\right)$$
$$z = 1$$

28.

$$\frac{2}{3} = \frac{4j}{6}$$
$$\left(\frac{6}{4}\right)\frac{2}{3} = \left(\frac{6}{4}\right)\frac{4}{6}j$$
$$1 = j$$

29.

$$\frac{3}{4} = \frac{3}{8}x - \frac{3}{2}$$
$$+\frac{2}{3} \quad +\frac{3}{2}$$
$$\frac{9}{4} = \frac{3}{8}x$$
$$\left(\frac{8}{3}\right)\frac{9}{4} = \left(\frac{8}{3}\right)\frac{3}{8}x$$
$$6 = x$$

30.

$$\frac{1}{5} - \frac{x}{5} = -\frac{2}{5}$$
$$-\frac{1}{5} \quad -\frac{1}{5}$$
$$-\frac{x}{5} = -\frac{3}{5}$$
$$(-5)\left(-\frac{x}{5}\right) = (-5)\left(-\frac{3}{5}\right)$$
$$x = 3$$

31. Let w represent the number of weeks.

$$245 = 125 + 15w$$
$$-125 \quad -125$$
$$120 = 15w$$
$$\frac{120}{15} = \frac{15w}{15}$$
$$8 = w$$

Jennifer needs to save for another 8 weeks to buy the bike.

32. C = Casey's portion; A = Amanda's portion $= 2C$

$$C + 2C + 2(2) = 34$$
$$3C + 4 = 34$$
$$-4 \quad -4$$
$$3C = 30$$
$$\frac{3C}{3} = \frac{30}{3}$$
$$C = 10$$

Casey paid \$10. Amanda paid \$20.

33. b = the number of brothers

$$7.50b + 2(7.50) = 30$$
$$7.5b + 15 = 30$$
$$-15 \quad -15$$
$$7.5b = 15$$
$$\frac{7.5b}{7.5} = \frac{15}{7.5}$$
$$b = 2$$

Marissa has 2 brothers.

34.

$$5s - 7 > -42$$
$$+7 \quad +7$$
$$5s > -35$$
$$\frac{5s}{5} > \frac{-35}{5}$$
$$s > -7$$

Appendix: Essential Algebra Review

Review 13.1 / Solving Equations and Inequalities, Part 1

35. $\frac{b}{2} + 3 < 9$

$-3 \quad -3$

$\frac{b}{2} < 6$

$(2)\frac{b}{2} < (2)6$

$b < 12$

36. $-2q + 5 \geq 19$

$-5 \quad -5$

$-2q \geq 14$

$\frac{-2q}{-2} \leq \frac{14}{-2}$

$q \leq -7$

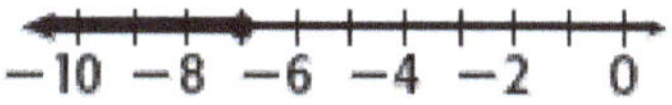

37. $-8c - 11 \leq 13$

$+11 \quad +11$

$-8c \leq 24$

$\frac{-8c}{-8} \geq \frac{24}{-8}$

$c \geq -3$

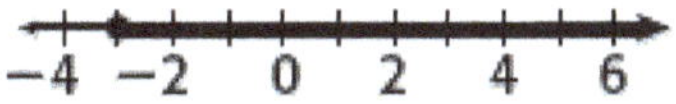

38. $\frac{y}{-4} + 6 > 10$

$-6 \quad -6$

$\frac{y}{-4} > 4$

$(-4)\frac{y}{-4} < (-4)4$

$y < -16$

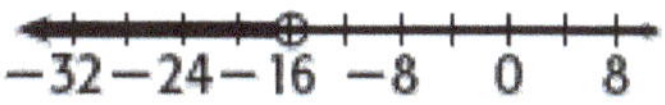

39. $\frac{x}{9} - 5 \leq -8$

$+5 \quad +5$

$\frac{x}{9} \leq -3$

$(9)\frac{x}{9} \leq (9)-3$

$x \leq -27$

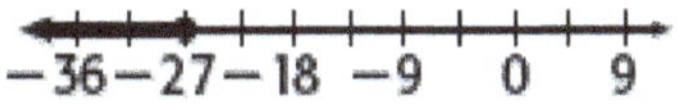

40. $\frac{r}{-2} - 9 > -14$

$+9 \quad +9$

$\frac{r}{-2} > -5$

$(-2)\frac{r}{-2} < (-2)-5$

$r < 10$

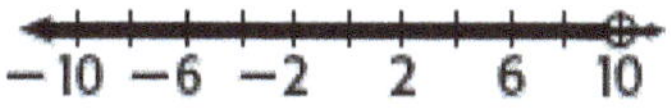

41. $44 \geq 13j + 18$

$-18 \quad -18$

$26 \geq 13j$

$\frac{26}{13} \geq \frac{13j}{13}$

$2 \geq j$

42. $\frac{d}{13} - 12 > 27$

$+12 \quad +12$

$\frac{d}{13} > 39$

$(13)\frac{d}{13} > (13)39$

$d > 507$

43. Let x represent the number of bagels Rico can buy.

$1 + 0.65x \leq 5$

$-1 \quad -1$

$0.065x \leq 4$

$\frac{0.65x}{0.65} \leq \frac{4}{0.65}$

$x \leq 6.15$

Rico can buy at most 6 bagels.

44. Let x represent the average amount each member should still raise.

$500 + 35x \geq 1200$

$-500 \quad -500$

$35x \geq 700$

$\frac{35x}{35} \geq \frac{700}{35}$

$x \geq 20$

On average, each member should raise at least $20.

Appendix: Essential Algebra Review

Review 13.1 / Solving Equations and Inequalities, Part 2

1. $b + 18 + 3b = 74$

$$\begin{aligned} 4b + 18 &= 74 \\ -18 \quad & \quad -18 \\ \hline 4b &= 56 \\ \frac{4b}{4} &= \frac{56}{4} \\ b &= 14 \end{aligned}$$

2. $10x - 3 - 2x = 4$

$$\begin{aligned} 8x - 3 &= 4 \\ +3 \quad & \quad +3 \\ \hline 8x &= 7 \\ \frac{8x}{8} &= \frac{7}{8} \\ x &= \frac{7}{8} \end{aligned}$$

3. $18w - 10 - 6w = 50$

$$\begin{aligned} 12w - 10 &= 50 \\ +10 \quad & \quad +10 \\ \hline 12w &= 60 \\ \frac{12w}{12} &= \frac{60}{12} \\ w &= 5 \end{aligned}$$

4. $19 = 5n + 7 - 3n$

$$\begin{aligned} 19 &= 2n + 7 \\ -7 \quad & \quad -7 \\ \hline 12 &= 2n \\ \frac{12}{2} &= \frac{2n}{2} \\ 6 &= n \end{aligned}$$

5. $-27 = -3p + 15 - 3p$

$$\begin{aligned} -27 &= -6p + 15 \\ -15 \quad & \quad -15 \\ \hline -42 &= -6p \\ \frac{-42}{-6} &= \frac{-6p}{-6} \\ 7 &= p \end{aligned}$$

6. $-x - 8 + 14x = -34$

$$\begin{aligned} 13x - 8 &= -34 \\ +8 \quad & \quad +8 \\ \hline 13x &= -26 \\ \frac{13x}{13} &= \frac{-26}{13} \\ x &= -2 \end{aligned}$$

7. $2(x + 4) + 6 = 22$

$$\begin{aligned} 2x + 8 + 6 &= 22 \\ 2x + 14 &= 22 \\ -14 \quad & \quad -14 \\ \hline 2x &= 8 \\ \frac{2x}{2} &= \frac{8}{2} \\ x &= 4 \end{aligned}$$

8. $1 - 3(n + 5) = -8$

$$\begin{aligned} 1 - 3n - 15 &= -8 \\ -3n - 14 &= -8 \\ +14 \quad & \quad +14 \\ \hline -3n &= 6 \\ \frac{-3n}{-3} &= \frac{6}{-3} \\ n &= -2 \end{aligned}$$

9. $4.3 - 1.4(p + 7) = -9.7$

$$\begin{aligned} 4.3 - 1.4p - 9.8 &= -9.7 \\ -5.5 - 1.4p &= -9.7 \\ +5.5 \quad & \quad +5.5 \\ \hline -1.4p &= -4.2 \\ \frac{-1.4p}{-1.4} &= \frac{-4.2}{-1.4} \\ p &= 3 \end{aligned}$$

10. $1.8 + 6n - 3.2 = 7.6$

$$\begin{aligned} -1.4 + 6n &= 7.6 \\ +1.4 \quad & \quad +1.4 \\ \hline 6n &= 9 \\ \frac{6n}{6} &= \frac{9}{6} \\ n &= 1.5 \end{aligned}$$

11. $0 = 9\left(k - \frac{2}{3}\right) + 33$

$$\begin{aligned} 0 &= 9k - 6 + 33 \\ 0 &= 9k + 27 \\ -27 \quad & \quad -27 \\ \hline -27 &= 9k \\ \frac{-27}{9} &= \frac{9k}{9} \\ -3 &= k \end{aligned}$$

12. $6(t - 2) - 76 = -142$

$$\begin{aligned} 6t - 12 - 76 &= -142 \\ 6t - 88 &= -142 \\ +88 \quad & \quad +88 \\ \hline 6t &= -54 \\ \frac{6t}{6} &= \frac{-54}{6} \\ t &= -9 \end{aligned}$$

Appendix: Essential Algebra Review

Review 13.1 / Solving Equations and Inequalities, Part 2

13. Let x represent the number of laps Karen ran. Then $3x$ represents the number Abby ran, and $\frac{3x+4}{7}$ represents the number Jill ran, which equals 1.

$$\frac{3x+4}{7} = 1$$
$$(7)\frac{3x+4}{7} = (7)1$$
$$3x + 4 = 7$$
$$-4 \quad -4$$
$$3x = 3$$
$$\frac{3x}{3} = \frac{3}{3}$$
$$x = 1$$

14.
$$7a - 17 = 4a + 1$$
$$-4a \quad -4a$$
$$3a - 17 = 1$$
$$+17 \quad +17$$
$$3a = 18$$
$$\frac{3a}{3} = \frac{18}{3}$$
$$a = 6$$

15.
$$2b - 5 = 8b + 1$$
$$-2b \quad -2b$$
$$-5 = 6b + 1$$
$$-1 \quad -1$$
$$-6 = 6b$$
$$\frac{-6}{6} = \frac{6b}{6}$$
$$-1 = b$$

16.
$$4x - 2 = 3x + 4$$
$$-3x \quad -3x$$
$$x - 2 = 4$$
$$+2 \quad +2$$
$$x = 6$$

17.
$$2x - 5 = 4x - 1$$
$$-2x \quad -2x$$
$$-5 = 2x - 1$$
$$+1 \quad +1$$
$$-4 = 2x$$
$$\frac{-4}{2} = \frac{2x}{2}$$
$$-2 = x$$

18.
$$8x - 2 = 3x + 12.25$$
$$-3x \quad -3x$$
$$5x - 2 = 12.25$$
$$+2 \quad +2$$
$$5x = 14.25$$
$$\frac{5x}{5} = \frac{14.25}{5}$$
$$x = 2.85$$

19.
$$5x + 2 = 3x$$
$$-5x \quad -5x$$
$$2 = -2x$$
$$\frac{2}{-2} = \frac{-2x}{-2}$$
$$-1 = x$$

20.
$$3c - 5 = 2c + 5$$
$$-2c \quad -2c$$
$$c - 5 = 5$$
$$+5 \quad +5$$
$$c = 10$$

21.
$$-17 - 2x = 6 - x$$
$$+2x \quad +2x$$
$$-17 = 6 + x$$
$$-6 \quad -6$$
$$-23 = x$$

22.
$$3(t - 1) = 9 + t$$
$$3(t) + 3(-1) = 9 + t$$
$$3t - 3 = 9 + t$$
$$-t \quad -t$$
$$2t - 3 = 9$$
$$+3 \quad +3$$
$$2t = 12$$
$$\frac{2t}{2} = \frac{12}{2}$$
$$t = 6$$

Appendix: Essential Algebra Review

Review 13.1 / Solving Equations and Inequalities, Part 2

23.
$$\begin{aligned} 5 - x - 2 &= 3 + 4x + 5 \\ 3 - x &= 4x + 8 \\ +x \quad & \quad +x \\ 3 &= 5x + 8 \\ -8 \quad & \quad -8 \\ -5 &= 5x \\ \frac{-5}{5} &= \frac{5x}{5} \\ -1 &= x \end{aligned}$$

24.
$$\begin{aligned} 2(x + 4) &= 3(x - 2) \\ 2(x) + 2(4) &= 3(x) + 3(-2) \\ 2x + 8 &= 3x - 6 \\ -2x \quad & \quad -2x \\ 8 &= x - 6 \\ +6 \quad & \quad +6 \\ 14 &= x \end{aligned}$$

25.
$$\begin{aligned} 3m - 10 &= 2(4m - 5) \\ 3m - 10 &= 2(4m) + 2(-5) \\ 3m - 10 &= 8m - 10 \\ -3m \quad & \quad -3m \\ -10 &= 5m - 10 \\ +10 \quad & \quad +10 \\ 0 &= 5m \\ \frac{0}{5} &= \frac{5m}{5} \\ 0 &= m \end{aligned}$$

26.
$$\begin{aligned} 5 - (n - 4) &= 3(n + 2) \\ 5 - 1(n) - 1(-4) &= 3(n) + 3(2) \\ 5 - n + 4 &= 3n + 6 \\ 9 - n &= 3n + 6 \\ +n \quad & \quad +n \\ 9 &= 4n + 6 \\ -6 \quad & \quad -6 \\ 3 &= 4n \\ \frac{3}{4} &= \frac{4n}{4} \\ \frac{3}{4} &= n \end{aligned}$$

27.
$$\begin{aligned} 6(x + 7) - 20 &= 6x \\ 6(x) + 6(7) - 20 &= 6x \\ 6x + 42 - 20 &= 6x \\ 6x + 22 &= 6x \\ -6x \quad & \quad -6x \\ 22 &= 0 \ ✗ \end{aligned}$$

no solution

28.
$$\begin{aligned} 8(x + 1) &= 4x - 8 \\ 8(x) + 8(1) &= 4x - 8 \\ 8x + 8 &= 4x - 8 \\ -4x \quad & \quad -4x \\ 4x + 8 &= -8 \\ -8 \quad & \quad -8 \\ 4x &= -16 \\ \frac{4x}{4} &= \frac{-16}{4} \\ x &= -4 \end{aligned}$$

29.
$$\begin{aligned} x - 4 - 3x &= -2x - 3 - 1 \\ -2x - 4 &= -2x - 4 \\ +2x \quad & \quad +2x \\ -4 &= -4 \ ✓ \end{aligned}$$

all real numbers

30.
$$\begin{aligned} -2(x + 2) &= -2x + 1 \\ -2(x) - 2(2) &= -2x + 1 \\ -2x - 4 &= -2x + 1 \\ +2x \quad & \quad +2x \\ -4 &= 1 \ ✗ \end{aligned}$$

no solution

31.
$$\begin{aligned} 2(x + 4) - 5 &= 2x + 3 \\ 2(x) + 2(4) - 5 &= 2x + 3 \\ 2x + 8 - 5 &= 2x + 3 \\ 2x + 3 &= 2x + 3 \\ +2x \quad & \quad +2x \\ 3 &= 3 \ ✓ \end{aligned}$$

all real numbers

Appendix: Essential Algebra Review

Review 13.1 / Solving Equations and Inequalities, Part 2

32. a. Let w represent the number of weeks.

$$\begin{aligned} 150 + 2w &= 195 - w \\ +w &\quad\;\; +w \\ 150 + 3w &= 195 \\ -150 &\quad -150 \\ 3w &= 45 \\ \frac{3w}{3} &= \frac{45}{3} \\ w &= 15 \end{aligned}$$

Justin and Tyson will weigh the same amount in 15 weeks.

b. $195 - w$
$= 195 - 15$
$= 180$ lb

33. $y = -2x - 1$
$y = 2x + 3$
The expressions $-2x - 1$ and $2x + 3$ both equal y, so they equal each other.
Solve the equation $-2x - 1 = 2x + 3$ to find x.

$$\begin{aligned} -2x - 1 &= 2x + 3 \\ +2x &\quad +2x \\ -1 &= 4x + 3 \\ -3 &\quad\;\; -3 \\ -4 &= 4x \\ \frac{-4}{4} &= \frac{4x}{4} \\ x &= -1 \end{aligned}$$

To find y, substitute -1 for x in one of the original equations.
$y = -2x - 1 = -2(-1) - 1 = 2 - 1 = 1$
The solution is $(-1, 1)$.

34. $y = 3x - 6$
$y = x + 2$
The expressions $3x - 6$ and $x + 2$ both equal y, so they equal each other.
Solve the equation $3x - 6 = x + 2$ to find x.

$$\begin{aligned} 3x - 6 &= x + 2 \\ -x &\quad -x \\ 2x - 6 &= 2 \\ +6 &\quad +6 \\ 2x &= 8 \\ \frac{2x}{2} &= \frac{8}{2} \\ x &= 4 \end{aligned}$$

To find y, substitute 4 for x in one of the original equations.
$y = 3x - 6 = 3(4) - 6 = 12 - 6 = 6$
The solution is $(4, 6)$.

35. $y = 5x - 3$
$y = -3x + 13$
The expressions $5x - 3$ and $-3x + 13$ both equal y, so they equal each other.
Solve the equation $5x - 3 = -3x + 13$ to find x.

$$\begin{aligned} 5x - 3 &= -3x + 13 \\ +3x &\quad +3x \\ 8x - 3 &= 13 \\ +3 &\quad +3 \\ 8x &= 16 \\ \frac{8x}{8} &= \frac{16}{8} \\ x &= 2 \end{aligned}$$

To find y, substitute 2 for x in one of the original equations.
$y = 5x - 3 = 5(2) - 3 = 10 - 3 = 7$
The solution is $(2, 7)$.

36. $y = x + 6$
$y = -2x - 12$
The expressions $x + 6$ and $-2x - 12$ both equal y, so they equal each other.
Solve the equation $x + 6 = -2x - 12$ to find x.

$$\begin{aligned} x + 6 &= -2x - 12 \\ +2x &\quad +2x \\ 3x + 6 &= -12 \\ -6 &\quad -6 \\ 3x &= -18 \\ \frac{3x}{3} &= \frac{-18}{3} \\ x &= -6 \end{aligned}$$

To find y, substitute -6 for x in one of the original equations.
$y = x + 6 = -6 + 6 = 0$
The solution is $(-6, 0)$.

37. $y = 3x - 1$
$y = -2x + 9$
The expressions $3x - 1$ and $-2x + 9$ both equal y, so they equal each other.
Solve the equation $3x - 1 = -2x + 9$ to find x.

$$\begin{aligned} 3x - 1 &= -2x + 9 \\ +2x \quad &\quad +2x \\ 5x - 1 &= 9 \\ +1 \quad &\quad +1 \\ 5x &= 10 \\ \frac{5x}{5} &= \frac{10}{5} \\ x &= 2 \end{aligned}$$

To find y, substitute 2 for x in one of the original equations.
$y = 3x - 1 = 3(2) - 1 = 6 - 1 = 5$
The solution is (2, 5).

38. $y = -2x - 6$
$y = 3x + 29$
The expressions $-2x - 6$ and $3x + 29$ both equal y, so they equal each other.
Solve the equation $-2x - 6 = 3x + 29$ to find x.

$$\begin{aligned} -2x - 6 &= 3x + 29 \\ +2x \quad &\quad +2x \\ -6 &= 5x + 29 \\ -29 \quad &\quad -29 \\ -35 &= 5x \\ \frac{-35}{5} &= \frac{5x}{5} \\ x &= -7 \end{aligned}$$

To find y, substitute -7 for x in one of the original equations.
$y = -2x - 6 = -2(-7) - 6 = 14 - 6 = 8$
The solution is (−7, 8).

39. $3x + 3y = 15$ $\qquad$ $3x - 6y = -12$
These two equations can be reduced to:

$$\begin{aligned} x - 2y &= -4 \\ +2y \quad &\quad +2y \\ x &= -4 + 2y \end{aligned}$$

$$\begin{aligned} x + y &= 5 \\ -y \quad &\quad -y \\ x &= 5 - y \\ 5 - y &= -4 + 2y \\ +y \quad &\quad +y \\ 5 &= -4 + 3y \\ +4 \quad &\quad +4 \\ 9 &= 3y \\ \frac{9}{3} &= \frac{3y}{3} \\ y &= 3 \\ x &= 5 - y \\ &= 5 - 3 = 2 \end{aligned}$$

The solution is (2, 3).

40.

$$\begin{aligned} 2x + y &= 11 \\ -y \quad &\quad -y \\ 2x &= 11 - y \\ \frac{2x}{2} &= \frac{11}{2} - \frac{y}{2} \\ x &= \frac{11}{2} - \frac{y}{2} \end{aligned}$$

$$\begin{aligned} -x + 2y &= 2 \\ -2y \quad &\quad -2y \\ -x &= 2 - 2y \\ x &= 2y - 2 \end{aligned}$$

$$\begin{aligned} \frac{11}{2} - \frac{y}{2} &= 2y - 2 \\ 2\left(\frac{11}{2} - \frac{y}{2}\right) &= 2(2y - 2) \\ 2\left(\frac{11}{2}\right) - 2\left(\frac{y}{2}\right) &= 4y - 4 \\ 11 - y &= 4y - 4 \\ +y \quad &\quad +y \\ 11 &= 5y - 4 \\ +4 \quad &\quad +4 \\ 15 &= 5y \\ \frac{15}{5} &= \frac{5y}{5} \\ y &= 3 \\ x &= 2y - 2 \\ &= 2(3) - 2 = 6 - 2 = 4 \end{aligned}$$

The solution is (4, 3).

41. $4x + 3y = 13$ $\quad y = 5x - 2$

$$4x + 3y - 4x = 13 - 4x$$
$$3y = 13 - 4x$$
$$\frac{3y}{3} = \frac{13}{3} - \frac{4}{3}x$$
$$y = \frac{13}{3} - \frac{4}{3}x$$
$$\frac{13}{3} - \frac{4}{3}x = 5x - 2$$
$$3\left(\frac{13}{3} - \frac{4}{3}x\right) = 3(5x - 2)$$
$$3\left(\frac{13}{3}\right) - 3\left(\frac{4}{3}x\right) = 15x - 6$$
$$13 - 4x = 15x - 6$$
$$+4x \quad +4x$$
$$13 = 19x - 6$$
$$+6 \quad +6$$
$$19 = 19x$$
$$\frac{19}{19} = \frac{19x}{19}$$
$$x = 1$$
$$y = 5x - 2$$
$$= 5(1) - 2 = 5 - 2 = 3$$

The solution is (1, 3).

42. $5x - 9y = 11$ $\quad 3x + 7y = 19$

$$+9y \quad +9y \qquad\qquad -7y \quad -7y$$
$$5x = 11 + 9y \qquad\qquad 3x = 19 - 7y$$
$$\frac{5x}{5} = \frac{11}{5} + \frac{9y}{5} \qquad\qquad \frac{3x}{3} = \frac{19}{3} - \frac{7y}{3}$$
$$x = \frac{11}{5} + \frac{9y}{5} \qquad\qquad x = \frac{19}{3} - \frac{7y}{3}$$
$$\frac{11}{5} + \frac{9y}{5} = \frac{19}{3} - \frac{7y}{3}$$
$$15\left(\frac{11}{5} + \frac{9y}{5}\right) = 15\left(\frac{19}{3} - \frac{7y}{3}\right)$$
$$15\left(\frac{11}{5}\right) + 15\left(\frac{9y}{5}\right) = 15\left(\frac{19}{3}\right) - 15\left(\frac{7y}{3}\right)$$
$$33 + 27y = 95 - 35y$$
$$+35y \quad +35y$$
$$33 + 62y = 95$$
$$-33 \quad -33$$
$$62y = 62$$
$$\frac{62y}{62} = \frac{62}{62}$$
$$y = 1$$
$$x = \frac{11}{5} + \frac{9y}{5}$$
$$= \frac{11}{5} + \frac{9(1)}{5} = \frac{11}{5} + \frac{9}{5} = \frac{20}{5} = 4$$

The solution is (4, 1).

43. $12x + 18y = 30$ $\quad 4x - 13y = 67$

$$-18y \quad -18y \qquad\qquad +13y \quad +13y$$
$$12x = 30 - 18y \qquad\qquad 4x = 67 + 13y$$
$$\frac{12x}{12} = \frac{30}{12} - \frac{18y}{12} \qquad\qquad \frac{4x}{4} = \frac{67}{4} + \frac{13y}{4}$$
$$x = \frac{5}{2} - \frac{3y}{2} \qquad\qquad x = \frac{67}{4} + \frac{13y}{4}$$
$$\frac{5}{2} - \frac{3y}{2} = \frac{67}{4} + \frac{13y}{4}$$
$$4\left(\frac{5}{2} - \frac{3y}{2}\right) = 4\left(\frac{67}{4} + \frac{13y}{4}\right)$$
$$10 - 6y = 67 + 13y$$
$$\frac{-57}{19} = \frac{19y}{19}$$
$$y = -3$$
$$x = \frac{5}{2} - \frac{3y}{2}$$
$$= \frac{5}{2} - \frac{3(-3)}{2} = \frac{5}{2} - \frac{-9}{2} = \frac{5}{2} + \frac{9}{2}$$
$$= \frac{14}{2} = 7$$

The solution is (7, −3).

44. $-14x - 11y = 97$ $\quad -12y + 11x = 27$

$$+11y \quad +11y \qquad\qquad +12y \quad +12y$$
$$-14x = 97 + 11y \qquad\qquad 11x = 27 + 12y$$
$$\frac{-14x}{-14} = \frac{97}{-14} - \frac{11y}{14} \qquad\qquad \frac{11x}{11} = \frac{27}{11} + \frac{12y}{11}$$
$$x = -\frac{97}{14} - \frac{11y}{14} \qquad\qquad x = \frac{27}{11} + \frac{12y}{11}$$
$$-\frac{97}{14} - \frac{11y}{14} = \frac{27}{11} + \frac{12y}{11}$$
$$154\left(-\frac{97}{14} - \frac{11y}{14}\right) = 154\left(\frac{27}{11} + \frac{12y}{11}\right)$$
$$-1067 - 121y = 378 + 168y$$
$$\frac{-1445}{289} = \frac{289y}{289}$$
$$y = -5$$
$$x = \frac{27}{11} + \frac{12(-5)}{11}$$
$$= \frac{27}{11} + \frac{-60}{11} = \frac{-33}{11} = -3$$

The solution is (−3, −5).

Appendix: Essential Algebra Review

Graphing and the Coordinate Plane

The Coordinate Plane

1. II
2. IV
3. III
4. I

5-8.

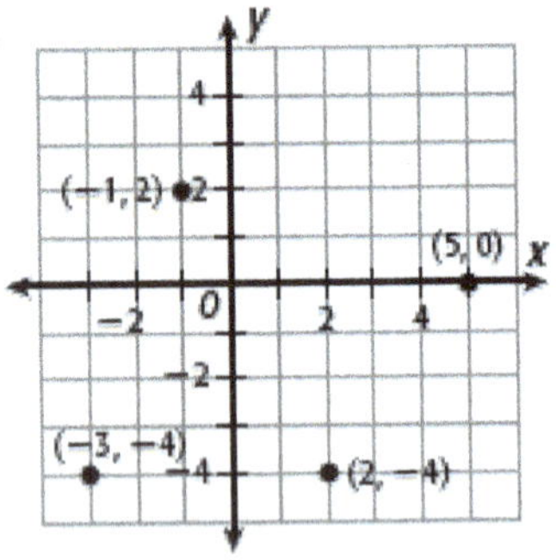

5. (−1, 2)
   Start at the origin. Move 1 unit left and 2 units up.
6. (2, −4)
   Start at the origin. Move 2 units right and 4 units down.
7. (−3, −4)
   Start at the origin. Move 3 units left and 4 units down.
8. (5, 0)
   Start at the origin. Move 5 units right.
9. Start at the origin. Point *J* is 6 units right and 3 units down.
   (6, −3)
10. Start at the origin. Point *P* is 3 units left and 2 units up.
    (−3, 2)
11. Start at the origin. Point *S* is 4 units left.
    (−4, 0)
12. Start at the origin. Point *M* is 5 units right.
    (5, 0)

Appendix: Essential Algebra Review

Graphing and the Coordinate Plane

Equations, Tables, and Graphs

1. $g = 15m$
   Possible answer:

m	0	1	2	3	4
g	0	15	30	45	60

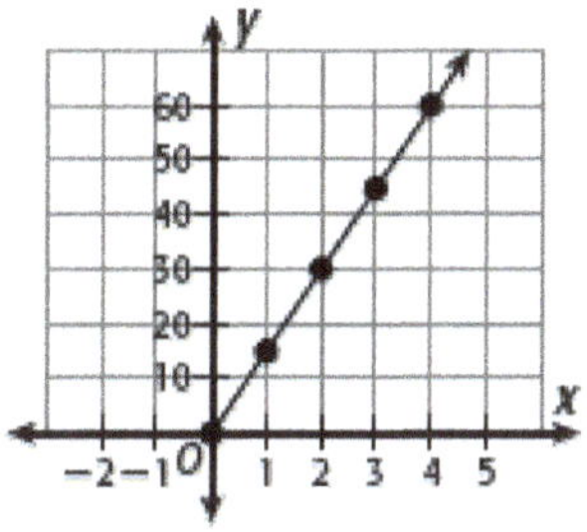

2. $y = x + 3$

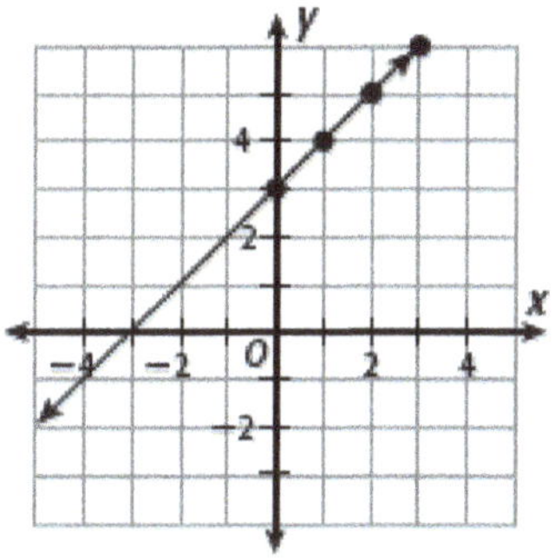

3. $y = 3x$
   Possible answer:

x	−2	−1	0	1	2
y	−6	−3	0	3	6

Appendix: Essential Algebra Review

Graphing and the Coordinate Plane

Graphing Using Intercepts

1. The x-intercept is -5. The y-intercept is 1.
2. The x-intercept is 2. The y-intercept is -4.
3. The x-intercept is -3. The y-intercept is -2.
4.
$$\begin{aligned} 2x - 4y &= 4 \\ 2x - 4(0) &= 4 \\ 2x - 0 &= 4 \\ 2x &= 4 \\ \frac{2x}{2} &= \frac{4}{2} \\ x &= 2 \end{aligned}$$

The x-intercept is 2.

$$\begin{aligned} 2x - 4y &= 4 \\ 2(0) - 4y &= 4 \\ 0 - 4y &= 4 \\ -4y &= 4 \\ \frac{-4y}{-4} &= \frac{4}{-4} \\ y &= -1 \end{aligned}$$

The y-intercept is -1.

5.
$$\begin{aligned} -2y &= 3x - 6 \\ -3x \quad & \quad -3x \\ -3x - 2y &= -6 \end{aligned}$$

$$\begin{aligned} -3x - 2y &= -6 \\ -3x - 2(0) &= -6 \\ -3x - 0 &= -6 \\ -3x &= -6 \\ \frac{-3x}{-3} &= \frac{-6}{-3} \\ x &= 2 \end{aligned}$$

The x-intercept is 2.

$$\begin{aligned} -3x - 2y &= -6 \\ -3(0) - 2y &= -6 \\ 0 - 2y &= -6 \\ -2y &= -6 \\ \frac{-2y}{-2} &= \frac{-6}{-2} \\ y &= 3 \end{aligned}$$

The y-intercept is 3.

6.
$$\begin{aligned} 4y + 5x &= 2y - 3x + 16 \\ -2y \quad & \quad -2y \\ 2y + 5x &= -3x + 16 \\ +3x \quad & \quad +3x \\ 2y + 8x &= 16 \\ 8x + 2y &= 16 \end{aligned}$$

$$\begin{aligned} 8x + 2y &= 16 \\ 8x + 2(0) &= 16 \\ 8x + 0 &= 16 \\ 8x &= 16 \\ \frac{8x}{8} &= \frac{16}{8} \\ x &= 2 \end{aligned}$$

The x-intercept is 2.

$$\begin{aligned} 8x + 2y &= 16 \\ 8(0) + 2y &= 16 \\ 0 + 2y &= 16 \\ 2y &= 16 \\ \frac{2y}{2} &= \frac{16}{2} \\ y &= 8 \end{aligned}$$

The y-intercept is 8.

Appendix: Essential Algebra Review

Review 13.2 / Graphing and the Coordinate Plane

1. I

2. III

3. IV

4. II

5–8.

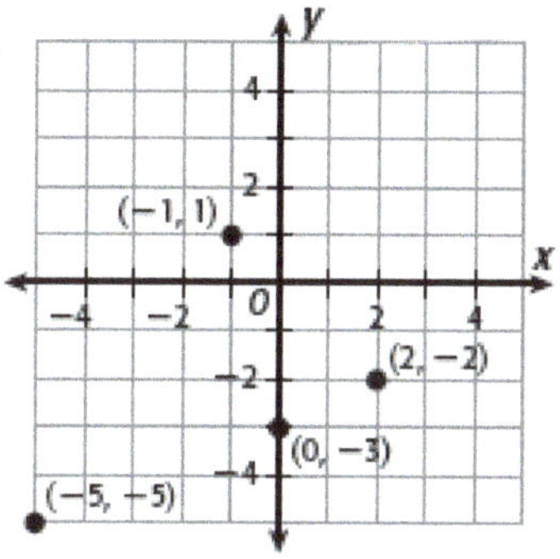

5. (−1, 1)
Start at the origin. Move 1 unit left and 1 unit up.

6. (2, −2)
Start at the origin. Move 2 units right and 2 units down.

7. (−5, −5)
Start at the origin. Move 5 units left and 5 units down.

8. (0, −3)
Start at the origin. Move 3 units down.

9. Start at the origin. Point *Q* is 4 units left and 4 units up.
(−4, 4)

10. Start at the origin. Point *V* is 5 units right and 6 units down.
(5, −6)

11. Start at the origin. Point *R* is 5 units left and 4 units down.
(−5, −4)

12. Start at the origin. Point *P* is 4 units down.
(0, −4)

13. Start at the origin. Point *S* is 5 units right and 6 units up.
(5, 6)

14. Start at the origin. Point *L* is 2 units right and 2 units up.
(2, 2)

15. Possible answer:

s	0	1	2	3	4
h	0	−0.5	−1.0	−1.5	−2.0

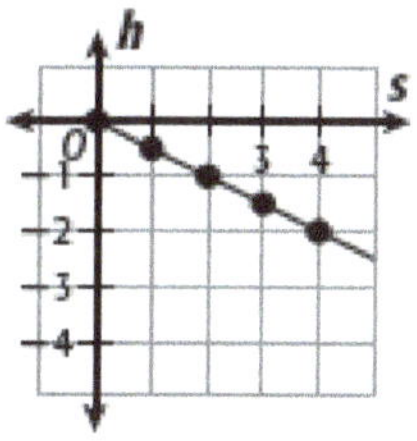

16. $y = 12 - x$

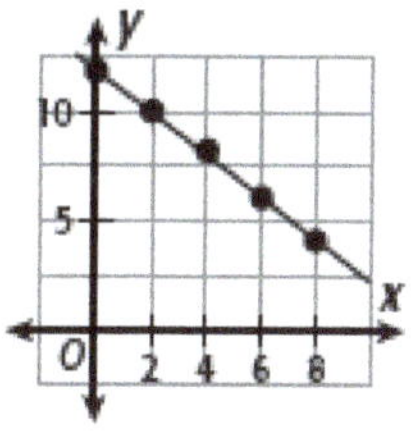

Appendix: Essential Algebra Review

Review 13.2 / Graphing and the Coordinate Plane

17. $y = x + 1$

x	-3	-2	-1	0	1
y	-2	-1	0	1	2

18. The x-intercept is -1. The y-intercept is 3.

19. The x-intercept is -5. The y-intercept is -1.

20. The x-intercept is -4. The y-intercept is 2.

21.

$$\begin{aligned} 6x + 3y &= 12 \\ 6x + 3(0) &= 12 \\ 6x + 0 &= 12 \\ 6x &= 12 \\ \frac{6x}{6} &= \frac{12}{6} \\ x &= 2 \end{aligned}$$

The x-intercept is 2.

$$\begin{aligned} 6x + 3y &= 12 \\ 6(0) + 3y &= 12 \\ 0 + 3y &= 12 \\ 3y &= 12 \\ \frac{3y}{3} &= \frac{12}{3} \\ y &= 4 \end{aligned}$$

The y-intercept is 4.

22.

$$\begin{aligned} 4y - 8 &= 2x \\ -2x \quad & \quad -2x \\ 4y - 2x - 8 &= 0 \\ +8 \quad & \quad +8 \\ 4y - 2x &= 8 \\ -2x + 4y &= 8 \end{aligned}$$

$$\begin{aligned} -2x + 4y &= 8 \\ -2x + 4(0) &= 8 \\ -2x + 0 &= 8 \\ -2x &= 8 \\ \frac{-2x}{-2} &= \frac{8}{-2} \\ x &= -4 \end{aligned}$$

The x-intercept is -4.

$$\begin{aligned} -2x + 4y &= 8 \\ -2(0) + 4y &= 8 \\ 0 + 4y &= 8 \\ \frac{4y}{4} &= \frac{8}{4} \\ y &= 2 \end{aligned}$$

The y-intercept is 2.

23.

$$\begin{aligned} -2y + x &= 2y - 8 \\ -2y \quad & \quad -2y \\ -4y + x &= -8 \\ x - 4y &= -8 \end{aligned}$$

$$\begin{aligned} x - 4y &= -8 \\ x - 4(0) &= -8 \\ x - 0 &= -8 \\ x &= -8 \end{aligned}$$

The x-intercept is -8.

$$\begin{aligned} x - 4y &= -8 \\ 0 - 4y &= -8 \\ -4y &= -8 \\ \frac{-4y}{-4} &= \frac{-8}{-4} \\ y &= 2 \end{aligned}$$

The y-intercept is 2.

24.

$$\begin{aligned} 4x + y &= 8 \\ 4x + 0 &= 8 \\ 4x &= 8 \\ \frac{4x}{4} &= \frac{8}{4} \\ x &= 2 \end{aligned}$$

The x-intercept is 2.

$$\begin{aligned} 4x + y &= 8 \\ 4(0) + y &= 8 \\ 0 + y &= 8 \\ y &= 8 \end{aligned}$$

The y-intercept is 8.

25.

$$\begin{aligned} y - 3x &= -15 \\ 0 - 3x &= -15 \\ -3x &= -15 \\ \frac{-3x}{-3} &= \frac{-15}{-3} \\ x &= 5 \end{aligned}$$

The x-intercept is 5.

$$\begin{aligned} y - 3x &= -15 \\ y - 3(0) &= -15 \\ y - 0 &= -15 \\ y &= -15 \end{aligned}$$

The y-intercept is -15.

26.

$$\begin{aligned} 2x + y &= 10x - 1 \\ -10x \quad & \quad -10x \\ -8x + y &= -1 \end{aligned}$$

$$\begin{aligned} -8x + y &= -1 \\ -8x + 0 &= -1 \\ -8x &= -1 \\ \frac{-8x}{-8} &= \frac{-1}{-8} \\ x &= \frac{1}{8} \end{aligned}$$

The x-intercept is $\frac{1}{8}$.

$$\begin{aligned} -8x + y &= -1 \\ -8(0) + y &= -1 \\ 0 + y &= -1 \\ y &= -1 \end{aligned}$$

The y-intercept is -1.

Appendix: Essential Algebra Review

Review 13.2 / Graphing and the Coordinate Plane

27. a.

x	0	3	6	9	12
$f(x) = 300 - 25x$	300	225	150	75	0

Bass Population

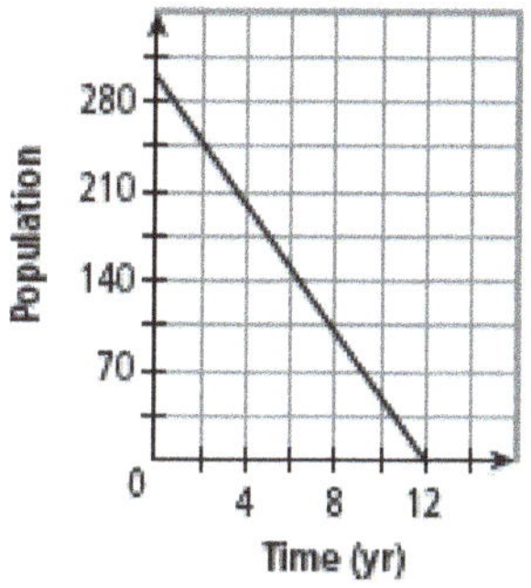

The x-intercept is 12. The y-intercept is 300.

b. x-intercept: time when bass population is 0
y-intercept: number of bass originally in the lake

28. a.

x	0	5	15	20	25
$f(x) = 5 - \frac{1}{5}x$	5	4	2	1	0

5K Race

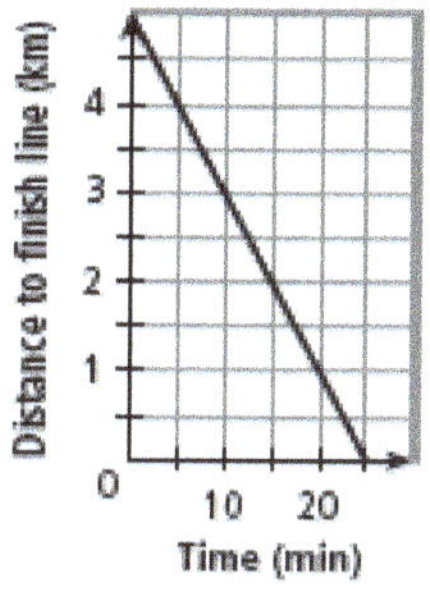

The x-intercept is 25. The y-intercept is 5.

b. x-intercept: total time to run the race (when the distance to the finish line is 0)
y-intercept: total length of the race (when time is 0)

29. x-intercept:

$4x - 6y = 12$
$4x - 6(0) = 12$
$4x = 12$
$\frac{4x}{4} = \frac{12}{4}$
$x = 3$

y-intercept:

$4x - 6y = 12$
$4(0) - 6y = 12$
$-6y = 12$
$\frac{-6y}{-6} = \frac{12}{-6}$
$y = -2$

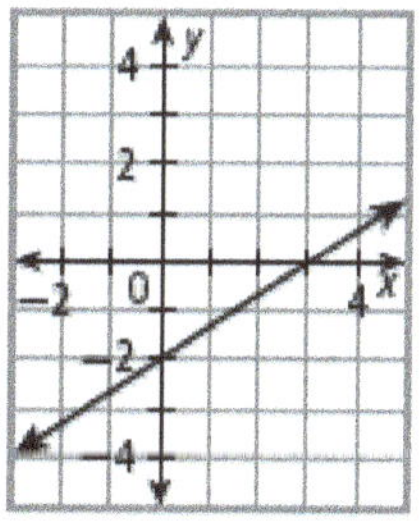

30. x-intercept:

$2x + 3y = 18$
$2x + 3(0) = 18$
$2x = 18$
$\frac{2x}{2} = \frac{18}{2}$
$x = 9$

y-intercept:

$2x + 3y = 18$
$2(0) + 3y = 18$
$3y = 18$
$\frac{3y}{3} = \frac{18}{3}$
$y = 6$

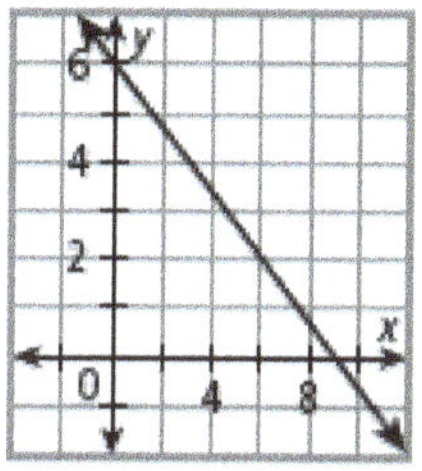

Appendix: Essential Algebra Review

Review 13.2 / Graphing and the Coordinate Plane

31. *x*-Intercept:

$$\frac{1}{2}x - 4y = 4$$
$$\frac{1}{2}x - 4(0) = 4$$
$$\frac{1}{2}x = 4$$
$$2\left(\frac{1}{2}x\right) = 2(4)$$
$$x = 8$$

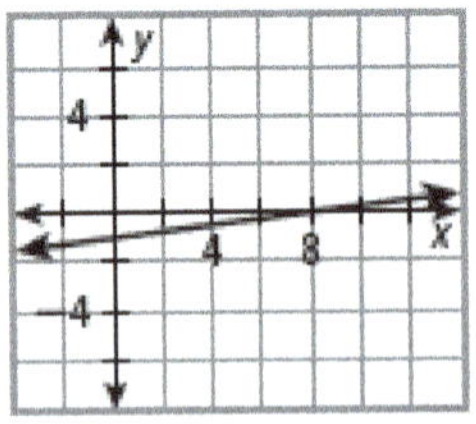

***y*-Intercept:**

$$\frac{1}{2}x - 4y = 4$$
$$\frac{1}{2}(0) - 4y = 4$$
$$-4y = 4$$
$$\frac{-4y}{-4} = \frac{4}{-4}$$
$$y = -1$$

32. *x*-Intercept:

$$y - x = -1$$
$$0 - x = -1$$
$$-x = -1$$
$$-1(-x) = -1(-1)$$
$$x = 1$$

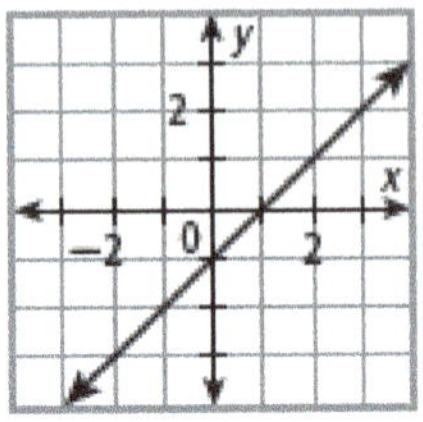

***y*-Intercept:**

$$y - x = -1$$
$$y - 0 = -1$$
$$y = -1$$

33. *x*-Intercept:

$$5x + 3y = 15$$
$$5x + 3(0) = 15$$
$$5x = 15$$
$$\frac{5x}{5} = \frac{15}{5}$$
$$x = 3$$

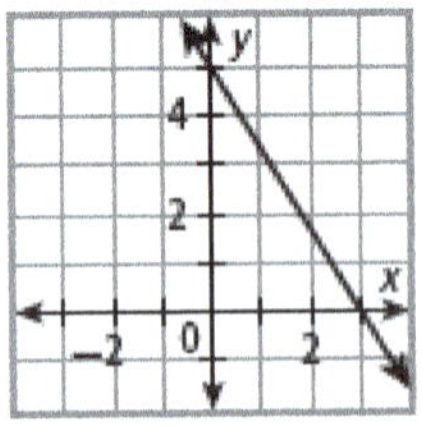

***y*-Intercept:**

$$5x + 3y = 15$$
$$5(0) + 3y = 15$$
$$3y = 15$$
$$\frac{3y}{3} = \frac{15}{3}$$
$$y = 5$$

34. *x*-Intercept:

$$x - 3y = -1$$
$$x - 3(0) = -1$$
$$x = -1$$

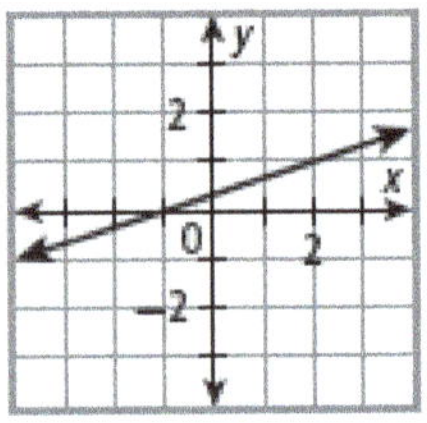

***y*-Intercept:**

$$x - 3y = -1$$
$$0 - 3y = -1$$
$$-3y = -1$$
$$\frac{-3y}{-3} = \frac{-1}{-3}$$
$$y = \frac{1}{3}$$

Appendix: Essential Algebra Review

Ratio and Proportion

Rates, Ratios, and Proportions

1. $\frac{x}{\$64} = \frac{3}{4}$

   $x = \frac{3}{4} \times \$64$

   $x = \$48$

2. $\frac{x}{341 \text{ trillion}} = \frac{2}{1}$

   $x = \frac{2}{1} \times 341 \text{ trillion}$

   $= 682 \text{ trillion}$

3. $\frac{x}{1} = \frac{2000}{40}$

   $x = 50 \text{ rotations/s}$

4. $\frac{x}{1} = \frac{224{,}988}{12}$

   $x = 18{,}749 \text{ lb/cow}$

5. $\frac{x}{1} = \frac{4.5}{9}$

   $x = 0.5 \text{ m/s}$

6. $4\frac{1}{2}\text{/h} \cdot \frac{1 \text{ h}}{60 \text{ min}}$

   $= 0.075 \text{ page/min}$

7. $\frac{18 \text{ ft}}{2 \text{ s}} \cdot \frac{3600 \text{ s}}{1 \text{ h}} \cdot \frac{1 \text{ mi}}{5280 \text{ ft}}$

   $\approx 6.14 \text{ mi/h}$

8. $\frac{125 \text{ yards}}{1 \text{ tablespoon}} \cdot \frac{256 \text{ tablespoons}}{1 \text{ gallon}} \cdot \frac{1 \text{ mi}}{1760 \text{ yards}}$

   $= 18 \text{ mi/gal}$

9. $\frac{3}{z} = \frac{1}{8}$

   $1(z) = 3(8)$

   $z = 24$

10. $\frac{x}{3} = \frac{1}{5}$

    $5(x) = 1(3)$

    $5x = 3$

    $x = \frac{3}{5}$

11. $\frac{b}{4} = \frac{3}{2}$

    $2(b) = 4(3)$

    $2b = 12$

    $b = 6$

12. $\frac{f+3}{12} = \frac{7}{2}$

$2(f+3) = 12(7)$

$2f + 6 = 84$

$\underline{\quad -6 \quad} \; \underline{-6}$

$2f = 78$

$f = 39$

13. $\frac{-1}{5} = \frac{3}{2d}$

$-1(2d) = 5(3)$

$-2d = 15$

$d = -7.5$

14. $\frac{3}{14} = \frac{s-2}{21}$

$14(s-2) = 21(3)$

$14s - 28 = 63$

$\underline{\quad +28} \; \underline{+28}$

$14s = 91$

$s = 6.5$

15. $\frac{-4}{9} = \frac{7}{x}$

$-4(x) = 7(9)$

$-4x = 63$

$x = -15.75$

16. $\frac{3}{s-2} = \frac{1}{7}$

$1(s-2) = 3(7)$

$s - 2 = 21$

$\underline{\quad +2} \; \underline{+2}$

$s = 23$

17. $\frac{10}{h} = \frac{52}{13}$

$52(h) = 10(13)$

$52h = 130$

$h = 2.5$

18. Let h be the height of the Altar Stone.

$\frac{3}{5} = \frac{h}{4.9}$

$5(h) = 4.9(3)$

$5h = 14.7$

$h = 2.94\ m$

Appendix: Essential Algebra Review

Ratio and Proportion

Applications of Proportions

1. $\frac{3}{\$45} = \frac{6}{t}$

$3 \cdot t = 6 \cdot 45$

$3t = 270$

$\frac{3t}{3} = \frac{270}{3}$

$t = \$90$

2. $\frac{2}{\$14} = \frac{m}{\$42}$

$14 \cdot m = 2 \cdot 42$

$14m = 84$

$\frac{14m}{14} = \frac{84}{14}$

$m = 6$

3. $\frac{12}{9} = \frac{n}{3}$

$9 \cdot n = 12 \cdot 3$

$9n = 36$

$\frac{9n}{9} = \frac{36}{9}$

$n = 4$

4. $\frac{t}{5} = \frac{28}{20}$

$20 \cdot t = 5 \cdot 28$

$20t = 140$

$\frac{20t}{20} = \frac{140}{20}$

$t = 7$

5. $\frac{1}{c} = \frac{6}{12}$

$6 \cdot c = 1 \cdot 12$

$6c = 12$

$\frac{6c}{6} = \frac{12}{6}$

$c = 2$

6. $\frac{6}{7} = \frac{30}{b}$

$6 \cdot b = 7 \cdot 30$

$6b = 210$

$\frac{6b}{6} = \frac{210}{6}$

$b = 35$

7. $\frac{\$0.75}{0.5\,\text{mi}} = \frac{x}{17.5\,\text{mi}}$

$0.5x = 0.75 \cdot 17.5$

$0.5x = 13.125$

$\frac{0.5x}{0.5} = \frac{13.125}{0.5}$

$x = \$26.25$

Appendix: Essential Algebra Review

Ratio and Proportion

Finding and Using Percents

1. $p \cdot 91 = 45$
$p = \frac{45}{91}$
$p \approx 0.495$
$p \approx 49.5\%$

2. $p \cdot 1270 = 375$
$p = \frac{375}{1270}$
$p \approx 0.295$
$p \approx 29.5\%$

3. $p \cdot 240 = 180$
$p = \frac{180}{240}$
$p = \frac{3}{4}$
$p = 0.75$
$p = 75\%$

4. $p \cdot 186 = 75$
$p = \frac{75}{186}$
$p \approx 0.403$
$p \approx 40.3\%$

5. $2 \cdot 80\% = 2 \cdot 0.8 = 1.6$ miles
Bailey walked 1.6 miles.

6. $18 \cdot 175\% = 18 \cdot 1.75 = 31.5$ pages
Madison's paper is 31.5 pages long.

7. $p \cdot 109{,}600{,}000 = 19{,}508{,}800$
$p = \frac{19{,}508{,}800}{109{,}600{,}000}$
$p \approx 0.178$
$p \approx 17.8\%$
17.8% of America watched *CSI* this week.

8. $6.9 = 11\frac{1}{2}\% \cdot n$
$6.9 = 0.115 \cdot n$
$\frac{6.9}{0.115} = n$
$60 = n$
6.9 is $11\frac{1}{2}\%$ of 60

9. $92 = 66\frac{2}{3}\% \cdot n$
$92 = \frac{2}{3} \cdot n$
$92 \cdot \frac{3}{2} = n$
$\frac{276}{2} = n$
$138 = n$
92 is $66\frac{2}{3}\%$ of 138

10. $12\% \cdot n = 20$
$0.12 \cdot n = 20$
$n = \frac{20}{0.12}$
$n = 166.7$
12% of 166.7 is 20

11. $30\% \cdot n = 96$
$0.30 \cdot n = 96$
$n = \frac{96}{0.30}$
$n = 320$
30% of 320 is 96

12. 32% of $7.4 = 0.32 \cdot 7.4 \approx 2.4$ oz.
2.4 ounces

13. $155\% \cdot n = 23.25$
$1.55 \cdot n = 23.25$
$n = \frac{23.25}{1.55}$
$n = 15$
The actual height of the pole is 15 feet.

Appendix: Essential Algebra Review

Review 13.3 / Ratio and Proportion

1. $\frac{6}{x} = \frac{1}{9}$
$1(x) = 6(9)$
$x = 54$ in.

2. $\frac{x}{12{,}000} = \frac{3}{500}$
$500(x) = 12{,}000(3)$
$500x = 36{,}000$
$x = 72$

3. $\frac{25 \text{ lb}}{4 \text{ gal}} = 6.25 \text{ lb/gal}$

4. $\frac{\$6058.50}{15 \text{ oz}} = \$403.90/\text{oz}$

5. $\frac{11.9 \text{ ft}}{3 \text{ days}}$
$= \frac{11.9 \text{ ft}}{3 \text{ days}} \cdot \frac{1 \text{ day}}{24 \text{ h}} \cdot \frac{12 \text{ in.}}{1 \text{ ft}}$
≈ 1.98 in./h
11.9 ft in 3 days is roughly equivalent to 12 ft in 72 h, or $\frac{12}{72} = \frac{1}{6}$ ft/h. There are 12 in. in 1 ft, so this speed is equivalent to $\frac{12}{6} = 2$ in./h. So 1.98 in./h is reasonable.

6. 694 m/s
$= \frac{694 \text{ m}}{1 \text{ s}} \cdot \frac{3600 \text{ s}}{1 \text{ h}} \cdot \frac{1 \text{ km}}{1000 \text{ m}}$
$= 2498.4$ km/h

7. $\frac{v}{6} = \frac{1}{2}$
$2(v) = 6(1)$
$\frac{2v}{2} = \frac{6}{2}$
$v = 3$

8. $\frac{2}{5} = \frac{4}{y}$
$2(y) = 4(5)$
$\frac{2y}{2} = \frac{20}{2}$
$y = 10$

9. $\frac{2}{h} = \frac{-5}{6}$
$(-5)(h) = 2(6)$
$\frac{(-5)h}{(-5)} = \frac{12}{(-5)}$
$h = -2.4$

10. $\frac{3}{10} = \frac{b+7}{20}$
$10(b+7) = 20(3)$
$10b + 70 = 60$
$\underline{\quad - 70 \quad - 70}$
$10b = -10$
$\frac{10b}{10} = \frac{-10}{10}$
$b = -1$

11. $\frac{5t}{9} = \frac{1}{2}$
$2(5t) = 1(9)$
$\frac{10t}{10} = \frac{9}{10}$
$t = 0.9$

12. $\frac{2}{3} = \frac{6}{q-4}$
$2(q-4) = 3(6)$
$2q - 8 = 18$
$\underline{\quad + 8 \quad + 8}$
$2q = 26$
$\frac{2q}{2} = \frac{26}{2}$
$q = 13$

13. $\frac{x}{8} = \frac{7.5}{20}$

$20(x) = 8(7.5)$

$\frac{20x}{20} = \frac{60}{20}$

$x = 3$

14. $\frac{3}{k} = \frac{45}{18}$

$45(k) = 3(18)$

$\frac{45k}{45} = \frac{54}{45}$

$k = 1.2$

15. $\frac{6}{a} = \frac{15}{17}$

$15(a) = 6(17)$

$\frac{15a}{15} = \frac{102}{15}$

$a = 6.8$

16. $\frac{9}{2} = \frac{5}{x+1}$

$9(x+1) = 2(5)$

$9x + 9 = 10$

$\underline{\quad -9} \quad \underline{-9}$

$9x \quad = \quad 1$

$\frac{9x}{9} = \frac{1}{9}$

$x = \frac{1}{9}$

17. $\frac{3}{5} = \frac{x}{100}$

$5(x) = 3(100)$

$\frac{5x}{5} = \frac{300}{5}$

$x = 60$

18. $\frac{38}{19} = \frac{n-5}{20}$

$19(n-5) = 20(38)$

$19n - 95 = 760$

$\underline{\quad +95} \quad \underline{+95}$

$19n \quad = 855$

$\frac{19n}{19} = \frac{855}{19}$

$n = 45$

19. $\frac{0.3 \text{ mm}}{3 \text{ cm}}$

$= \frac{0.3 \text{ mm}}{3 \text{ cm}} \cdot \frac{1 \text{ cm}}{10 \text{ mm}}$

$= \frac{0.3 \text{ mm}}{30 \text{ mm}} = \frac{1}{100}$

So, the scale must be 1:100.

20. $\frac{3}{\$18} = \frac{12}{\text{m}}$

$3 \cdot \text{m} = 12 \cdot 18$

$3\text{m} = 216$

$\frac{3\text{m}}{3} = \frac{216}{3}$

$\text{m} = \$72$

21. $\frac{3}{\$27} = \frac{\text{c}}{\$54}$

$3 \cdot 54 = 27 \cdot \text{c}$

$162 = 27\text{c}$

$\frac{162}{27} = \frac{27\text{c}}{27}$

$6 \text{ lbs} = \text{c}$

22. $\frac{3}{2} = \frac{24}{d}$

$3 \cdot d = 2 \cdot 24$

$3d = 48$

$\frac{3d}{3} = \frac{48}{3}$

$d = 16$

23. $\frac{p}{40} = \frac{3}{8}$
$8 \cdot p = 40 \cdot 3$
$8p = 120$
$\frac{8p}{8} = \frac{120}{8}$
$p = 15$

24. $\frac{6}{14} = \frac{x}{7}$
$14 \cdot x = 6 \cdot 7$
$14x = 42$
$\frac{14x}{14} = \frac{42}{14}$
$x = 3$

25. $\frac{5}{p} = \frac{7}{77}$
$7 \cdot p = 5 \cdot 77$
$7p = 385$
$\frac{7p}{7} = \frac{385}{7}$
$p = 55$

26. $\frac{\text{20 min of study}}{\text{60 min of class}} = \frac{x \text{ min of study}}{\text{90 min of class}}$
$60 \cdot x = 20 \cdot 90$
$60x = 1{,}800$
$\frac{60x}{60} = \frac{1{,}800}{60}$
$x = 30$
30 min

27. 16% of 70
$\frac{16}{100} = \frac{n}{70}$
$1{,}120 = 100n$
$11.2 = n$
16% of 70 is 11.2.

28. 118% of 3000
$\frac{118}{100} = \frac{n}{3000}$
$354{,}000 = 100n$
$3540 = n$
118% of 3000 is 3540.

29. 5% of 58
$\frac{5}{100} = \frac{n}{58}$
$290 = 100n$
$2.9 = n$
5% of 58 is 2.9.

30. 103% of 50
$\frac{103}{100} = \frac{n}{50}$
$5{,}150 = 100n$
$51.5 = n$
103% of 50 is 51.5.

31. 86% of 19,396
$0.86 \cdot 19{,}396 = 16{,}680.56$
About 16,681 bulldogs were registered.

32. 56 is 140% of what number?
$\frac{140}{100} = \frac{56}{n}$
$140n = 5600$
$n = 40$
56 is 140% of 40.

33. 45 is 20% of what number?
$45 = 0.20 \cdot n$
$225 = n$
45 is 20% if 225.

34. 18 is 15% of what number?
$18 = 0.15 \cdot n$
$120 = n$
18 is 15% of 120.

35. What percent of 60 is 40?
$40 = n \cdot 60$
$0.\overline{6} = n$
40 is $66\frac{2}{3}$% of 60.

36. What percent of 48 is 16?
$\frac{n}{100} = \frac{16}{48}$
$48n = 1600$
$n = 33\frac{1}{3}$
16 is $33\frac{1}{3}$% of 48.

37. What percent of 6 is 18?

$18 = n \cdot 6$

$3 = n$

18 is 300% of 6.

38. $2.10 \cdot n = 147$

$n = \frac{147}{2.1}$

$n = 70$

210% of 70 is 147.

39. Restate the problem.

\$0.21 is what percent of \$3?

$\frac{n}{100} = \frac{\$0.21}{\$3.00}$

$3n = 21$

$n = 7$

The sales tax rate is 7%.

Appendix: Essential Algebra Review

Operations with Polynomials and Factoring

Adding and Subtracting Polynomials

1. $\left(5n^3 + 3n + 6\right) + \left(18n^3 + 9\right)$
$= \left(5n^3 + 18n^3\right) + 3n + (6 + 9)$
$= 23n^3 + 3n + 15$

2. $\left(3.7q^2 - 8q + 3.7\right) + \left(4.3q^2 - 2.9q + 1.6\right)$
$= \left(3.7q^2 + 4.3q^2\right) + (-8q - 2.9q) + (3.7 + 1.6)$
$= 8q^2 - 10.9q + 5.3$

3. $(-3x + 12) + \left(9x^2 + 2x - 18\right)$
$= 9x^2 + (-3x + 2x) + (12 - 18)$
$= 9x^2 - x - 6$

4. $\left(9x^4 + x^3\right) + \left(2x^4 + 6x^3 - 8x^4 + x^3\right)$
$= \left(9x^4 + 2x^4 - 8x^4\right) + \left(x^3 + 6x^3 + x^3\right)$
$= 3x^4 + 8x^3$

5. $m\angle ABD = \left(8a^2 - 2a + 5\right) + (7a + 4)$
$= 8a^2 + (-2a + 7a) + (5 + 4)$
$= 8a^2 + 5a + 9$

6. $-\left(0.32r^3 + 0.19r^4\right) = -0.32r^3 - 0.19r^4$

7. $-\left(9y^2 - 6x^2\right) = -9y^2 + 6x^2$

8. $-(-4 + w - 2z) = 4 - w + 2z$

9. $\left(6c^4 + 8c + 6\right) - \left(2c^4\right)$
$= \left(6c^4 + 8c + 6\right) + \left(-2c^4\right)$
$= \left(6c^4 - 2c^4\right) + 8c + 6$
$= 4c^4 + 8c + 6$

10. $\left(16y^2 - 8y + 9\right) - \left(6y^2 - 2y + 7y\right)$
$= \left(16y^2 - 8y + 9\right) + \left(-6y^2 + 2y - 7y\right)$
$= \left(16y^2 - 6y^2\right) + (-8y + 2y - 7y) + 9$
$= 10y^2 - 13y + 9$

11. $(2r + 5) - (5r - 6)$
$= (2r + 5) + (-5r + 6)$
$= (2r - 5r) + (5 + 6)$
$= -3r + 11$

12. $\left(-7k^2 + 3\right) - \left(2k^2 + 5k - 1\right)$
$= \left(-7k^2 + 3\right) + \left(-2k^2 - 5k + 1\right)$
$= \left(-7k^2 - 2k^2\right) + (-5k) + (3 + 1)$
$= -9k^2 - 5k + 4$

13. $BC = (15x - y) - (8x + 3y - 1)$
$= 15x - y - 8x - 3y + 1$
$= 15x - 8x - y - 3y + 1$
$= 7x - 4y + 1$

Appendix: Essential Algebra Review

Operations with Polynomials and Factoring

Multiplying and Dividing Monomials

1. $(2x^2)(7x^4)$
$= (2 \cdot 7)(x^2 \cdot x^4)$
$= 14x^6$

2. $(-5mn^3)(4m^2n^2)$
$= (-5 \cdot 4)(m \cdot m^2)(n^3 \cdot n^2)$
$= -20m^3n^5$

3. $(6rs^2)(s^3t^2)\left(\frac{1}{2}r^4t^3\right)$
$= \left(6 \cdot \frac{1}{2}\right)(r \cdot r^4)(s^2 \cdot s^3)(t^2 \cdot t^3)$
$= 3r^5s^5t^5$

4. $\left(\frac{1}{3}a^5\right)(12a)$
$= \left(\frac{1}{3} \cdot 12\right)(a^5 \cdot a)$
$= 4a^6$

5. $(-3x^4y^2)(-7x^3y)$
$= (-3 \cdot (-7))(x^4 \cdot x^3)(y^2 \cdot y)$
$= 21x^7y^3$

6. $(-2pq^3)(5p^2q^2)(-3q^4)$
$= (-2 \cdot 5 \cdot (-3))(p \cdot p^2)(q^3 \cdot q^2 \cdot q^4)$
$= 30p^3q^9$

7. $\dfrac{15n^5}{5n^2} = \dfrac{15}{5} \cdot \dfrac{n^5}{n^2} = 3n^{5-2} = 3n^3$

8. $\dfrac{-4x^7}{2x^6} = \dfrac{-4}{2} \cdot \dfrac{x^7}{x^6} = -2x^{7-6} = -2x^1 = -2x$

9. $\dfrac{18a^2b^8}{12ab^5} = \dfrac{18}{12} \cdot \dfrac{a^2}{a} \cdot \dfrac{b^8}{b^5} = \dfrac{3}{2}a^{2-1}b^{8-5} = \dfrac{3}{2}ab^3$

10. $\dfrac{24s^{10}t^{12}}{18s^5} = \dfrac{24}{18} \cdot \dfrac{s^{10}}{s^5} \cdot t^{12} = \dfrac{4}{3}s^{10-5}t^{12} = \dfrac{4}{3}s^5t^{12}$

11. $\dfrac{42a^4c^7}{7a^4c^2} = \dfrac{42}{7} \cdot \dfrac{a^4}{a^4} \cdot \dfrac{c^7}{c^2} = 6a^{4-4}c^{7-2}$
$= 6a^0c^5 = 6 \cdot 1 \cdot c^5 = 6c^5$

12. $\dfrac{16m^5p}{8m^3p} = \dfrac{16}{8} \cdot \dfrac{m^5}{m^3} \cdot \dfrac{p}{p} = 2m^{5-3}p^{1-1}$
$= 2m^2p^0 = 2m^2$

13. $(5a)^3 = (5)^3(a)^3 = 125a^3$

14. $(9m^3)^2 = (9)^2(m^3)^2 = 81(m^{3 \cdot 2}) = 81m^6$

15. $(6x^2y^5)^3 = (6)^3(x^2)^3(y^5)^3$
$= 216(x^{2 \cdot 3})(y^{5 \cdot 3}) = 216x^6y^{15}$

16. $(-7p^5q^3)^2 = (-7)^2(p^5)^2(q^3)^2$
$= 49(p^{5 \cdot 2})(q^{3 \cdot 2}) = 49(p^{10})(q^6) = 49p^{10}q^6$

17. $(2m^3n^2)^5 = (2)^5(m^3)^5(n^2)^5$
$= 32(m^{3 \cdot 5})(n^{2 \cdot 5}) = 32m^{15}n^{10}$

18. $(-3x^4yz)^4 = (-3)^4(x^4)^4(y)^4(z)^4$
$= 81(x^{4 \cdot 4})(y^4)(z^4) = 81x^{16}y^4z^4$

Multiplying Monomials by Polynomials

1. $4(x^2 + 2x + 1)$
$= 4(x^2) + 4(2x) + 4(1)$
$= 4x^2 + 8x + 4$

2. $3ab(2a^2 + 3b^3)$
$= 3ab(2a^2) + 3ab(3b^3)$
$= (3 \cdot 2)(a \cdot a^2)(b) + (3 \cdot 3)(a)(b \cdot b^3)$
$= 6a^3b + 9ab^4$

3. $2a^3b(3a^2b + ab^2)$
$= 2a^3b(3a^2b) + 2a^3b(ab^2)$
$= (2 \cdot 3)(a^3 \cdot a^2)(b \cdot b) + (2)(a^3 \cdot a)(b \cdot b^2)$
$= 6a^5b^2 + 2a^4b^3$

4. $-3x(x^2 - 4x + 6)$
$= -3x(x^2) - 3x(-4x) - 3x(6)$
$= -3x^3 + 12x^2 - 18x$

5. $5x^2y(2xy^3 - y)$
$= 5x^2y(2xy^3) + 5x^2y(-y)$
$= (5 \cdot 2)(x^2 \cdot x)(y \cdot y^3) + (5 \cdot (-1))(x^2)(y \cdot y)$
$= 10x^3y^4 - 5x^2y^2$

6. $5m^2n^3 \cdot mn^2(4m - n)$
$= (5)(m^2 \cdot m)(n^3 \cdot n^2)(4m - n)$
$= 5m^3n^5(4m - n)$
$= 5m^3n^5(4m) + 5m^3n^5(-n)$
$= (5 \cdot 4)(m^3 \cdot m)(n^5) + (5 \cdot (-1))(m^3)(n^5 \cdot n)$
$= 20m^4n^5 - 5m^3n^6$

7. a. $A = \ell w$
$= (2x - 3)(x)$
$= 2x(x) - 3(x)$
$= 2x^2 - 3x$
The area is represented by $2x^2 - 3x$.

b. $A = 2x^2 - 3x$
$= 2(4)^2 - 3(4)$
$= 2(16) - 3(4)$
$= 32 - 12 = 20$
The area is 20 in^2.

Appendix: Essential Algebra Review

Operations with Polynomials and Factoring

Multiplying Binomials

1. $(x+1)(x-2)$
$= x(x) + x(-2) + 1(x) + 1(-2)$
$= x^2 - 2x + x - 2$
$= x^2 - x - 2$

2. $(y-3)(y-5)$
$= y(y) + y(-5) - 3(y) - 3(-5)$
$= y^2 - 5y - 3y + 15$
$= y^2 - 8y + 15$

3. $(4a^3 - 2b)(a - 3b^2)$
$= 4a^3(a) + 4a^3(-3b^2) - 2b(a) - 2b(-3b^2)$
$= 4a^2 - 2ab - 12a^3b^2 + 6b^3$

4. $(m^2 - 2mn)(3mn + n^2)$
$= m^2(3mn) + m^2(n^2) - 2mn(3mn) - 2mn(n^2)$
$= 3m^3n + m^2n^2 - 6m^2n^2 - 2mn^3$
$= 3m^3n - 5m^2n^2 - 2mn^3$

5. Area of big □ $(x+3)^2 = (x+3)(x+3) = x^2 + 3x + 3x + 9 = x^2 + 6x + 9$
Area of small □ $(x+1)^2 = (x+1)(x+1) = x^2 + x + x + 1 = x^2 + 2x + 1$
Total area = area of big □ + area of small □
$= (x^2 + 6x + 9) + (x^2 + 2x + 1)$
$= x^2 + 6x + 9 + x^2 + 2x + 1$
$= x^2 + x^2 + 6x + 2x + 9 + 1$
$= 2x^2 + 8x + 10$
The area of the figure is $2x^2 + 8x + 10$.

6. $(x+7)^2 = (x+7)(x+7)$
$= x^2 + 7x + 7x + 49$
$= x^2 + 14x + 49$

7. $(2+x)^2 = (2+x)(2+x)$
$= 4 + 2x + 2x + x^2$
$= x^2 + 4x + 4$

8. $(5x+9)^2 = (5x+9)(5x+9)$
$= 25x^2 + 45x + 45x + 81$
$= 25x^2 + 90x + 81$

9. $(2a+7b)^2 = (2a+7b)(2a+7b)$
$= 4a^2 + 14ab + 14ab + 49b^2$
$= 4a^2 + 28ab + 49b^2$

10. $(x-6)^2 = (x-6)(x-6)$
$= x^2 - 6x - 6x + 36$
$= x^2 - 12x + 36$

11. $(2x-1)^2 = (2x-1)(2x-1)$
$= 4x^2 - 2x - 2x + 1$
$= 4x^2 - 4x + 1$

Appendix: Essential Algebra Review

Operations with Polynomials and Factoring

Multiplying Binomials

12. $(8-x)^2 = (8-x)(8-x)$
$= 64 - 8x - 8x + x^2$
$= 64 - 16x + x^2$
$= x^2 - 16x + 64$

13. $(6p-q)^2 = (6p-q)(6p-q)$
$= 36p^2 - 6qp - 6qp + q^2$
$= 36p^2 - 12qp + q^2$

14. $(7a-2b)^2 = (7a-2b)(7a-2b)$
$= 49a^2 - 14ab - 14ab + 4b^2$
$= 49a^2 - 28ab + 4b^2$

15. $(x-5)(x+5) = x^2 + 5x - 5x - 25$
$= x^2 - 25$

16. $(x+6)(x-6) = x^2 - 6x + 6x - 36$
$= x^2 - 36$

17. $(5x+1)(5x-1) = 25x^2 - 5x + 5x - 1$
$= 25x^2 - 1$

18. $(2x^2+3)(2x^2-3) = 4x^4 - 6x^2 + 6x^2 - 9$
$= 4x^2 - 9$

19. $(9-x^3)(9+x^3) = 81 + 9x^3 - 9x^3 - x^6$
$= 81 - x^6$

20. $(2x-5y)(2x+5y) = 4x^2 + 10xy - 10xy - 25y^2$
$= 4x^2 - 25y^2$

Appendix: Essential Algebra Review

Operations with Polynomials and Factoring

Factoring with the GCF

1. $6x^2 = 2 \cdot 3 \cdot x \cdot x$
   $5x^2 = 5 \cdot x \cdot x$
   The GCF of $6x^2$ and $5x^2$ is x^2.

2. $15y^3 = 3 \cdot 5 \cdot y \cdot y \cdot y$
   $-20y = -1 \cdot 2 \cdot 2 \cdot 5 \cdot y$
   The GCF of $15y^3$ and $-20y$ is $5y$.

3. $13q^4 = 13 \cdot q \cdot q \cdot q \cdot q$
   $2p^2 = 2 \cdot p \cdot p$
   The GCF of $13q^4$ and $2p^2$ is 1.

4. $15a = 3 \cdot 5 \cdot a$
   $5a^2 = 5 \cdot a \cdot a$
   The GCF of $15a$ and $5a^2$ is $5a$.
   $15a - 5a^2$
   $3(5a) - a(5a)$
   $5a(3 - a)$

5. $10g^3 = 2 \cdot 5 \cdot g \cdot g \cdot g$
   $3g = 3 \cdot g$
   The GCF of $10g^3$ and $3g$ is g.
   $10g^3 - 3g$
   $10g^2(g) - 3(g)$
   $g(10g^2 - 3)$

6. $35x = 5 \cdot 7 \cdot x$
   $42 = 6 \cdot 7$
   The GCF of $35x$ and 42 is 7.
   $-35x + 42$
   $-5x(7) + 6(7)$
   $7(-5x + 6)$

7. $4x^2 = 2 \cdot 2 \cdot x \cdot x$
   $6x = 2 \cdot 3 \cdot x$
   The GCF of $4x^2$ and $6x$ is $2x$.
   $-(4x^2 + 6x)$
   $-\left(2x(2x) + 3(2x)\right)$
   $-2x(2x + 3)$

8. $12h^4 = 2 \cdot 2 \cdot 3 \cdot h \cdot h \cdot h \cdot h$
   $8h^2 = 2 \cdot 2 \cdot 2 \cdot h \cdot h$
   $6h = 2 \cdot 3 \cdot h$
   The GCF of $12h^4$, $8h^2$ and $6h$ is $2h$.
   $12h^4 + 8h^2 - 6h$
   $6h^3(2h) + 4h(2h) - 3(2h)$
   $2h(6h^3 + 4h - 3)$

9. $3x^2 = 3 \cdot x \cdot x$
   $9x = 3 \cdot 3 \cdot x$
   $3 = 3$
   The GCF of $3x^2$, $9x$ and 3 is 3.
   $3x^2 - 9x + 3$
   $x^2(3) - 3x(3) + 1(3)$
   $3(x^2 - 3x + 1)$

10. $9m^2 = 3 \cdot 3 \cdot m \cdot m$
    $m = m$
    The GCF of $9m^2$ and m is m.
    $9m^2 + m$
    $9m(m) + 1(m)$
    $m(9m + 1)$

11. $14n^3 = 2 \cdot 7 \cdot n \cdot n \cdot n$
    $7n = 7 \cdot n$
    $7n^2 = 7 \cdot n \cdot n$
    The GCF of $14n^3$, $7n$ and $7n^2$ is $7n$.
    $14n^3 + 7n + 7n^2$
    $2n^2(7n) + 1(7n) + n(7n)$
    $7n(2n^2 + 1 + n)$

12. $36f = 2 \cdot 2 \cdot 3 \cdot 3 \cdot f$
    $18f^2 = 2 \cdot 3 \cdot 3 \cdot f \cdot f$
    $3 = 3$
    The GCF of $36f$, $18f^2$ and 3 is 3.
    $36f + 18f^2 + 3$
    $12f(3) + 6f^2(3) + 1(3)$
    $3(12f + 6f^2 + 1)$

Appendix: Essential Algebra Review

Operations with Polynomials and Factoring

Factoring with the GCF

13. $15b^2 = 3 \cdot 5 \cdot b \cdot b$
$7b \quad = 7 \cdot b$
The GCF of $15b^2$ and $7b$ is b.
$-15b^2 + 7b$
$-15b(b) + 7(b)$
$b(-15b + 7)$

14. $-16t^2 + 320t$
$-t(16t) + 20t(16)$
$16(-6t + 20)$
Using the factored form of the expression can tell you when the rocket will land again.

15. $5(m - 2) - m(m - 2)$
$(5 - m)(m - 2)$

16. $2b(b + 3) + 5(b + 3)$
$(2b + 5)(b + 3)$

17. $4(x - 3) - x(y + 2)$
Cannot be factored

Appendix: Essential Algebra Review

Operations with Polynomials and Factoring

Factoring $x^2 + bx + c$

1. $x^2 + 13x + 36$

$(x + 1)(x + 36) = x^2 + 37x + 36$ ✗

$(x + 2)(x + 18) = x^2 + 20x + 36$ ✗

$(x + 3)(x + 12) = x^2 + 15x + 36$ ✗

$(x + 4)(x + 9) = x^2 + 13x + 36$ ✓

The factors of $x^2 + 13x + 36$ are $(x + 4)$ and $(x + 9)$.

$x^2 + 13x + 36 = (x + 4)(x + 9)$

2. $x^2 + 11x + 24$

$(x + 1)(x + 24) = x^2 + 25x + 24$ ✗

$(x + 2)(x + 12) = x^2 + 14x + 24$ ✗

$(x + 3)(x + 8) = x^2 + 11x + 24$ ✓

The factors of $x^2 + 11x + 24$ are $(x + 3)$ and $(x + 8)$.

$x^2 + 11x + 24 = (x + 3)(x + 8)$

3. $x^2 + 14x + 40$

$(x + 1)(x + 40) = x^2 + 41x + 40$ ✗

$(x + 2)(x + 20) = x^2 + 22x + 40$ ✗

$(x + 4)(x + 10) = x^2 + 14x + 40$ ✓

The factors of $x^2 + 14x + 40$ are $(x + 4)$ and $(x + 10)$.

$x^2 + 14x + 40 = (x + 4)(x + 10)$

4. $x^2 + 4x + 3$

Factors of 3	Sum
1 and 3	4 ✓

$(x + 1)(x + 3)$

5. $x^2 + 10x + 16$

Factors of 16	Sum
1 and 16	17 ✗
2 and 8	10 ✓

$(x + 2)(x + 8)$

6. $x^2 + 15x + 44$

Factors of 44	Sum
1 and 44	45 ✗
2 and 22	24 ✗
4 and 11	15 ✓

$(x + 4)(x + 11)$

7. $x^2 - 7x + 6$

Factors of 6	Sum
−1 and −6	−7 ✓

$(x - 1)(x - 6)$

8. $x^2 - 9x + 14$

Factors of 14	Sum
−1 and −14	−15 ✗
−2 and −7	−9 ✓

$(x - 2)(x - 7)$

9. $x^2 - 11x + 24$

Factors of 24	Sum
−1 and −24	−25 ✗
−2 and −12	−14 ✗
−3 and −8	−11 ✓
−4 and −6	−10 ✓

$(x - 3)(x - 8)$

10. $x^2 + 6x - 7$

Factors of −7	Sum
−1 and 7	6 ✓

$(x - 7)(x + 1)$

11. $x^2 + 6x - 27$

Factors of −27	Sum
−1 and 27	26 ✗
−3 and 9	6 ✓

$(x - 3)(x + 9)$

12. $x^2 + x - 30$

Factors of −30	Sum
−1 and 30	29 ✗
−2 and 15	13 ✗
−3 and 10	7 ✗
−5 and 6	1 ✓

$(x - 5)(x + 6)$

13. $x^2 - x - 2$

Factors of −2	Sum
1 and −2	−1 ✓

$(x + 1)(x - 2)$

14. $x^2 - 3x - 18$

Factors of −18	Sum
1 and −17	−16 ✗
2 and −9	−7 ✗
3 and −6	−3 ✓

$(x + 3)(x - 6)$

15. $x^2 - 4x - 45$

Factors of −45	Sum
1 and −45	−44 ✗
3 and −15	−12 ✗
5 and −9	−4 ✓

$(x + 5)(x - 9)$

Appendix: Essential Algebra Review

Operations with Polynomials and Factoring

Factoring $ax^2 + bx + c$

1. $2x^2 + 9x + 10$
   $(1x + 10)(2x + 1) = 2x^2 + 21x + 10$ ✗
   $(1x + 5)(2x + 2) = 2x^2 + 12x + 10$ ✗
   $(1x + 2)(2x + 5) = 2x^2 + 9x + 10$ ✓
   The factors of $2x^2 + 9x + 10$ are $(x + 2)$ and $(2x + 5)$.
   $2x^2 + 9x + 10 = (x + 2)(2x + 5)$

2. $5x^2 + 31x + 6$
   $(1x + 6)(5x + 1) = 5x^2 + 31x + 6$ ✓
   The factors of $5x^2 + 31x + 6$ are $(x + 6)$ and $(5x + 1)$.
   $5x^2 + 31x + 6 = (x + 6)(5x + 1)$

3. $5x^2 + 7x - 6$
   $(1x - 6)(5x + 1) = 5x^2 - 29x - 6$ ✗
   $(1x - 3)(5x + 2) = 5x^2 - 13x - 6$ ✗
   $(1x - 2)(5x + 3) = 5x^2 - 7x - 6$ ✗
   $(1x + 2)(5x - 3) = 5x^2 + 7x - 6$ ✓
   The factors of $5x^2 + 7x - 6$ are $(x + 2)$ and $(5x - 3)$.
   $5x^2 + 7x - 6 = (x + 2)(5x - 3)$

4. $6x^2 + 37x + 6$
   $(1x + 6)(6x + 1) = 6x^2 + 37x + 6$ ✓
   The factors of $6x^2 + 37x + 6$ are $(x + 6)$ and $(6x + 1)$.
   $6x^2 + 37x + 6 = (x + 6)(6x + 1)$

5. $3x^2 - 14x - 24$
   $(1x - 24)(3x + 1) = 3x^2 - 71x - 24$ ✗
   $(1x - 12)(3x + 2) = 3x^2 - 34x - 24$ ✗
   $(1x - 8)(3x + 3) = 3x^2 - 21x - 24$ ✗
   $(1x - 6)(3x + 4) = 3x^2 - 14x - 24$ ✓
   The factors of $3x^2 - 14x - 24$ are $(x - 6)$ and $(3x + 4)$.
   $3x^2 - 14x - 24 = (x - 6)(3x + 4)$

6. $6x^2 + x - 2$
   $(1x - 2)(6x + 1) = 6x^2 - 11x - 2$ ✗
   $(1x - 1)(6x + 2) = 6x^2 - 4x - 2$ ✗
   $(1x + 1)(6x - 2) = 6x^2 + 4x - 2$ ✗
   $(1x + 2)(6x - 1) = 6x^2 + 11x - 2$ ✗
   $(2x - 2)(3x + 1) = 6x^2 - 4x - 2$ ✗
   $(2x - 1)(3x + 2) = 6x^2 + x - 2$ ✓
   The factors of $6x^2 + x - 2$ are $(2x - 1)$ and $(3x + 2)$.
   $6x^2 + x - 2 = (2x - 1)(3x + 2)$.

7. $5x^2 + 11x + 2$

Factors of 5	Factors of 2	Outer + Inner	
1 and 5	1 and 2	$1(2) + 5(1) = 7$	✗
1 and 5	2 and 1	$1(1) + 5(2) = 11$	✓

$(x + 2)(5x + 1)$

8. $2x^2 + 11x + 5$

Factors of 2	Factors of 5	Outer + Inner	
1 and 2	1 and 5	$1(5) + 2(1) = 7$	✗
1 and 2	5 and 1	$1(1) + 2(5) = 11$	✓

$(x + 5)(2x + 1)$

9. $4x^2 - 9x + 5$

Factors of 4	Factors of 5	Outer + Inner	
1 and 4	−1 and −5	$1(-5) + 4(-1) = -9$	✓

$(x - 1)(4x - 5)$

10. $2y^2 - 11y + 14$

Factors of 2	Factors of 14	Outer + Inner	
1 and 2	−1 and −14	$1(-14) + 2(-1) = -16$	✗
1 and 2	−14 and −1	$1(-1) + 2(-14) = -29$	✗
1 and 2	−2 and −7	$1(-7) + 2(-2) = -11$	✓

$(2y - 7)(y - 2)$

11. $5x^2 + 9x + 4$

Factors of 5	Factors of 4	Outer + Inner	
1 and 5	1 and 4	$1(4) + 5(1) = 9$	✓

$(x + 1)(5x + 4)$

12. $3x^2 + 7x + 2$

Factors of 3	Factors of 2	Outer + Inner	
1 and 3	1 and 2	$1(2) + 3(1) = 5$	✗
1 and 3	2 and 1	$1(1) + 3(2) = 7$	✓

$(x + 2)(3x + 1)$

13. $4a^2 + 8a - 5$

Factors of 4	Factors of −5	Outer + Inner	
1 and 4	1 and −5	$1(-5) + 4(1) = -1$	✗
1 and 4	−1 and 5	$1(5) + 4(-1) = 1$	✗
1 and 4	5 and −1	$1(-1) + 4(5) = 19$	✗
1 and 4	−5 and 1	$1(1) + 4(-5) = -19$	✗
2 and 2	1 and −5	$2(-5) + 2(1) = -8$	✗
2 and 2	−1 and 5	$2(5) + 2(-1) = 8$	✓

$(2a - 1)(2a + 5)$

14. $15x^2 + 4x - 3$

Factors of 15	Factors of −3	Outer + Inner
1 and 15	1 and −3	$1(-3) + 15(1) = 12$ ✗
1 and 15	−1 and 3	$1(3) + 15(-1) = -12$ ✗
1 and 15	3 and −1	$1(-1) + 15(3) = 44$ ✗
1 and 15	−3 and 1	$1(1) + 15(-3) = -44$ ✗
3 and 5	1 and −3	$3(-3) + 5(1) = -4$ ✗
3 and 5	−1 and 3	$3(3) + 5(-1) = 4$ ✓

$(3x - 1)(5x + 3)$

15. $2x^2 + x - 6$

Factors of 2	Factors of −6	Outer + Inner
1 and 2	1 and −6	$1(-6) + 2(1) = -4$ ✗
1 and 2	−1 and 6	$1(6) + 2(-1) = 4$ ✗
1 and 2	2 and −3	$1(-3) + 2(2) = 1$ ✓

$(x + 2)(2x - 3)$

16. $6n^2 - 11n - 10$

Factors of 6	Factors of −10	Outer + Inner
1 and 6	1 and −10	$1(-10) + 6(1) = -4$ ✗
1 and 6	−1 and 10	$1(10) + 6(-1) = 4$ ✗
1 and 6	2 and −5	$1(-5) + 6(2) = 7$ ✗
1 and 6	−2 and 5	$1(5) + 6(-2) = -7$ ✗
1 and 6	5 and −2	$1(-2) + 6(5) = 28$ ✗
1 and 6	−5 and 2	$1(2) + 6(-5) = -28$ ✗
1 and 6	10 and −1	$1(-1) + 6(10) = 59$ ✗
1 and 6	−10 and 1	$1(1) + 6(-10) = -59$ ✗
2 and 3	2 and −5	$2(-5) + 3(2) = -4$ ✗
2 and 3	−2 and 5	$2(5) + 3(-2) = 4$ ✗
2 and 3	5 and −2	$2(-2) + 3(5) = 11$ ✗
2 and 3	−5 and 2	$2(2) + 3(-5) = -11$ ✓

$(2n - 5)(3n + 2)$

17. $10x^2 - 9x - 1$

Factors of 10	Factors of −1	Outer + Inner
1 and 10	1 and −1	$1(-1) + 10(1) = 9$ ✗
1 and 10	−1 and 1	$1(1) + 10(-1) = -9$ ✓

$(x - 1)(10x + 1)$

18. $7x^2 - 3x - 10$

Factors of 7	Factors of −10	Outer + Inner
1 and 7	1 and −10	$1(-10) + 7(1) = -3$ ✓

$(x + 1)(7x - 10)$

19. $-2x^2 + 5x + 12$

$-1(2x^2 - 5x - 12)$

Factors of 2	Factors of −12	Outer + Inner
1 and 2	1 and −12	$1(-12) + 2(1) = -10$ ✗
1 and 2	−1 and 12	$1(12) + 2(-1) = 10$ ✗
1 and 2	2 and −6	$1(-6) + 2(2) = -2$ ✗
1 and 2	−2 and 6	$1(6) + 2(-2) = 2$ ✗
1 and 2	3 and −4	$1(-4) + 2(3) = 2$ ✗
1 and 2	−3 and 4	$1(4) + 2(-3) = -2$ ✗
1 and 2	4 and −3	$1(-3) + 2(4) = 5$ ✗
1 and 2	−4 and 3	$1(3) + 2(-4) = -5$ ✓

$= -1(2x + 3)(x - 4)$

20. $-4n^2 - 16n + 9$

$-1(4n^2 + 16n - 9)$

Factors of 4	Factors of −9	Outer + Inner
1 and 4	1 and −9	$1(-9) + 4(1) = -5$ ✗
1 and 4	−1 and 9	$1(9) + 4(-1) = 5$ ✗
1 and 4	3 and −3	$1(-3) + 4(3) = 9$ ✗
1 and 4	−3 and 3	$1(3) + 4(-3) = -9$ ✗
1 and 4	9 and −1	$1(-1) + 4(9) = 35$ ✗
1 and 4	−9 and 1	$1(1) + 4(-9) = -35$ ✗
2 and 2	1 and −9	$2(-9) + 2(1) = -16$ ✗
2 and 2	−1 and 9	$2(9) + 2(-1) = 16$ ✓

$-1(2n - 1)(2n + 9)$

21. $-5x^2 + 7x + 6$

$-1(5x^2 - 7x - 6)$

Factors of 5	Factors of −6	Outer + Inner
1 and 5	1 and −6	$1(-6) + 5(1) = -1$ ✗
1 and 5	−1 and 6	$1(6) + 5(-1) = 1$ ✗
1 and 5	2 and −3	$1(-3) + 5(2) = 7$ ✗
1 and 5	−2 and 3	$1(3) + 5(-2) = -7$ ✓

$-1(x - 2)(5x + 3)$

22. $-6x^2 + 13x - 2$

$-1(6x^2 - 13x + 2)$

Factors of 6	Factors of 2	Outer + Inner
1 and 6	−1 and −2	$1(-2) + 6(-1) = -8$ ✗
1 and 6	−2 and −1	$1(-1) + 6(-2) = -13$ ✓

$-1(x - 2)(6x - 1)$

23. $-4x^2 - 8x + 5$

$-1(4x^2 + 8x - 5)$

Factors of 4	Factors of −5	Outer + Inner	
1 and 4	1 and −5	$1(-5) + 4(1) = -1$	✗
1 and 4	−1 and 5	$1(5) + 4(-1) = 1$	✗
1 and 4	5 and −1	$1(-1) + 4(5) = 19$	✗
1 and 4	−5 and 1	$1(1) + 4(-5) = -19$	✗
2 and 2	1 and −5	$2(-5) + 2(1) = -8$	✗
2 and 2	−1 and 5	$2(5) + 2(-1) = 8$	✓

$-1(2x - 1)(2x + 5)$

24. $-5x^2 + x + 18$

$-1(5x^2 - x - 18)$

Factors of 5	Factors of −18	Outer + Inner	
1 and 5	1 and −18	$1(-18) + 5(1) = -13$	✗
1 and 5	−1 and 18	$1(18) + 5(-1) = 13$	✗
1 and 5	2 and −9	$1(-9) + 5(2) = 1$	✗
1 and 5	−2 and 9	$1(9) + 5(-2) = -1$	✓

$-1(x - 2)(5x + 9)$

Appendix: Essential Algebra Review

Review 13.4 / Operations with Polynomials and Factoring, Part 1

1. $(2t^2 - 8t) + (8t^2 + 9t)$
$= (2t^2 + 8t^2) + (-8t + 9t)$
$= 10t^2 + t$

2. $(-7x^2 - 2x + 3) + (4x^2 - 9x)$
$= (-7x^2 + 4x^2) + (-2x - 9x) + 3$
$= -3x^2 - 11x + 3$

3. $(x^5 - x) + (x^4 + x)$
$= (x^5 + x^4) + (-x + x)$
$= x^5 + x^4$

4. $(-2z^3 + z + 2z^3 + z) + (3z^3 - 5z^2)$
$= (-2z^3 + 2z^3 + 3z^3) + (-5z^2) + (z + z)$
$= 3z^3 - 5z^2 + 2z$

5. $(t^3 + 8t^2) - (3t^3)$
$= (t^3 + 8t^2) + (-3t^3)$
$= (t^3 - 3t^3) + 8t^2$
$= -2t^3 + 8t^2$

6. $(3x^2 - x) - (x^2 + 3x - x)$
$= (3x^2 - x) + (-x^2 - 3x + x)$
$= (3x^2 - x^2) + (-x - 3x + x)$
$= 2x^2 - 3x$

7. $(5m + 3) - (6m^3 - 2m^2)$
$= (5m + 3) + (-6m^3 + 2m^2)$
$= -6m^3 + 2m^2 + 5m + 3$

8. $(3s^2 + 4s) - (-10s^2 + 6s)$
$= (3s^2 + 4s) + (10s^2 - 6s)$
$= (3s^2 + 10s^2) + (4s - 6s)$
$= 13s^2 - 2s$

9. width $= (6w^2 + 8) - 2(w^2 - 3w + 2)$
$= (6w^2 + 8) + (-2(w^2) - 2(-3w) - 2(2))$
$= (6w^2 + 8) + (-2w^2 + 6w - 4)$
$= (6w^2 - 2w^2) + 6w + (8 - 4)$
$= 4w^2 + 6w + 4$

10. $P = 2\ell + 2w$
$= 2(4a + 3b) + 2(7a - 2b)$
$= 2(4a) + 2(3b) + 2(7a) + 2(-2b)$
$= 8a + 6b + 14a - 4b$
$= 8a + 14a + 6b - 4b$
$= 22a - 2b$

11. $(3x^2)(8x^5)$
$= (3 \cdot 8)(x^2 \cdot x^5)$
$= 24x^7$

12. $(-2r^3s^4)(6r^2s)$
$= (-2 \cdot 6)(r^3 \cdot r^2)(s^4 \cdot s)$
$= -12r^5s^5$

13. $(15xy^2)\left(\frac{1}{3}x^2z^3\right)(y^3z^4)$
$= \left(15 \cdot \frac{1}{3}\right)(x \cdot x^2)(y^2 \cdot y^3)(z^3 \cdot z^4)$
$= 5x^3y^5z^7$

14. $(-2a^3)(-5a)$
$= (-2 \cdot (-5))(a^3 \cdot a)$
$= 10a^4$

15. $(6x^3y^2)(-2x^2y)$
$= (6 \cdot (-2))(x^3 \cdot x^2)(y^2 \cdot y)$
$= -12x^5y^3$

16. $(-3a^2b)(-2b^3)(-a^3b^2)$
$= (-3 \cdot (-2) \cdot (-1))(a^2 \cdot a^3)(b \cdot b^3 \cdot b^2)$
$= -6a^5b^6$

Appendix: Essential Algebra Review

Review 13.4 / Operations with Polynomials and Factoring, Part 1

17. $(7x^2)(xy^5)(2x^3y^2)$
$= (7 \cdot 2)(x^2 \cdot x \cdot x^3)(y^5 \cdot y^2)$
$= 14x^6y^7$

18. $(-4a^3bc^2)(a^3b^2c)(3ab^4c^5)$
$= (-4 \cdot 3)(a^3 \cdot a^3 \cdot a)(b \cdot b^2 \cdot b^4)(c^2 \cdot c \cdot c^5)$
$= -12a^7b^7c^8$

19. $(12mn^2)(2m^2n)(mn)$
$= (12 \cdot 2)(m \cdot m^2 \cdot m)(n^2 \cdot n \cdot n)$
$= 24m^4n^4$

20. $\frac{35x^9}{7x^7} = \frac{35}{7} \cdot \frac{x^9}{x^7} = 5x^{9-7} = 5x^2$

21. $\frac{20n^{12}}{5n^4} = \frac{20}{5} \cdot \frac{n^{12}}{n^4} = 4n^{12-4} = 4n^8$

22. $\frac{8pq^8}{4pq^5} = \frac{8}{4} \cdot \frac{p}{p} \cdot \frac{q^8}{q^5} = 2p^{1-1}q^{8-5} = 2p^0q^3 = 2q^3$

23. $\frac{12a^5b^4}{8a^4b^{11}} = \frac{12}{8} \cdot \frac{a^5}{a^4} \cdot \frac{b^4}{b^{11}}$
$= \frac{3}{2} \cdot \frac{\cancel{aaaa}a}{\cancel{aaaa}} \cdot \frac{\cancel{bbbb}}{\cancel{bbbb}bbbbbbb}$
$= \frac{3}{2} \cdot \frac{a}{1} \cdot \frac{1}{bbbbbbb} = \frac{3}{2} \cdot a \cdot \frac{1}{b^7} = \frac{3a}{2b^7}$

24. $\frac{15x^2y^7}{10xy^2} = \frac{15}{10} \cdot \frac{x^2}{x} \cdot \frac{y^7}{y^2} = \frac{3}{2}x^{2-1}y^{7-2}$
$= \frac{3}{2}x^1y^5 = \frac{3}{2}xy^5$

25. $\frac{-25a^{10}b^3}{5a^3} = \frac{-25}{5} \cdot \frac{a^{10}}{a^3} \cdot b^3 = -5a^{10-3}b^3 = -5a^7b^3$

26. $2y\ ^{8} = 2\ ^{8}\ y\ ^{8} = 256y^8$

27. $7x^3\ ^{3} = 7\ ^{3}\ x^3\ ^{3} = 343\ x^{3\cdot3} = 343x^9$

28. $4b^2\ ^{4} = 4\ ^{4}\ b^2\ ^{4} = 256\ b^{2\cdot4} = 256b^8$

29. $5x^4y^5\ ^{3} = 5\ ^{3}\ x^4\ ^{3}\ y^5\ ^{3} = 125\ x^{4\cdot3}\ y^{5\cdot3}$
$= 125\ x^{12}\ y^{15} = 125x^{12}y^{15}$

30. $2a^3b^{10}\ ^{5} = 2\ ^{5}\ a^3\ ^{5}\ b^{10}\ ^{5} = 32\ a^{3\cdot5}\ b^{10\cdot5}$
$= 32\ a^{15}\ b^{50} = 32a^{15}b^{50}$

31. $-2p^3r\ ^{4} = -2\ ^{4}\ p^3\ ^{4}\ r\ ^{4} = 16\ p^{3\cdot4}\ r\ ^{4}$
$= 16\ p^{12}\ r^4 = 16p^{12}r^4$

32. $9s(s+6)$
$= 9s(s) + 9s(6)$
$= 9s^2 + 54s$

33. $9(2x^2 - 5x)$
$= 9(2x^2) + 9(-5x)$
$= 18x^2 - 45x$

34. $3x(9x^2 - 4x)$
$= 3x(9x^2) + 3x(-4x)$
$= 27x^3 - 12x^2$

35. $3(2x^2 + 5x + 4)$
$= 3(2x^2) + 3(5x) + 3(4)$
$= 6x^2 + 15x + 12$

36. $5s^2t^3(2s - 3t^2)$
$= 5s^2t^3(2s) + 5s^2t^3(-3t^2)$
$= (5 \cdot 2)(s^2 \cdot s)(t^3) + (5 \cdot (-3))(s^2)(t^3 \cdot t^2)$
$= 10s^3t^3 - 15s^2t^5$

Appendix: Essential Algebra Review

Review 13.4 / Operations with Polynomials and Factoring, Part 1

37. $x^2y^3 \cdot 5x^2y(6x + y^2)$
$= (5)(x^2 \cdot x^2)(y^3 \cdot y)(6x + y^2)$
$= 5x^4y^4(6x + y^2)$
$= 5x^4y^4(6x) + 5x^4y^4(y^2)$
$= (5 \cdot 6)(x^4 \cdot x)(y^4) + (5)(x^4)(y^4 \cdot y^2)$
$= 30x^5y^4 + 5x^4y^6$

38. $-5x(2x^2 - 3x - 1)$
$= -5x(2x^2) - 5x(-3x) - 5x(-1)$
$= -10x^3 + 15x^2 + 5x$

39. $-2a^2b^3(3ab^2 - a^2b)$
$= -2a^2b^3(3ab^2) - 2a^2b^3(-a^2b)$
$= (-2 \cdot 3)(a^2 \cdot a)(b^3 \cdot b^2) - (2 \cdot -1)(a^2 \cdot a^2)(b^3 \cdot b)$
$= -6a^3b^5 + 2a^4b^4$

40. $-7x^3y \cdot x^2y^2(2x - y)$
$= (-7)(x^3 \cdot x^2)(y \cdot y^2)(2x - y)$
$= -7x^5y^3(2x - y)$
$= -7x^5y^3(2x) - 7x^5y^3(-y)$
$= (-7 \cdot 2)(x^5 \cdot x)(y^3) + (-7 \cdot (-1))(x^5)(y^3 \cdot y)$
$= -14x^6y^3 + 7x^5y^4$

41. a. $A = \ell w$
$= (x + 3)(x)$
$= x(x) + 3(x)$
$= x^2 + 3x$
The area is represented by $x^2 + 3x$.

b. $A = x^2 + 3x$
$= (5)^2 + 3(5)$
$= 25 + 15 = 40$
The area is 40 ft^2.

42. $(x + 5)(x - 3)$
$= x(x) + x(-3) + 5(x) + 5(-3)$
$= x^2 - 3x + 5x - 15$
$= x^2 + 2x - 15$

43. $(5x - 2)(x + 3)$
$= 5x(x) + 5x(3) - 2(x) - 2(3)$
$= 5x^2 + 15x - 2x - 6$
$= 5x^2 + 13x - 6$

44. $(5x + 2)(2x - 1)$
$= 5x(2x) + 5x(-1) + 2(2x) + 2(-1)$
$= 10x^2 - 5x + 4x - 2$
$= 10x^2 - x - 2$

45. $(x - 1)(x - 2)$
$= x(x) + x(-2) - 1(x) - 1(-2)$
$= x^2 - 2x - x + 2$
$= x^2 - 3x + 2$

46. $(x - 8)(7x + 4)$
$= x(7x) + x(4) - 8(7x) - 8(4)$
$= 7x^2 + 4x - 56x - 32$
$= 7x^2 - 52x - 32$

47. $(2x + 7)(3x + 7)$
$= 2x(3x) + 2x(7) + 7(3x) + 7(7)$
$= 6x^2 + 14x + 21x + 49$
$= 6x^2 + 35x + 49$

48. $x + 3\ ^2 = x + 3 \quad x + 3$
$= x^2 + 3x + 3x + 9$
$= x^2 + 6x + 9$

Appendix: Essential Algebra Review

Review 13.4 / Operations with Polynomials and Factoring, Part 1

49. $2+3x\ ^{2} = \ 2+3x \quad 2+3x$

$= 4+6x+6x+9x^{2}$

$= 4+12x+9x^{2}$

50. $r^{2}+5t\ ^{2} = \ r^{2}+5t \quad r^{2}+5t$

$= r^{4}+5r^{2}t+5r^{2}t+25t^{2}$

$= r^{4}+10r^{2}t+25t^{2}$

51. $s^{2}-7\ ^{2} = \ s^{2}-7 \quad s^{2}-7$

$= s^{4}-7s^{2}-7s^{2}+49$

$= s^{4}-14s^{2}+49$

52. $2c-d^{3}\ ^{2} = \ 2c-d^{3} \quad 2c-d^{3}$

$= 4c^{2}-2cd^{3}-2cd^{3}+d^{6}$

$= 4c^{2}-4cd^{3}+d^{6}$

53. $1-x^{2}\ ^{2} = \ 1-x^{2} \quad 1-x^{2}$

$= 1-x^{2}-x^{2}+x^{4}$

$= 1-2x^{2}+x^{4}$

54. $a-10 \quad a+10 = a^{2}+10a-10a-100$

$= a^{2}-100$

55. $5a^{2}+9 \quad 5a^{2}-9 = 25a^{4}-45a^{2}+45a^{2}-81$

$= 25a^{4}-81$

56. $x^{3}+y^{2} \quad x^{3}-y^{2} = x^{6}-x^{3}y^{2}+x^{3}y^{2}-y^{4}$

$= x^{6}-y^{4}$

57. $A = \pi r^{2}$

$= \pi \ \ x+4\ ^{2}$

$= \pi \ \ x+4 \quad x+4$

$= \pi \ \ x^{2}+8x+16$

$= \pi x^{2}+8\pi x+16\pi$

Appendix: Essential Algebra Review

Review 13.4 / Operations with Polynomials and Factoring, Part 2

1. $8a^2 = 2 \cdot 2 \cdot 2 \cdot a \cdot a$
$11 = 1 \cdot 11$
The GCF of $8a^2$ and 11 is 1.

2. $9s = 3 \cdot 3 \cdot s$
$63s^3 = 3 \cdot 3 \cdot 7 \cdot s \cdot s \cdot s$
The GCF of $9s$ and $63s^3$ is $9s$.

3. $-64n^4 = -1 \cdot 2 \cdot 2 \cdot 2 \cdot 2 \cdot 2 \cdot 2 \cdot n \cdot n \cdot n \cdot n$
$24n^2 = 2 \cdot 2 \cdot 2 \cdot 3 \cdot n \cdot n$
The GCF of $-64n^4$ and $24n^2$ is $8n^2$.

4. $9y^2 = 3 \cdot 3 \cdot y \cdot y$
$45y = 3 \cdot 3 \cdot 5 \cdot y$
The GCF of $9y^2$ and $45y$ is $9y$.
$9y^2 + 45y$
$y(9y) + 5(9y)$
$9y(y + 5)$

5. $36d^3 = 2 \cdot 2 \cdot 3 \cdot 3 \cdot d \cdot d \cdot d$
$24 = 2 \cdot 2 \cdot 2 \cdot 3$
The GCF of $36d^3$ and 24 is 12.
$36d^2 + 24$
$3d^3(12) + 2(12)$
$12(3d^3 + 2)$

6. $14x^4 = 2 \cdot 7 \cdot x \cdot x \cdot x \cdot x$
$5x^2 = 5 \cdot x \cdot x$
The GCF of $14x^4$ and $5x^2$ is x^2.
$-14x^4 + 5x^2$
$-14x^2(x^2) + 5(x^2)$
$x^2(-14x^2 + 5)$

7. $15f = 3 \cdot 5 \cdot f$
$10f^2 = 2 \cdot 5 \cdot f \cdot f$
The GCF of $15f$ and $10f^2$ is $5f$.
$-15f - 10f^2$
$3(-5f) + 2f(-5f)$
$-5f(3 + 2f)$

8. $4d^4 = 2 \cdot 2 \cdot d \cdot d \cdot d \cdot d$
$d^3 = d \cdot d \cdot d$
$3d^2 = 3 \cdot d \cdot d$
The GCF of $4d^4$, d^3, and $3d^2$ is d^2.
$-4d^4 + d^3 - 3d^2$
$-4d^2(d^2) + d(d^2) - 3(d^2)$
$d^2(-4d^2 + d - 3)$

9. $14x^3 = 2 \cdot 7 \cdot x \cdot x \cdot x$
$63x^2 = 3 \cdot 3 \cdot 7 \cdot x \cdot x$
$7x = 7 \cdot x$
The GCF of $14x^2$, $63x^2$ and $7x$ is $7x$.
$14x^3 + 63x^2 - 7x$
$2x^2(7x) + 9x(7x) - 1(7x)$
$7x(2x^2 + 9x - 1)$

10. $21c^2 = 3 \cdot 7 \cdot c \cdot c$
$14c = 2 \cdot 7 \cdot c$
The GCF of $21c^2$ and $14c$ is $7c$.
$21c^2 + 14c$
$3c(7c) + 2(7c)$
$7c(3c + 2)$

11. $33d^3 = 3 \cdot 11 \cdot d \cdot d \cdot d$
$22d = 2 \cdot 11 \cdot d$
$11 = 11$
The GCF of $33d^3$, $22d$ and 11 is 11.
$33d^3 + 22d + 11$
$3d^3(11) + 2d(11) + 1(11)$
$11(3d^3 + 2d + 1)$

12. $5g^3 = 5 \cdot g \cdot g \cdot g$
$15g^2 = 3 \cdot 5 \cdot g \cdot g$
The GCF of $5g^3$ and $15g^2$ is $5g^2$.
$-5g^3 - 15g^2$
$g(-5g^2) + 3(-5g^2)$
$-5g^2(g + 3)$

13. $S = P + Prt$
$= 1(P) + rt(P)$
$= P(1 + rt)$
$P(1 + rt)$ is the factored expression for $P + Prt$.

14. Cannot be factored

Appendix: Essential Algebra Review

Review 13.4 / Operations with Polynomials and Factoring, Part 2

15. $-4x(x+2)+9(x+2)$
$(x+2)(-4x+9)$

16. $6y(y-7)+(y-7)$
$6y(y-7)+1(y-7)$
$(6y+1)(y-7)$

17. $a(x-3)+2b(x-3)$
$(a+2b)(x-3)$

18. $-3(2+b)+4b(b+2)$
$(-3+4b)(b+2)$

19. $5(3x-2)+x(3x-2)$
$(5+x)(3x-2)$

20. $x^2+13x+30$
$(x+1)(x+30)=x^2+31x+30$ ✗
$(x+2)(x+15)=x^2+17x+30$ ✗
$(x+3)(x+10)=x^2+13x+30$ ✓
The factors of $x^2+13x+30$ are $(x+3)$ and $(x+10)$.
$x^2+13x+30=(x+3)(x+10)$

21. $x^2+11x+28$
$(x+1)(x+28)=x^2+29x+28$ ✗
$(x+2)(x+14)=x^2+16x+28$ ✗
$(x+4)(x+7)=x^2+11x+28$ ✓
The factors of $x^2+11x+28$ are $(x+4)$ and $(x+7)$.
$x^2+11x+28=(x+4)(x+7)$

22. $x^2+16x+48$
$(x+1)(x+48)=x^2+49x+48$ ✗
$(x+2)(x+24)=x^2+26x+48$ ✗
$(x+3)(x+16)=x^2+19x+48$ ✗
$(x+4)(x+12)=x^2+16x+48$ ✓
The factors of $x^2+16x+48$ are $(x+4)$ and $(x+12)$.
$x^2+16x+48=(x+4)(x+12)$

23. $x^2+12x+11$

Factors of 11	Sum
1 and 11	12 ✓

$(x+1)(x+11)$

24. $x^2+16x+28$

Factors of 28	Sum
1 and 28	29 ✗
2 and 14	16 ✓

$(x+2)(x+14)$

25. $x^2+15x+36$

Factors of 36	Sum
1 and 36	37 ✗
2 and 18	20 ✗
3 and 12	15 ✓

$(x+3)(x+12)$

26. x^2-6x+5

Factors of 5	Sum
−1 and −5	−6 ✓

$(x-1)(x-5)$

27. $x^2-9x+18$

Factors of 18	Sum
−1 and −18	−19 ✗
−2 and −9	−11 ✗
−3 and −6	−9 ✓

$(x-3)(x-6)$

28. $x^2-12x+32$

Factors of 32	Sum
−1 and −32	−33 ✗
−2 and −16	−18 ✗
−4 and −8	−12 ✓

$(x-4)(x-8)$

29. x^2+x-12

Factors of −12	Sum
−1 and 12	11 ✗
−2 and 6	4 ✗
−3 and 4	1 ✓

$(x-3)(x+4)$

30. $x^2+4x-21$

Factors of −21	Sum
−1 and 21	20 ✗
−3 and 7	4 ✓

$(x-3)(x+7)$

Appendix: Essential Algebra Review

Review 13.4 / Operations with Polynomials and Factoring, Part 2

31. $x^2 + 9x - 36$

Factors of −36	Sum
−1 and 36	35 ✗
−2 and 18	16 ✗
−3 and 12	9 ✓

$(x - 3)(x + 12)$

32. $x^2 - 12x - 13$

Factors of −13	Sum
1 and −13	−12 ✓

$(x + 1)(x - 13)$

33. $x^2 - 10x - 24$

Factors of −24	Sum
1 and −24	−23 ✗
2 and −12	−10 ✓

$(x + 2)(x - 12)$

34. $x^2 - 2x - 35$

Factors of −35	Sum
1 and −35	−33 ✗
5 and −7	−2 ✓

$(x + 5)(x - 7)$

35. $n^2 - 12n - 45$

Factors of −45	Sum
1 and −45	−44 ✗
3 and −15	−12 ✓

$(n + 3)(n - 15)$

n	$n^2 - 12n - 45$
0	$0^2 - 12(0) - 45 = -45$
1	$1^2 - 12(1) - 45 = -56$
2	$2^2 - 12(2) - 45 = -65$
3	$3^2 - 12(3) - 45 = -72$
4	$4^2 - 12(4) - 45 = -77$

n	$(n + 3)(n - 15)$
0	$(0 + 3)(0 - 15) = -45$
1	$(1 + 3)(1 - 15) = -56$
2	$(2 + 3)(2 - 15) = -65$
3	$(3 + 3)(3 - 15) = -72$
4	$(4 + 3)(4 - 15) = -77$

36. $9x^2 + 9x + 2$

$(1x + 2)(9x + 1) = 9x^2 + 19x + 2$ ✗

$(1x + 1)(9x + 2) = 9x^2 + 11x + 2$ ✗

$(3x + 2)(3x + 1) = 9x^2 + 9x + 2$ ✓

The factors of $9x^2 + 9x + 2$ are $(3x + 2)$ and $(3x + 1)$.

$9x^2 + 9x + 2 = (3x + 2)(3x + 1)$

37. $2x^2 + 7x + 5$

$(1x + 5)(2x + 1) = 2x^2 + 11x + 5$ ✗

$(1x + 1)(2x + 5) = 2x^2 + 7x + 5$ ✓

The factors of $2x^2 + 7x + 5$ are $(x + 1)$ and $(2x + 5)$.

$2x^2 + 7x + 5 = (x + 1)(2x + 5)$

38. $10d^2 + 17d + 7$

$(1d + 7)(10d + 1) = 10d^2 + 71d + 7$ ✗

$(1d + 1)(10d + 7) = 10d^2 + 17d + 7$ ✓

The factors of $10d^2 + 17d + 7$ are $(d + 1)$ and $(10d + 7)$.

$10d^2 + 17d + 7 = (d + 1)(10d + 7)$

Appendix: Essential Algebra Review

Review 13.4 / Operations with Polynomials and Factoring, Part 2

39. $6x^2 + 14x + 4$

$(1x + 4)(6x + 1) = 6x^2 + 25x + 4$ ✗

$(1x + 2)(6x + 2) = 6x^2 + 14x + 4$ ✓

The factors of $6x^2 + 14x + 4$ are $(x + 2)$ and $(6x + 2)$.

$6x^2 + 14x + 4 = 2(3x + 1)(x + 2)$

40. $6x^2 - 13x + 6$

$(1x - 6)(6x - 1) = 6x^2 - 37x + 6$ ✗

$(1x - 3)(6x - 2) = 6x^2 - 20x + 6$ ✗

$(1x - 2)(6x - 3) = 6x^2 - 15x + 6$ ✗

$(1x - 1)(6x - 6) = 6x^2 - 12x + 6$ ✗

$(2x - 6)(3x - 1) = 6x^2 - 20x + 6$ ✗

$(2x - 3)(3x - 2) = 6x^2 - 13x + 6$ ✓

The factors of $6x^2 - 13x + 6$ are $(2x - 3)$ and $(3x - 2)$.

$6x^2 - 13x + 6 = (2x - 3)(3x - 2)$

41. $5x^2 + 9x - 18$

$(1x + 18)(5x - 1) = 5x^2 + 89x - 18$ ✗

$(1x + 9)(5x - 2) = 5x^2 + 43x - 18$ ✗

$(1x + 6)(5x - 3) = 5x^2 + 27x - 18$ ✗

$(1x + 3)(5x - 6) = 5x^2 + 9x - 18$ ✓

The factors of $5x^2 + 9x - 18$ are $(x + 3)$ and $(5x - 6)$.

$5x^2 + 9x - 18 = (x + 3)(5x - 6)$

42. $6x^2 + 23x + 7$

Factors of 6	Factors of 7	Outer + Inner	
1 and 6	1 and 7	$1(7) + 6(1) = 13$	✗
1 and 6	7 and 1	$1(1) + 6(7) = 43$	✗
2 and 3	1 and 7	$2(7) + 3(1) = 17$	✗
2 and 3	7 and 1	$2(1) + 3(7) = 23$	✓

$(2x + 7)(3x + 1)$

43. $10n^2 - 17n + 7$

Factors of 10	Factors of 7	Outer + Inner
1 and 10	−1 and −7	$1(-7) + 10(-1) = -17$ ✓

$(n - 1)(10n - 7)$

44. $3x^2 + 11x + 6$

Factors of 3	Factors of 6	Outer + Inner	
1 and 3	1 and 6	$1(6) + 3(1) = 9$	✗
1 and 3	6 and 1	$1(1) + 3(6) = 19$	✗
1 and 3	2 and 3	$1(3) + 3(2) = 9$	✗
1 and 3	3 and 2	$1(2) + 3(3) = 11$	✓

$(x + 3)(3x + 2)$

45. $7x^2 + 15x + 2$

Factors of 7	Factors of 2	Outer + Inner	
1 and 7	1 and 2	$1(2) + 7(1) = 9$	✗
1 and 7	2 and 1	$1(1) + 7(2) = 15$	✓

$(x + 2)(7x + 1)$

46. $3n^2 + 4n + 1$

Factors of 3	Factors of 1	Outer + Inner	
1 and 3	1 and 1	$1(1) + 3(1) = 4$	✓

$(n + 1)(3n + 1)$

47. $3x^2 - 19x + 20$

Factors of 3	Factors of 20	Outer + Inner
1 and 3	−1 and −20	$1(-20) + 3(-1) = -23$ ✗
1 and 3	−20 and −1	$1(-1) + 3(-20) = -61$ ✗
1 and 3	−2 and −10	$1(-10) + 3(-2) = -16$ ✗
1 and 3	−10 and −2	$1(-2) + 3(-10) = -32$ ✗
1 and 3	−4 and −5	$1(-5) + 3(-4) = -17$ ✗
1 and 3	−5 and −4	$1(-4) + 3(-5) = -19$ ✓

$(x - 5)(3x - 4)$

48. $6x^2 + 11x + 4$

Factors of 6	Factors of 4	Outer + Inner	
1 and 6	1 and 4	$1(4) + 6(1) = 10$	✗
1 and 6	4 and 1	$1(1) + 6(4) = 25$	✗
1 and 6	2 and 2	$1(2) + 6(2) = 14$	✗
2 and 3	1 and 4	$2(4) + 3(1) = 11$	✓

$(2x + 1)(3x + 4)$

Appendix: Essential Algebra Review

Review 13.4 / Operations with Polynomials and Factoring, Part 2

49. $4x^2 - 31x + 21$

Factors of 4	Factors of 21	Outer + Inner
1 and 4	−1 and −21	$1(-21) + 4(-1) = -25$ ✗
1 and 4	−21 and −1	$1(-1) + 4(-21) = -85$ ✗
1 and 4	−3 and −7	$1(-7) + 4(-3) = -19$ ✗
1 and 4	−7 and −3	$1(-3) + 4(-7) = -31$ ✓

$(x - 7)(4x - 3)$

50. $10x^2 + 31x + 15$

Factors of 10	Factors of 15	Outer + Inner
1 and 10	1 and 15	$1(15) + 10(1) = 25$ ✗
1 and 10	15 and 1	$1(1) + 10(15) = 151$ ✗
1 and 10	3 and 5	$1(5) + 10(3) = 35$ ✗
1 and 10	5 and 3	$1(3) + 10(5) = 53$ ✗
2 and 5	1 and 15	$2(15) + 5(1) = 35$ ✗
2 and 5	15 and 1	$2(1) + 5(15) = 77$ ✗
2 and 5	3 and 5	$2(5) + 5(3) = 25$ ✗
2 and 5	5 and 3	$2(3) + 5(5) = 31$ ✓

$(2x + 5)(5x + 3)$

51. $2n^2 - 7n - 4$

Factors of 2	Factors of −4	Outer + Inner
1 and 2	1 and −4	$1(-4) + 2(1) = -2$ ✗
1 and 2	−1 and 4	$1(4) + 2(-1) = 2$ ✗
1 and 2	2 and −2	$1(-2) + 2(2) = 2$ ✗
1 and 2	−2 and 2	$1(2) + 2(-2) = -2$ ✗
1 and 2	4 and −1	$1(-1) + 2(4) = 7$ ✗
1 and 2	−4 and 1	$1(1) + 2(-4) = -7$ ✓

$(n - 4)(2n + 1)$

52. $3x^2 - 4x - 15$

Factors of 3	Factors of −15	Outer + Inner
1 and 3	1 and −15	$1(-15) + 3(1) = -12$ ✗
1 and 3	−1 and 15	$1(15) + 3(-1) = 12$ ✗
1 and 3	3 and −5	$1(-5) + 3(3) = 4$ ✗
1 and 3	−3 and 5	$1(5) + 3(-3) = -4$ ✓

$(x - 3)(3x + 5)$

53. $3n^2 - n - 4$

Factors of 3	Factors of −4	Outer + Inner
1 and 3	1 and −4	$1(-4) + 3(1) = -1$ ✓

$(n + 1)(3n - 4)$

54. $-4x^2 - 4x + 15$

$-1(4x^2 + 4x - 15)$

Factors of 4	Factors of −15	Outer + Inner
1 and 4	1 and −15	$1(-15) + 4(1) = -11$ ✗
1 and 4	−1 and 15	$1(15) + 4(-1) = 11$ ✗
1 and 4	3 and −5	$1(-5) + 4(3) = 7$ ✗
1 and 4	−3 and 5	$1(5) + 4(-3) = -7$ ✗
1 and 4	5 and −3	$1(-3) + 4(5) = 17$ ✗
1 and 4	−5 and 3	$1(3) + 4(-5) = -17$ ✗
1 and 4	15 and −1	$1(-1) + 4(15) = 59$ ✗
1 and 4	−15 and 1	$1(1) + 4(-15) = -59$ ✗
2 and 2	1 and −15	$2(-15) + 2(1) = -28$ ✗
2 and 2	−1 and 15	$2(15) + 2(-1) = 28$ ✗
2 and 2	3 and −5	$2(-5) + 2(3) = -4$ ✗
2 and 2	−3 and 5	$2(5) + 2(-3) = 4$ ✓

$-1(2x - 3)(2x + 5)$

55. $-3x^2 + 16x - 16$

$-1(3x^2 - 16x + 16)$

Factors of 3	Factors of 16	Outer + Inner
1 and 3	−1 and −16	$1(-16) + 3(-1) = -19$ ✗
1 and 3	−16 and −1	$1(-1) + 3(-16) = -49$ ✗
1 and 3	−2 and −8	$1(-8) + 3(-2) = -14$ ✗
1 and 3	−8 and −2	$1(-2) + 2(-8) = -18$ ✗
1 and 3	−4 and −4	$1(-4) + 3(-4) = -16$ ✓

$-1(x - 4)(3x - 4)$

56. $-3x^2 - x + 2$

$-1(3x^2 + x - 2)$

Factors of 3	Factors of −2	Outer + Inner
1 and 3	1 and −2	$1(-2) + 3(1) = 1$ ✓

$-1(x + 1)(3x - 2)$

Appendix: Essential Algebra Review

Radicals and Solving Quadratics

Radical Expressions

1. $\sqrt{81} = 9$

2. $\sqrt{\dfrac{98}{2}} = \sqrt{49}$
   $= 7$

3. $\sqrt{(a+7)^2} = |a+7|$

4. $\sqrt{180} = \sqrt{(36)5}$
   $= \sqrt{36}\sqrt{5}$
   $= 6\sqrt{5}$

5. $\sqrt{40} = \sqrt{(4)10}$
   $= \sqrt{4}\sqrt{10}$
   $= 2\sqrt{10}$

6. $\sqrt{648} = \sqrt{(324)2}$
   $= 18\sqrt{2}$

7. $\sqrt{m^5n^3} = \sqrt{m^5}\sqrt{n^3}$
   $= \sqrt{m^4}\sqrt{n^2}\sqrt{mn}$
   $= m^2n\sqrt{mn}$

8. $\sqrt{32x^4y^3} = \sqrt{16(2)x^4y^2y}$
   $= 4x^2y\sqrt{2y}$

9. $\sqrt{200a^2b} = \sqrt{(2)100a^2b}$
   $= \sqrt{100a^2}\sqrt{2b}$
   $= 10a\sqrt{2b}$

10. $\sqrt{\dfrac{17}{25}} = \dfrac{\sqrt{17}}{\sqrt{25}}$
    $= \dfrac{\sqrt{17}}{5}$

11. $\sqrt{\dfrac{7}{16}} = \dfrac{\sqrt{7}}{\sqrt{16}}$
    $= \dfrac{\sqrt{7}}{4}$

12. $\sqrt{\dfrac{6}{49}} = \dfrac{\sqrt{6}}{\sqrt{49}}$
    $= \dfrac{\sqrt{6}}{7}$

13. $\sqrt{\dfrac{b}{c^2}} = \dfrac{\sqrt{b}}{\sqrt{c^2}}$
    $= \dfrac{\sqrt{b}}{c}$

14. $\sqrt{\dfrac{4x^2}{36x}} = \sqrt{\dfrac{x}{9}}$
    $= \dfrac{\sqrt{x}}{\sqrt{9}}$
    $= \dfrac{\sqrt{x}}{3}$

15. $\sqrt{\dfrac{7a^4}{9a^3}} = \sqrt{\dfrac{7a}{9}}$
    $= \dfrac{\sqrt{7a}}{\sqrt{9}}$
    $= \dfrac{\sqrt{7a}}{3}$

16. $\sqrt{\dfrac{108}{49}} = \dfrac{\sqrt{(36)3}}{\sqrt{49}}$
    $= \dfrac{\sqrt{36}\sqrt{3}}{7}$
    $= \dfrac{6\sqrt{3}}{7}$

Appendix: Essential Algebra Review

Radicals and Solving Quadratics

Radical Expressions

17. $\sqrt{\frac{204}{25}} = \frac{\sqrt{(4)51}}{\sqrt{25}}$
$= \frac{\sqrt{4}\sqrt{51}}{5}$
$= \frac{2\sqrt{51}}{5}$

18. $\sqrt{\frac{512}{81}} = \frac{\sqrt{256(2)}}{\sqrt{81}}$
$= \frac{\sqrt{256}\sqrt{2}}{\sqrt{81}}$
$= \frac{16\sqrt{2}}{9}$

19. $\sqrt{\frac{1}{36x^2}} = \frac{\sqrt{1}}{\sqrt{36}\sqrt{x^2}}$
$= \frac{1}{6x}$

20. $\sqrt{\frac{50x^2}{169}} = \frac{\sqrt{(25)2x^2}}{\sqrt{169}}$
$= \frac{\sqrt{25}\sqrt{2}\sqrt{x^2}}{\sqrt{169}}$
$= \frac{5x\sqrt{2}}{13}$

21. $\sqrt{\frac{72x^7}{4x^4}} = \sqrt{18x^3}$
$= \sqrt{9}\sqrt{x^2}\sqrt{2x}$
$= 3x\sqrt{2x}$

22. $c = \sqrt{a^2 + b^2}$
$= \sqrt{20^2 + 25^2}$
$= \sqrt{25(16 + 25)}$
$= \sqrt{25}\sqrt{41}$
$= 5\sqrt{41}$ mi ≈ 32 mi

Appendix: Essential Algebra Review

Radicals and Solving Quadratics

Solving Quadratic Equations by Factoring

1. $(x+2)(x-8)=0$
$x+2=0$ or $x-8=0$
$x=-2$ or $x=8$
The solutions are -2 and 8.

2. $(x-6)(x-5)=0$
$x-6=0$ or $x-5=0$
$x=6$ or $x=5$
The solutions are 6 and 5.

3. $(x+7)(x+9)=0$
$x+7=0$ or $x+9=0$
$x=-7$ or $x=-9$
The solutions are -7 and -9.

4. $(x)(x-1)=0$
$x=0$ or $x-1=0$
$x=1$
The solutions are 0 and 1.

5. $(x)(x+11)=0$
$x=0$ or $x+11=0$
$x=-11$
The solutions are 0 and -11.

6. $(3x+2)(4x-1)=0$
$3x+2=0$ or $4x-1=0$
$3x=-2$ or $4x=1$
$x=-\frac{2}{3}$ or $x=\frac{1}{4}$
The solutions are $-\frac{2}{3}$ and $\frac{1}{4}$.

7. $x^2+4x-12=0$
$(x+6)(x-2)=0$
$x+6=0$ or $x-2=0$
$x=-6$ or $x=2$
The solutions are -6 and 2.

8. $x^2-8x-9=0$
$(x+1)(x-9)=0$
$x+1=0$ or $x-9=0$
$x=-1$ or $x=9$
The solutions are -1 and 9.

9. $x^2-5x+6=0$
$(x-2)(x-3)=0$
$x-2=0$ or $x-3=0$
$x=2$ or $x=3$
The solutions are 2 and 3.

10.
$$\begin{aligned} x^2-3x &= 10 \\ -10 \quad &\ \ -10 \\ x^2-3x-10 &= 0 \\ (x+2)(x-5) &= 0 \end{aligned}$$
$x+2=0$ or $x-5=0$
$x=-2$ or $x=5$
The solutions are -2 and 5.

11.
$$\begin{aligned} x^2+10x &= -16 \\ +16 \quad &\ \ +16 \\ x^2+10x+16 &= 0 \\ (x+8)(x+2) &= 0 \end{aligned}$$
$x+8=0$ or $x+2=0$
$x=-8$ or $x=-2$
The solutions are -8 and -2.

12.
$$\begin{aligned} x^2+2x &= 15 \\ -15 \quad &\ \ -15 \\ x^2+2x-15 &= 0 \\ (x+5)(x-3) &= 0 \end{aligned}$$
$x+5=0$ or $x-3=0$
$x=-5$ or $x=3$
The solutions are -5 and 3.

13. $x^2-8x+16=0$
$(x-4)(x-4)=0$
$x-4=0$
$x=4$
The only solution is 4.

14.
$$\begin{aligned} -3x^2 &= 18x+27 \\ +3x^2 \quad &\ \ +3x^2 \\ 0 &= 3x^2+18x+27 \\ 0 &= 3(x^2+6x+9) \\ 0 &= 3(x+3)(x+3) \end{aligned}$$
$3\neq 0$ or $x+3=0$
$x=-3$
The only solution is -3.

15.
$$\begin{aligned} x^2+36 &= 12x \\ -12x \quad &\ \ -12x \\ x^2-12x+36 &= 0 \\ (x-6)(x-6) &= 0 \end{aligned}$$
$x-6=0$
$x=6$
The only solution is 6.

Solving Quadratic Equations by Factoring

16. $x^2 + 14x + 49 = 0$

$(x + 7)(x + 7) = 0$

$x + 7 = 0$

$x = -7$

The only solution is -7.

17. $x^2 - 16x + 64 = 0$

$(x - 8)(x - 8) = 0$

$x - 8 = 0$

$x = 8$

The only solution is 8.

18.

$$\begin{aligned} 2x^2 + 6x &= -18 \\ +18 &\quad +18 \\ 2x^2 + 6x + 18 &= 0 \\ 2(x^2 + 3x + 9) &= 0 \end{aligned}$$

Since $2 \neq 0$ and since $y = x^2 + 3x + 9$ opens upward and has its vertex at $(-1.5, 6.75)$, it does not intersect the x-axis. So there are no real solutions.

19. $h = -16t^2 + 14t + 2$

$0 = -16t^2 + 14t + 2$

$0 = -2(8t^2 - 7t - 1)$

$0 = -2(8t + 1)(t - 1)$

$-2 \neq 0,\ 8t + 1 = 0$ or $t - 1 = 0$

$8t = -1$ or $t = 1$

$t = -\frac{1}{8}$ ✗

The beanbag takes 1 s to reach the ground.

Appendix: Essential Algebra Review

Radicals and Solving Quadratics

Solving Quadratic Equations by Using Square Roots

1. $x^2 = 225$

$x = \pm\sqrt{225}$

$x = \pm 15$

The solutions are 15 and −15.

2. $x^2 = 49$

$x = \pm\sqrt{49}$

$x = \pm 7$

The solutions are 7 and −7.

3. $x^2 = -100$

$x \neq \pm\sqrt{100}$

There are no real solutions.

4. $x^2 = 400$

$x = \pm\sqrt{400}$

$x = \pm 20$

The solutions are 20 and −20.

5. $-25 = x^2$

$\pm\sqrt{25} \neq x$

There are no real solutions.

6. $36 = x^2$

$\pm\sqrt{36} = x$

$\pm 6 = x$

The solutions are 6 and −6.

7. $3x^2 - 75 = 0$

$+75 \quad +75$

$\frac{3x^2}{3} = \frac{75}{3}$

$x^2 = 25$

$x = \pm\sqrt{25}$

$x = \pm 5$

The solutions are 5 and −5.

8. $0 = 81x^2 - 25$

$+25 \quad + 25$

$\frac{25}{81} = \frac{81x^2}{81}$

$\frac{25}{81} = x^2$

$\pm\sqrt{\frac{25}{81}} = x$

$\pm\frac{5}{9} = x$

The solutions are $\frac{5}{9}$ and $-\frac{5}{9}$.

9. $49x^2 + 64 = 0$

$- 64 \quad -64$

$\frac{49x^2}{49} = \frac{-64}{49}$

$x^2 = -\frac{64}{49}$

$x \neq \pm\sqrt{\frac{64}{49}}$

There are no real solutions.

10. $16x^2 + 10 = 131$

$- 10 \quad -10$

$\frac{16x^2}{16} = \frac{121}{16}$

$x^2 = \frac{121}{16}$

$x = \pm\sqrt{\frac{121}{16}}$

$x = \pm\frac{11}{4}$

The solutions are $\frac{11}{4}$ and $-\frac{11}{4}$.

11. $0 = 4x^2 - 16$

$+16 \quad + 16$

$\frac{16}{4} = \frac{4x^2}{4}$

$4 = x^2$

$\pm\sqrt{4} = x$

$\pm 2 = x$

The solutions are 2 and −2.

Appendix: Essential Algebra Review

Radicals and Solving Quadratics

Solving Quadratic Equations by Using Square Roots

12.
$$100x^2 + 26 = 10$$
$$-26 \quad -26$$
$$\frac{100x^2}{100} = \frac{-16}{100}$$
$$x^2 = -\frac{16}{100}$$
$$x \neq \pm\sqrt{\frac{16}{100}}$$
There are no real solutions.

13.
$$3x^2 = 81$$
$$\frac{3x^2}{3} = \frac{81}{3}$$
$$x^2 = 27$$
$$x = \pm\sqrt{27}$$
$$x \approx \pm 5.20$$
The approximate solutions are 5.20 and −5.20.

14.
$$0 = x^2 - 60$$
$$+60 \quad +60$$
$$60 = x^2$$
$$\pm\sqrt{60} = x$$
$$x \approx 7.75$$
The approximate solutions are 7.75 and −7.75.

15.
$$100 - 5x^2 = 0$$
$$+5x^2 \quad +5x^2$$
$$\frac{100}{5} = \frac{5x^2}{5}$$
$$20 = x^2$$
$$\pm\sqrt{20} = x$$
$$x \approx \pm 4.47$$
The approximate solutions are 4.47 and −4.47.

16.
$$A = \ell w$$
$$170 = (3w)w$$
$$170 = 3w^2$$
$$\frac{170}{3} = \frac{3w^2}{3}$$
$$\frac{170}{3} = w^2$$
$$\pm\sqrt{\frac{170}{3}} = w$$
$$w \approx \pm 7.5$$
Negative numbers are not reasonable for length, so $w \approx 7.5$ is the only solution that makes sense. Therefore, the width of the rectangle is 7.5 m.

Appendix: Essential Algebra Review

Review 13.5 / Radicals and Solving Quadratics

1. $\sqrt{100} = 10$

2. $\sqrt{\frac{800}{2}} = \sqrt{400}$
$= 20$

3. $\sqrt{3^2 + 4^2} = \sqrt{25}$
$= 5$

4. $\sqrt{3 \cdot 27} = \sqrt{81}$
$= 9$

5. $\sqrt{a^4} = a^2$

6. $\sqrt{(x+1)^2} = |x+1|$

7. $\sqrt{(5-x)^2} = |5-x|$

8. $\sqrt{(x-3)^2} = |x-3|$

9. $\sqrt{125} = \sqrt{25(5)}$
$= \sqrt{25}\sqrt{5}$
$= 5\sqrt{5}$

10. $\sqrt{4000} = \sqrt{10(400)}$
$= 20\sqrt{10}$

11. $\sqrt{216a^2b^2} = \sqrt{36a^2b^2}\sqrt{6}$
$= 6ab\sqrt{6}$

12. $\sqrt{320r^2s^2} = \sqrt{(5)64r^2s^2}$
$= 8rs\sqrt{5}$

13. $\sqrt{\frac{15}{64}} = \frac{\sqrt{15}}{\sqrt{64}}$
$= \frac{\sqrt{15}}{8}$

14. $\sqrt{\frac{45}{4}} = \frac{\sqrt{9(5)}}{\sqrt{4}}$
$= \frac{\sqrt{9}\sqrt{5}}{2}$
$= \frac{3\sqrt{5}}{2}$

15. $\sqrt{\frac{64a^4}{4a^6}} = \sqrt{\frac{16}{a^2}}$
$= \frac{\sqrt{16}}{\sqrt{a^2}}$
$= \frac{4}{a}$

16. $\sqrt{\frac{14z^3}{9z^3}} = \sqrt{\frac{14}{9}}$
$= \frac{\sqrt{14}}{\sqrt{9}}$
$= \frac{\sqrt{14}}{3}$

17. $\sqrt{\frac{128}{81}} = \frac{\sqrt{128}}{\sqrt{81}}$
$= \frac{\sqrt{64(2)}}{\sqrt{81}}$
$= \frac{8\sqrt{2}}{9}$

18. $\sqrt{\frac{x^3}{y^6}} = \frac{\sqrt{x^2x}}{\sqrt{y^6}}$
$= \frac{x\sqrt{x}}{y^3}$

19. $\sqrt{\frac{150}{196x^2}} = \frac{\sqrt{6(25)}}{\sqrt{196x^2}}$
$= \frac{\sqrt{25}\sqrt{6}}{14x}$
$= \frac{5\sqrt{6}}{14x}$

20. $\sqrt{\frac{192s^3}{49s}} = \sqrt{\frac{64(3)s^2}{49}}$
$= \frac{\sqrt{64s^2}\sqrt{3}}{\sqrt{49}}$
$= \frac{8s\sqrt{3}}{7}$

21. $t = \sqrt{\frac{d}{16}}$
$= \sqrt{\frac{160}{16}}$
$= \sqrt{10}$ s; 3.2 s

22. $(x - 8)(x + 6) = 0$
$x - 8 = 0$ or $x + 6 = 0$
$x = 8$ or $x = -6$
The solutions are 8 and −6.

23. $(x + 4)(x + 7) = 0$
$x + 4 = 0$ or $x + 7 = 0$
$x = -4$ or $x = -7$
The solutions are −4 and −7.

24. $(x - 2)(x - 5) = 0$
$x - 2 = 0$ or $x - 5 = 0$
$x = 2$ or $x = 5$
The solutions are 2 and 5.

25. $(x - 9)(x) = 0$
$x - 9 = 0$ or $x = 0$
$x = 9$
The solutions are 9 and 0.

26. $(x)(x + 25) = 0$
$x = 0$ or $x + 25 = 0$
$x = -25$
The solutions are 0 and −25.

27. $(2x + 1)(3x - 1) = 0$
$2x + 1 = 0$ or $3x - 1 = 0$
$2x = -1$ or $3x = 1$
$x = -\frac{1}{2}$ or $x = \frac{1}{3}$
The solutions are $-\frac{1}{2}$ and $\frac{1}{3}$.

28. $x^2 + 8x + 15 = 0$
$(x + 5)(x + 3) = 0$
$x + 5 = 0$ or $x + 3 = 0$
$x = -5$ or $x = -3$
The solutions are −5 and −3.

29. $x^2 - 2x - 8 = 0$
$(x + 2)(x - 4) = 0$
$x + 2 = 0$ or $x - 4 = 0$
$x = -2$ or $x = 4$
The solutions are −2 and 4.

30. $x^2 - 4x + 3 = 0$
$(x - 1)(x - 3) = 0$
$x - 1 = 0$ or $x - 3 = 0$
$x = 1$ or $x = 3$
The solutions are 1 and 3.

31. $x^2 + 10x + 25 = 0$
$(x + 5)(x + 5) = 0$
$x + 5 = 0$
$x = -5$
The only solution is −5.

32. $x^2 - x = 12$
$\underline{-12} \quad \underline{-12}$
$x^2 - x - 12 = 0$
$(x + 3)(x - 4) = 0$
$x + 3 = 0$ or $x - 4 = 0$
$x = -3$ or $x = 4$
The solutions are −3 and 4.

33. $-x^2 = 4x + 4$
$\underline{+x^2} \quad \underline{+x^2}$
$0 = x^2 + 4x + 4$
$0 = (x + 2)(x + 2)$
$x + 2 = 0$
$x = -2$
The only solution appears to be −2.

34. $h = -16t^2 + 95t + 6$
$0 = -16t^2 + 95t + 6$
$0 = -1(16t^2 - 95t - 6)$
$0 = -1(16t + 1)(t - 6)$
$-1 \neq 0$, $16t + 1 = 0$ or $t - 6 = 0$
$16t = -1$ or $t = 6$
$t = -\frac{1}{16}$ ✘
It takes the flare 6 s to reach the ground.

35. $x^2 = 169$
$x = \pm\sqrt{169}$
$x = \pm 13$
The solutions are 13 and −13.

36. $x^2 = 25$
$x = \pm\sqrt{25}$
$x = \pm 5$
The solutions are 5 and −5.

Appendix: Essential Algebra Review

Review 13.5 / Radicals and Solving Quadratics

37. $x^2 = -36$

$x \neq \pm\sqrt{36}$

There are no real solutions.

38. $x^2 = 10{,}000$

$x = \pm\sqrt{10{,}000}$

$x = \pm 100$

The solutions are 100 and -100.

39. $-121 = x^2$

$\pm\sqrt{121} \neq x$

There are no real solutions.

40. $625 = x^2$

$\pm\sqrt{625} = x$

$\pm 25 = x$

The solutions are 25 and -25.

41. $4 - 81x^2 = 0$

$+81x^2 \quad +81x^2$

$\frac{4}{81} = \frac{81x^2}{81}$

$\frac{4}{81} = x^2$

$\pm\sqrt{\frac{4}{81}} = x$

$\pm\frac{2}{9} = x$

The solutions are $\frac{2}{9}$ and $-\frac{2}{9}$.

42. $-4x^2 - 49 = 0$

$+4x^2 \quad +4x^2$

$\frac{-49}{4} = \frac{4x^2}{4}$

$-\frac{49}{4} = x^2$

$\pm\sqrt{\frac{49}{4}} \neq x$

There are no real solutions.

43. $64x^2 - 5 = 20$

$+5 \quad +5$

$\frac{64x^2}{64} = \frac{25}{64}$

$x^2 = \frac{25}{64}$

$x = \pm\sqrt{\frac{25}{64}}$

$x = \pm\frac{5}{8}$

The solutions are $\frac{5}{8}$ and $-\frac{5}{8}$.

44. $9x^2 + 9 = 25$

$-9 \quad -9$

$\frac{9x^2}{9} = \frac{16}{9}$

$x^2 = \frac{16}{9}$

$x = \pm\sqrt{\frac{16}{9}}$

$x = \pm\frac{4}{3}$

The solutions are $\frac{4}{3}$ and $-\frac{4}{3}$.

45. $49x^2 + 1 = 170$

$-1 \quad -1$

$\frac{49x^2}{49} = \frac{169}{49}$

$x^2 = \frac{169}{49}$

$x = \pm\sqrt{\frac{169}{49}}$

$x = \pm\frac{13}{7}$

The solutions are $\frac{13}{7}$ and $-\frac{13}{7}$.

46. $81x^2 + 17 = 81$

$-17 \quad -17$

$\frac{81x^2}{81} = \frac{64}{81}$

$x^2 = \frac{64}{81}$

$x = \pm\sqrt{\frac{64}{81}}$

$x = \pm\frac{8}{9}$

The solutions are $\frac{8}{9}$ and $-\frac{8}{9}$.

47. $4x^2 = 88$

$$\frac{4x^2}{4} = \frac{88}{4}$$
$$x^2 = 22$$
$$x = \pm\sqrt{22}$$
$$x \approx \pm 4.69$$

The approximate solutions are 4.69 and −4.69.

48. $x^2 - 29 = 0$

$$\underline{\quad + 29 \quad +29}$$
$$x^2 = 29$$
$$x = \pm\sqrt{29}$$
$$x \approx \pm 5.39$$

The approximate solutions are 5.39 and −5.39.

49. $x^2 + 40 = 144$

$$\underline{\quad - 40 \quad -40}$$
$$x^2 = 104$$
$$x = \pm\sqrt{104}$$
$$x \approx \pm 10.20$$

The approximate solutions are 10.20 and −10.20.

50. $3x^2 - 84 = 0$

$$\underline{\quad + 84 \quad +84}$$
$$\frac{3x^2}{3} = \frac{84}{3}$$
$$x^2 = 28$$
$$x = \pm\sqrt{28}$$
$$x \approx \pm 5.29$$

The approximate solutions are 5.29 and −5.29.

51. $50 - x^2 = 0$

$$\underline{\quad + x^2 \quad +x^2}$$
$$50 = x^2$$
$$\pm\sqrt{50} = x$$
$$x \approx \pm 7.07$$

The approximate solutions are 7.07 and −7.07.

52. $2x^2 - 10 = 64$

$$\underline{\quad + 10 \quad +10}$$
$$\frac{2x^2}{2} = \frac{74}{2}$$
$$x^2 = 37$$
$$x = \pm\sqrt{37}$$
$$x \approx \pm 6.08$$

The approximate solutions are 6.08 and −6.08.

53.

$$h = -16t^2 + 600$$
$$0 = -16t^2 + 600$$
$$\underline{+16t^2 \quad +16t^2}$$
$$\frac{16t^2}{16} = \frac{600}{16}$$
$$t^2 = \frac{75}{2}$$
$$t = \pm\sqrt{\frac{75}{2}}$$
$$t \approx \pm 6.1$$

Negative numbers are not reasonable for time, so $t \approx 6.1$ is the only solution that makes sense. Therefore, it takes the sack 6.1 s to reach the ground.